THE GHOST IN THE UNIVERSE

THE GHOST IN THE UNIVERSE

God in Light of Modern Science

TANER EDIS

59 John Glenn Drive
Amherst, New York 14228-2197

Published 2002 by Prometheus Books

Inquiries should be addressed to
Prometheus Books
59 John Glenn Drive
Amherst, New York 14228–2197
VOICE: 716–691–0133, ext. 207
FAX: 716–564–2711
WWW.PROMETHEUSBOOKS.COM

06 05 04 03 02 5 4 3 2 1

Library of Congress Cataloging-in-Publication Data

Edis, Taner, 1967–
The ghost in the universe : God in light of modern science / Taner Edis.
p. cm.
Includes bibliographical references and index.
ISBN 1–57392–977–8 (alk. paper)
1. Religion and science. I. Title.

BL240.3 .E35 2002
215—dc21

2002018958

Printed in the United States of America on acid-free paper

Contents

ACKNOWLEDGMENTS 9

INTRODUCTION: DOES GOD EXIST? 11

Sophisticated Gods 13
A Road Map 16

1. MAKING SENSE OF GOD 21

Something for Nothing 21
A Necessary Being? 24
The Impossible God 31
A Religious Theory 35
The Great Programmer 39
Is God a Philosophical Problem? 44

2. LET THERE BE LIFE 51

Specially Created 51
Resisting Evolution 54
Bringing Back the Designer 59
Order from Chaos 64
Darwin in Mind 69
God After Darwin 74

3. THE GODS OF MODERN PHYSICS 83

Physics in the New Age 83
Egocentric Cosmology 87
The Big Banger 92
A Quantum Spirit 97
Life, the Universe, and Everything 103
An Unnecessary Hypothesis 107

4. HISTORY AND HOLY WRIT 115

Special Revelation 115
Yahweh's Promise 118
The Messenger of God 124
God's Empires 129
The Meanings of History 133
The End of Revelation 139

5. GOD INCARNATE 147

In Search of Jesus 147
An Apocalyptic Prophet 150
Mythmaking 155
A Miracle Worker 161
The Risen Lord 166
The Bad News 171

6. SIGNS AND WONDERS 179

Wondrous Phenomena 179
A Spiritual Science 182
Great Performances 188
Statistical Miracles 192
Soaring Spirits 198
Thinking Meat 203

7. OF MYSTICS AND MACHINES 211

Feeling the Spirit 211
Thirty-one Flavors of Ultimate Reality 215
Beyond the Brain? 220
The Limits of Language 226
Holy Reason 230
Accidental Reason 235

8. LEAPS OF FAITH 243

Exorcizing Doubt 243
Reason Reformed 247
Progress? What Progress? 251
Round in Circles 257
The Sun Also Rises 262
Universal Reason 267

9. THE KNOWLEDGE OF GOOD AND EVIL 275

Moral Certainty 275
High Weirdness by Theology 278
The Morality of a Social Animal 283
Talking Morals 289
Believing the Absurd 293
Beyond Pragmatism 299

CONCLUSION: THE GOD OF SONG AND STORY 307

Divine Falsehoods 309
New Stories 312

INDEX 315

Acknowledgments

Some years ago, in what must have been one of my less sane moments, I decided to begin writing a book about science, religion, and philosophy. I found writing it even more rewarding than I anticipated—I learned a lot, and changed my mind on not a few things. But the process was also more difficult than I imagined. Fortunately, at every step along the way, I was helped by a remarkable band of "previewers" who criticized my drafts. I was lucky enough to have scientists, philosophers, and historians correct me, and also to have people from all walks of life who forced me to write clearly. Though not all agree with what I ended up saying, and the inevitable errors are of course all mine, there is something of their thoughts in this book as well as mine.

My special thanks go out to Amy Bix, Gerald Huber, W. D. "Bill" Loughman, and Victor J. Stenger, whose influence pervades the whole book. I could not have done it without their guidance and support.

I also owe many thanks to Achilles Avraamides, Virginia Barnett, Thaddeus M. Cowan, Stevan Davies, Keith Douglas, Ronald Ebert, Peter Fimmel, Lois Frankel, Marcus Harwell, Earl Hautala, Mike Huben, Jeffrey Jay Lowder, Keith Parsons, Wolf Roder, Peter Smitt, John David Stone, David Ussery, Ibn Warraq, Edmund Weinmann, and Thomas J. Wheeler. This is a better book because of their comments.

I have run the Internet SKEPTIC discussion group (see www.csicop.org/bibliography) for many years, using it to test-drive ideas in the making. I have depended on its members for their patience and constructive skepticism.

Books like this are, in part, products of other books. I am grateful to the libraries of Johns Hopkins, Iowa State, Louisiana State, and Truman State Universities. I also found some lovely books in the public libraries of Ames, Iowa; Highland Park, Illinois; and Baton Rouge, Louisiana; not to mention used-book stores across the land.

Paul Kurtz and Steven L. Mitchell at Prometheus Books were most helpful, even with a book like this, which they found hard to categorize as either science or philosophy.

Finally, I'd like to thank my parents, who early on infected me with a love of books and a respect for learning; my partner, Amy Bix, for being unreasonably wonderful; and our cats, Wimsey and Poirot, for lying about doing nothing and so showing me the *real* meaning of life.

Introduction

Does God Exist?

I don't believe in God because I don't believe in Mother Goose.
—Clarence Darrow, *Speech at Toronto* (1930)

A childhood friend in Turkey once told me there was a large stone suspended in midair above the Arabian desert; this miracle testified to the truth of Islam and the power of God. I never discovered the origin of the story—it does not seem to be part of any real tradition—but somehow it stuck with me. The popular gods seem to be magical beings, surfacing in that which is mysterious, unfathomable by puny human devices. Through the eyes of religion, the world often looks like a fairy tale which is nonetheless true.

Every story must have a villain, or at least a pesky skeptic; I told my friend I didn't believe him. He was sure, however, that he had heard impeccable testimony to his suspended stone. It was not just a fairy tale. So our gods come with arguments to ward off skeptics. A businessman once told me to look around his office: everything had an owner. Everything *must* have an owner—including me, he said—and this owner was God. He might have had a verse from the Quran in mind: "Whosoever is in the heavens or the earth belongs to Him."[1] But maybe he was just being a businessman. Philosophers like to think everything has an ultimate cause, which is God; maybe it came naturally to a businessman to think of an ultimate owner.

Of course, there are more interesting arguments out there than stones hanging in the air, or everything having an owner. Or at least arguments with more footnotes. After all, plenty of accomplished scholars, scientists, and philosophers are convinced there is a God. They believe we inhabit a deeply personal reality; that beneath surface appearances, we can come to see a God who touches just about everything in our lives. The idea of God holds their picture of the world together, including fact claims, moral ideals, and feelings of devotion and dependence. Their religious beliefs give them community, a dramatic story in which to set their lives, a tradition in which they can talk about what they believe and how they act.[2] And so the existence of a God becomes not just another fact, but a reality fundamental to a whole way of life.

Such a God is a very attractive idea, since it promises to make sense of so much in our lives. Maybe there is a subtle magic in the world, even though God has a way of getting tangled up in superstitions like suspended stones. Still, there are skeptics. Some of us have come to believe that even the subtle gods and demons are fictions. God used to be an obvious reality in our cultures, no more a matter of dispute than the existence of trees. A few philosophers toyed with doubt, but mostly as a prelude to a metaphysical discourse on why God must exist. Today, godless infidels are commonplace, especially among people with a philosophical or scientific background. Even as a social force, religion is no longer what it used to be. The devout believe in very different versions of God, and many people are apathetic, accepting religion as part of their social background but nothing more.

So we have an argument on our hands. And a fascinating argument it is, precisely because a spiritual reality is not supposed to be an ordinary fact. We do not argue about whether there is a God the way we debate the Loch Ness monster—God is not just a possible entity in addition to others in our world; our very picture of the world is at stake. All our knowledge comes into the argument—natural science, history, psychology, even our theories about how we learn about the world.

I join the argument as a skeptic. Being so obsessed with the idea of God as to write a book has made me appreciate religion as an expression of human hopes, even—at its best—as a work of art. Yet I remain among the godless infidels. I do not think there is any spiritual reality over and above the material universe. Our world does not look like it was created for any purpose; indeed, I believe we understand it best when we see it as an accidental world. As far as I can tell, our world is not the manifestation of any deep principle—moral, metaphysical, or theological.

It is not obvious that the world is a godless, accidental place. This idea looks crazy from the perspective of many intellectual traditions, it goes against common sense, it even denies many claims which people accept as facts. So I have a long argument to make. Which is just as well; I don't trust knockdown disproofs of God any more than proofs. And I want to build up a naturalistic picture of the world as well as criticize religious claims. This means I will lean heavily on science, though not, I hope, without being aware of the frailties of our sciences. In this, at least, I will be in good company. Infidels have usually embraced modern science, while religious thinkers have often had to make excuses for why the world looks different than what our religions let us expect. With science, we have stumbled upon an excellent way of learning about the world, and the best of our scientific knowledge consistently undermines our hope that there is a God.

SOPHISTICATED GODS

How, then, do we argue about God, science, and the way to best understand the world? Our religions do not always encourage skeptical questions. Perhaps to the truly faithful, God must be a basic assumption which comes before everything else. On the other hand, religious people also speak of a God who is obvious to human reason, so that when we ask how we know there is a God, we meet with overwhelming proof. The Christian apostle Paul thought our world was clearly created and sustained by a divine power:

> For all that can be known of God lies plain before their eyes; indeed God himself has disclosed it to them. Ever since the world began his invisible attributes, that is to say his everlasting power and deity, have been visible to the eye of reason, in the things he has made.[3]

Apparently, although there may always be those perverse enough to deny God, there are plenty of signs in God's creation to convince the honest skeptic. The Quran speaks of these signs, telling us:

> It is God who raised the skies without support, as you can see, then assumed His throne, and enthralled the sun and the moon (so that) each runs to a predetermined course. He disposes all affairs, distinctly explaining every sign that you may be certain of the meeting with your Lord. . . . In these are signs for those who reflect.[4]

In fact, both Paul and Muhammad seem so certain of their Gods, they declare that dissent from so obvious a truth must be a result of willful blindness or obstinate rebellion.

Our long-standing religious traditions and philosophical theologies also present God as a fact we can argue about. We should at least come to see that a God exists and that reality is ultimately spiritual, even if our merely human reasoning can never encompass the full mystery of divinity. The Catholic catechism instructs us that reason and science never truly contradict faith, "because the things of the world and the things of faith derive from the same God."[5] God is real, something we can speak of in statements like "God is the most perfect being," "God created the universe," or "God brought the Israelites out of Egypt." It is not just a metaphor expressing the Israelites' joy at being released from slavery or the wonder we feel when gazing at the stars. A most perfect divine power is actually responsible for Jewish history and the night sky. The existence of a God is supposed to *explain* things.

Unfortunately, old-time religion runs into trouble here, since its God was obvious only in a world we can no longer believe in. Before we had an idea of astrophysics, we might have thought that God hung the stars on the firmament, maybe even completed the whole job in just six days a few thousands of years ago. Before modern science and history, we might have trusted scriptural testimony about Noah's flood or the plagues of Egypt. But today, conservative, magical, scripture-waving religion has become obviously false to the well-educated person.

Of course, old-time religion is still very popular. Throw a brick in a crowd, and chances are it will hit someone who believes the stories in the Bible or Quran. Turn on a TV or radio, and it will not take long to find a preacher who not only proclaims the faith of his fathers, but says his scriptures are free of all error. Some of them will even argue that God's Word miraculously anticipates modern knowledge. Apparently "Can you dispatch the lightning on a mission and have it answer you, 'I am ready'?"(Job 38:35) is an "anticipation of radio."[6] Henry M. Morris, a leading creationist, reads the Bible as a physics textbook:

> . . . Ecclesiastes 3:14: "I know that, whatsoever God doeth, it shall be for ever: nothing can be put to it, nor any thing taken from it: and God doeth it, that men should fear before him." This striking verse actually anticipates the great principle of energy conservation. . . . The only way of accounting for the infinite reservoirs of energy in the universe is "God doeth it."[7]

Popular Muslim apologists also like to claim the Quran exhibits knowledge far beyond what was possible in the seventh century, when it was revealed. For example, 51 Adh-Dhariyat 57, "We built the heavens by Our authority; We are the Lord of power and expanse," supposedly speaks of the expanding universe, while the seven heavens or skies mentioned in 78 An-Naba 86 and other verses refer to parallel universes or cosmic structural hierarchies: solar systems, galaxies, and so on.[8]

Now, I don't intend to spend time refuting such claims; they are too blatantly wrong, and God does not stand or fall depending on whether our scriptures know their physics. We have more sophisticated Gods, who are not so fragile as to fall apart when we discover an error or contradiction in a holy book. The God of most liberal religious people does not dictate inerrant scriptures; what we have are human records of encounters with divinity. This God does not make the world in a matter of days and then fashion the first human out of earth, but starts up creation with a cosmic explosion and then directs long ages of evolution to produce intelligent creatures. God is the ghost in the universe, not visible on the surface, but the source of the order, the meaning, the very reality of material existence. And we are as likely to encounter this God within, in mystical experiences, as in the glory of creation. Such versions of divinity are defended by scholars and philosophers, not televangelists. Skeptics must wrestle with these sophisticated Gods, not just the fundamentalists' Big Boss In The Sky.

However, the old-time God is important, at least as a starting point. After all, traditional ideas of God have important strengths. If the world was as we once imagined it to be, we could not honestly be scientific naturalists. We would best understand our world as a cosmic drama with God as the leading character. In traditional religion, God is not a vague, content-free idea. God takes on concrete meaning within a network of religious fact claims telling us our world is dependent on a divine power.

If we don't keep the virtues of the traditional God in mind, our new and improved version will risk becoming nonsense. This has happened before. When Judaism entered the Greek intellectual world, its beliefs often looked philosophically dubious. Sophisticated Jews retooled their ancestral religion just as Greeks allegorized their myths. If Zeus's rape of Ganymede was really a metaphor for how the soul was enraptured by its encounter with divine power, then also "Adam was revealed to stand for Reason and Eve for the labile part of the soul and sensory component of the mind which, if distracted by lower things (the snake), will pull even Reason down with her."[9] Many Christian and Muslim philosophers would eventually do the same; God became the remote incomprehensible One of the Neoplatonists, emanating lesser divine entities like Existence, Reason, and Soul, producing the "intelligible" realm of which our "sensible" world was a mere echo. This prevented a lot of embarrassment; if scripture talked about a humanlike God sitting on a throne, well, this was merely an allegory to be explored for its Platonic meaning. But within the metaphysical song and dance, the reality of God tended to fade away. I am inclined to think that adopting Platonism turned an honest mistake into something worse. Conservative theologians rightly want to make real claims, not produce infinitely flexible verbiage.

So a religion with a sophisticated God must still set forth a picture of the world, and show how its God is central to it all. Creation does not have to take place in six days, and prophets can record their religious experiences without acquiring immunity from error. But God still must somehow help explain our world. Perhaps a creative spirit pervades all physical existence, or there is a purposive pattern in human history. Perhaps paranormal and mystical experiences let us see into a fundamentally personal reality, and our moral behavior originates in a God. We can argue about claims like these, and if they turn out to be correct discover that our world is, underneath its material surface, dependent on a supernatural personality.

But this also means we cannot defend God by denying that modern knowledge is relevant to religious questions. Among believers of all stripes, it is a commonplace that science is unable to answer ultimate questions about human origins, meaning, and destiny. This is at best an overly narrow view of science. In fact, I shall argue that we can say quite a few things about such questions without the benefit of religion. We come from accidents, not design. Our lives have no cosmic meaning. And our destiny is dust, not immortality. Many of us find such answers profoundly unappealing. Nevertheless, I believe they are most likely correct.

A ROAD MAP

In a long argument, the fun is in the details. But since I am trying to draw a broad picture of what the world is like, I should offer a road map to where my argument will be going.

First of all, I will not follow the usual style of philosophical debate. Philosophical theists have a way of trying to establish the reality of God with only minimal reference to the world as we see it. For example, they may observe that there is some order in our world, or simply that something exists, and then say a God is the best explanation for these facts. They talk about necessary beings, basic beliefs, and Very Very Important Deep Questions. Atheist philosophers join the fray by saying the theists' proofs are actually very bad arguments, and they go on to argue there cannot be a God, again based on some very minimal, obvious facts. Their old favorite is the problem of evil: how can a benevolent deity who is presumably not afflicted with a warped sense of humor create a universe as nasty as ours?

Now, I think the atheists do a good job of discovering the flaws in philosophical arguments for God. But if God should be a concept fundamental to the way we understand the world, it seems strange that so much of the debate has so little to do with what we have learned about our world. Plus, I must admit to a prejudice; immersing myself in the philosophy of religion has convinced me there is something very wrong with traditional philosophy. The God of an ordinary believer is, I think, a mistake, but I have come to suspect the God of a philosopher is sheer confusion. So while I have much to say about traditionally philosophical issues and arguments made by philosophers, I try not to give philosophy center stage.

In chapter 1, I start out by looking at philosophical arguments about God. I pay special attention to the classical theistic attempts at rational proof, since while these do not succeed, they still are important for our concept of God. I also argue that philosophical arguments against God only refute overly ambitious theologies. Traditional philosophical analysis does not take us very far. I then try to outline how we can make sense of God by drawing a picture of the world in which God is the central actor. This, of course, will shift the debate away from philosophy and toward science.

Chapter 2 is about Darwinian evolution, in biology and beyond. The complexities of life do not require intelligent design; accidents and blind mechanisms do the trick. Not only old-fashioned creationism but also more liberal attempts to find a progressive guiding hand in biology get nowhere. In fact, even the most sophisticated arguments for the compatibility of Darwinian processes with divine action fail since random, uncaused variation and selection is basic to all creativity in nature, including that of the human brain.

Physics seems more fundamental, and so perhaps closer to God than biology. Modern physics features often in current arguments that science reveals a God. Chapter 3 is where I criticize claims that cosmology shows that the universe has been designed for us, that the big bang was an event of divine creation, and that quantum physics proves consciousness central to physical existence. I also point out how we can understand the laws of physics not as expressions of a divine will, but as frameworks for accidents. Physics fixes us more firmly in a purely natural world.

We usually believe not in a generic God that might emerge from natural science, but in the God of Abraham revealed by history, scripture, and prophets. In chapter 4, I explore Jewish and Muslim history, and argue that our religions are very human creations, reflecting particular historical accidents. The authors of our holy books gave us stories shaped by theology, not reliable accounts of our past. Historical knowledge is continuous with natural science, giving us no more support for supernatural revelation.

Christianity's turn comes in chapter 5. Jesus is still central to liberal Christian theology, but the little we know about the historical Jesus does not help God. We discover hints of an apocalyptic prophet, a teacher who started social experiments, even a magician or spirit-possessed healer—but no Risen Lord, not necessarily even someone who was uniquely close to a God. In the end, all we get out of history is that some people have religious experiences and that religions are in part built on supernatural interpretations of these experiences.

Miracles and paranormal events could be the kind of religious experience which would establish a supernatural realm. Chapter 6 is where I look at strange phenomena and hints of mind over matter which suggest the existence of a soul. But psychical research gives us nothing but dubious anecdotes and very small and uncertain effects coaxed into sight by statistics. The evidence supports an unmagical world, where our minds are a product of nothing but our brains, and there is no such thing as a spirit or soul.

Mystics often say their experiences, indescribable and beyond all concepts, directly acquaint them with a divine power. Some tie their visions to elaborate Platonic philosophies. In chapter 7, I argue that modern psychology and artificial intelligence research helps us explain mystical experience without supernatural realities beyond our brains. Spiritual interpretations of mysticism and philosophies like mathematical Platonism are all most likely mistaken. We do not learn about the world by direct contact with an ultimate reality.

If I do a good job, we will have come a long way toward exorcizing God from our explanations of the world. But I rely on a scientific style of argument, and postmodern philosophy thinks science is overrated. Perhaps I arbitrarily choose presuppositions that lead me to trust science, and a religious person can take God on faith the same way. So chapter 8 is here to defend science against postmodern philosophizing. We do not have transcendental assurances behind our ability to reason, and science is very much a social activity. But it still is our best way of learning about the world. God cannot be pulled out of a philosopher's hat.

Postmodern worries, however, expose the *social* weaknesses of skepticism. And since the natural world contains no binding moral principles, if there are such things, they may require a God. In chapter 9, I deny there is any moral reality beyond our interests and the social enterprises we are a part of. Morality is rooted in our biology, not a transcendent realm. However, we are still not quite sure how to sustain morally engaged communities without supernatural guarantees. So religions have practical advantages even in secularized societies.

My concluding musings follow. Though I think we have excellent intellectual reasons to disbelieve in God, I am also fascinated by the stories our religions tell. They may be fictions, but fiction can tell us a lot about ourselves. God does not belong in our explanations; but in the end, I still think we have a lot to learn from religion.

NOTES

1. 21 Al-Anbiya 19. I use the translation by Ahmed Ali, 1984. Ali interprets the Quran according to modern scientific and moral sensibilities, so I err on the side of giving the Quran the benefit of the doubt.

2. I am, of course, using a substantive definition of religion, requiring belief in some sort of divinity. This is almost always true, even beyond the Abrahamic religions which are my defining examples; see Stewart Elliott Guthrie, *Faces in the Clouds: A New Theory of Religion* (Oxford: Oxford University Press, 1993), pp. 19–20. In some contexts, it may be legitimate to call certain nonsupernatural

attitudes religious—for example, if someone identifies with Judaism as a tradition and culture without accepting God—but I will not do so in this book.

3. Rom. 1:19–20. I use the *Revised English Bible*, 1989, for the New Testament.

4. 13 Ar-Rad 2–3. See also 30 Ar-Rum 20–27.

5. *Catechism of the Catholic Church* (English translation, Boston: Libreria Editrice Vaticana, 1994), ¶159.

6. Harold L. Fickett Jr., *A Layman's Guide to Baptist Beliefs* (Grand Rapids, Mich.: Zondervan, 1965), p. 16. For the Tanakh or Old Testament, I use the Jewish Publication Society translation, 1985.

7. Henry M. Morris, *Biblical Creationism* (Grand Rapids, Mich.: Baker, 1993), p. 108.

8. The best English-language example of this sort of apologetics is Maurice Bucaille, *The Bible, the Qur'an and Science* (Paris: Seghers, 1982).

9. Paula Fredriksen, *From Jesus to Christ: The Origins of the New Testament Images of Jesus* (New Haven, Conn.: Yale University Press, 1988), p. 15.

Chapter One

Making Sense of God

Philosophy, n. *A route of many roads leading from nowhere to nothing.*

—Ambrose Bierce, *The Devil's Dictionary* (1911)

Many philosophers, it seems, love to think our world rests on deep metaphysical principles, conveniently discovered from our armchairs. The traditional favorite among such principles is, of course, God. Unfortunately, this God of the philosophers becomes a dumping ground for arbitrary metaphysical intuitions about Ultimate Things. Believers make a living cooking up convoluted arguments for God, while more skeptical philosophers keep busy poking holes in these proofs. We need a better way to make sense of God—we should be arguing about how a God might explain the world we live in, not about first causes or necessary beings.

SOMETHING FOR NOTHING

Our Gods do not always need elaborate theologies. Many a religious person conceives of God as a totally overwhelming, literally awesome power; the details can wait. Yet no religion is silent about its God. The Quran tells us, in 112 Al-Ikhlas, to

> Say "He is God; the one that is most unique; God the immanently indispensable. He has begotten no one, and is begotten of none. There is none comparable to Him."

In these three sentences we get a hint of mystery, of greatness beyond comparison, even an objection to those who speak of sons or daughters of God. We get phrases like "immanently indispensable," which can keep generations of theologians at work trying to explain what the text means. The faithful do not hope to fully grasp the divine reality, but they are sure a God created our universe and sustains our present existence. This God is entirely holy, supremely worthy of worship, powerful beyond imagination.

God, then, is first of all something incomparably grand and glorious. Our God must be singular; it must serve as the ultimate explanation, the foundation of all value, the guarantee that monotheistic faith is superior to its rivals. Pagan gods may be able to dispatch lightning bolts against offending mortals, but the true God must be all-powerful. This God must not only be good, but so good that even our slightest sin makes us worthless in comparison. We must respond to God in fear and trembling, and then we must come to love God and desire to do the divine will.

This is a beginning, but we need to be more specific. Metaphysicians rise to the challenge by producing a list of infinite attributes for God. Theologian Gerald Bray gives an example, derived from the writings of John of Damascus:

Time	(Beginning):	without beginning, uncreated, unbegotten
	(End):	imperishable, immortal, everlasting
Space:		infinite, uncircumscribed, boundless, of infinite power
Matter:		simple, uncompound, incorporeal, without flux
Quality:		passionless, unchangeable, unalterable, unseen[1]

Such a list makes God sound like a Near Eastern tyrant with his roll of absurdly inflated titles. Even so, it is a decent stab at describing an incomparably great reality. We can now argue about God's attributes. Few quibble about God being uncreated, but philosophers will cross swords over how unchangeable or simple God must be. Change seems inappropriate for an ultimate principle behind our world; on the other hand, it is hard to think of a God we pray to, a personal God who acts on the world, as eternally unchangeable. Unfortunately, these disputes tend to devolve into metaphysical gamesmanship. Bray, for example, suggests that the *essence* of God is simple and unchanging while its *person* embodies complexity and flux. "At the level of the person, which is the point at which we enter into relationship with God, Christians insist that there is a plurality in unity, which is not to be confused with the simplicity of God's impersonal essence. The result is that everything which belongs to God's fixed and immutable essence is mediated to us through the relationship we have with the persons."[2] Now we can have a tedious argument over whether this simplicity wrapped in plurality is double-talk or legitimately having things both ways.

Defining what we mean by an incomparably great being is obviously difficult, and it is not clear whether metaphysicians help us or dazzle us with impressive but empty words. However, traditional metaphysics is attractive not only because it might refine our idea of divinity, but because it promises to actually prove there is a God. God, being incomparably great, might not merely happen to be, but would exist because it *must* exist. In other words, we might be compelled to acknowledge the reality of a God just by armchair reasoning, provided our head is screwed on right. We would still have to fight over the full list of divine attributes, but that would be just a matter of clearing up the details.

Metaphysicians have long tried to get something for nothing this way. Of course, their God is not much more than a vague perfect necessary being. But they hope to prove that at least an incomparably great object of spiritual devotion exists. Revealed religion, bolstered by philosophical respectability, can then tell us more about our salvation. Catholic tradition, for example, embraces metaphysical proofs of God:

> Created in God's image and called to know and love him, the person who seeks God discovers certain ways of coming to know him. These are also called proofs for the existence of God, not in the sense of proofs in the natural sciences, but rather in the sense of "converging and convincing arguments," which allow us to attain certainty about the truth.
>
> These "ways" of approaching God from creation have a twofold point of departure: the physical world and the human person.[3]

Apparently philosophy tells us there is a God, since the world must have a cause which is utterly beyond material things. On top of this, human experience confirms the existence of a soul, which can only come from God. The sensible person believes in God—in the end, is perhaps even compelled to accept papal authority.

Some take this certainty about God very seriously indeed, claiming God's nonexistence is a contradiction in terms. This "ontological argument" mainly attracts philosophers, but the other classical proofs of God appeal to nonphilosophers as well. Cosmological arguments claim the material universe must be caused by an independent, self-sufficient power; design arguments say a creative intelligence must be responsible for the order in the universe. These seem compelling to many ordinary believers and not a few scientists.[4]

With the help of such arguments, philosophy and God lived in harmony for many centuries. Popular religion was perhaps too vulgar and superstitious in the philosopher's eye, and philosophers were too prone to fanciful reinterpretation of doctrine from the perspective of orthodoxy; however, pious and metaphysical intuitions usually managed to coexist. It was obvious that everything had to have a cause, and that order required a designer. When religious thinkers applied these intuitions to the natural world, they found it needed a first cause beyond the universe, a God who imposed a purposeful plan on the world.

Of course, there were always skeptics;[5] even devout philosophers who liked one proof but thought another was worthless. After all, if our intuitions rest on truly universal principles, we have to wonder what caused the First Cause, and who designed the Designer. The usual answer is that God, as a necessary and self-sufficient being, is an exception to the rule. But now, it begins to look like all God does is preserve our pet metaphysical principles by taking their loose ends, knotting them together, and sweeping the whole mess under a carpet called Ultimate Mystery.

Especially after the European Enlightenment, philosophers began to subject the classical proofs to withering criticism. Unbelieving philosophers are common today, and the typical atheist argument starts by ripping apart the traditional proofs of God. Furthermore, modern science made us more skeptical about grandiose metaphysical schemes. After Darwin, we learned that intricate adaptive order did not require supernatural design; quantum physics showed us a random subatomic realm in which events took place without any cause. The classical proofs began to look like failures.

Today, many theologians agree that the old proofs do not work. As Gerald Bray admits, "the so-called 'proofs' for the existence of God can be used only as supporting evidence for a belief which is already held for other reasons."[6] Believers have to rely on revelation—God's self-disclosure—not the assurances of philosophers. But the old proofs remain important for our concept of God. God still has to be a self-sufficient being who is somehow the creating and sustaining cause of the world. So arguments for God continue to be informed by the classical proofs, even when they rely on empirical evidence or religious experience. It might even be said that proofs of God enjoy an immortality only truly bad ideas can aspire to. Something for nothing is too attractive a goal; philosophers will embark on that quest just like some inventors will always try to build perpetual-motion machines.

In fact, new and improved versions of the classical proofs are still going strong in the recent philosophy of religion. So the God of the philosophers, still alive if not brimming over with vitality, is a good starting point for us.

A NECESSARY BEING?

The most ambitious proof of God is the ontological argument, claiming to show that God's existence is logically necessary. It has many versions, but all boil down to saying God is defined to exist, therefore God exists.

Of course, scoffers are liable to say that, with that sort of reasoning, we can just as well declare that unicorns are defined to exist, and therefore unicorns exist. But metaphysicians keep hoping there is something special about *God* that will make the trick work.

The argument starts by defining God as an unsurpassable being. We should not even be able to conceive of anything greater—God should not just be the greatest being around, but the most perfect being possible. And obviously, an actual being is better than something which is a figment of our imagination. A perfect being must exist, since if it did not, it would not be perfect.[7] Having made God pop into existence by sheer force of logic, we now break out the champagne.

This comes too easily not to suspect smoke and mirrors. To begin with, "greatness" or "perfection" in an argument like this is quite arbitrary. We could replace godly perfection with unicorn-likeness and prove the existence of Mildred, the perfect unicorn. A horse, for example, is more unicorn-like than a rock. So we look for a maximally unicorn-like being. Obviously, an actual unicorn is more perfectly unicorn-like than a mere character in a fairy tale. So Mildred must exist.

Enlightenment philosophers, starting with Kant, pointed out that the mischief arises from taking "existence" to be a property we can attach to a concept.[8] So modern proofs no longer say God is defined to be perfect, perfection implies existence, hence God exists. They say God is defined to be perfect, perfection implies *necessary* existence, hence God exists.[9] This needs some translation. We can imagine a world with or without rocks or unicorns, and our brains will not overheat from the effort. God, being unsurpassably great, might be different. God might exist, but not like a rock which merely happens to be here. In our world, or any other world imaginable, God would exist because it would be impossible for such a perfect being not to exist. Indeed, it seems that if there is a God, it must exist out of necessity and not in a second-rate way like a rock. And if God is a necessary being, all sorts of wonders take place. A necessary being exists independently of other facts, so we need not worry about scientists casting doubt on the divine reality. The only way a necessary being can fail to exist is if it is impossible. All or nothing. Unless our idea of a perfect being is unintelligible or self-contradictory, God *must* exist.

Unfortunately, this sort of argument still leaves greatness quite arbitrary, so it fails to dispel Mildred. A perfect unicorn should also have a first-rate existence, it would seem. And since nothing but a metaphysical imagination can distinguish between an object which is real and one which is *necessarily* real, we can go ahead and define a maximally unicorn-like being to exist necessarily.[10] Just like God, if it is at all possible for Mildred to exist, then she *must* exist.

Maybe we should ignore Mildred. After all, a necessary unicorn is only a device to parody the proof. There is nothing wrong with Mildred in a formal, logical sense, but intuitively Mildred is ridiculous. God, on the other hand, is supposed to be eternal and self-sufficient, so perhaps we can legitimately call God logically necessary as well. In that case, the ontological argument expresses an intuition like:

(I1) It is possible—just barely possible, we need claim no more—that an unsurpassably great being exists. This being would be maximally excellent—omnipotent, morally perfect, and so on—and exist necessarily.

To some philosophers, (I1) seems true; and in that case, there must be a God. Richard Gale, however, points out other intuitions which seem at least as plausible as (I1), but are incompatible with (I1). For example:

(I2) It is possible—merely logically possible—that there could be a world in which morally unjustified evils take place.

If (I2) is true—and why not?—a necessary God is in trouble.[11] A necessary God would exist and be morally perfect in all imaginable worlds; therefore (I1) actually claims that morally unjustified evil is logically impossible.

There is more trouble ahead. Maybe an unsurpassably great, maximally excellent being makes no sense. Consider Mildred. Let us say the perfect unicorn must be so fleet of foot that nothing can beat her in a race. If Gerald, the pretty-good unicorn, could pass Mildred by half a neck, this would mean Mildred was not a maximally excellent unicorn after all. But this is no problem—maximum excellence means Mildred can move to any spot instantaneously, so while Gerald might tie with her, he can never go faster. If God's infinite attributes always corresponded to "intrinsic maxima" like Mildred's speed, there would also be no problem.[12] But now let us say Mildred should be unsurpassably strong, able to beat any conceivable Gerald in a tug-of-war. Interestingly, the situation is different than with a foot race: strength seems to have no intrinsic maximum, the way there is no such thing as a largest number. We can always imagine a Gerald who could best Mildred. And since God is supposed to be maximally great in just about everything, some of God's attributes are bound to run into trouble.[13] An infinitely powerful God is fine, but an unsurpassable God is dubious.

There is a way out. We can *define* God's attributes to be intrinsic maxima. Just as we can go beyond high school geometry and work with non-Euclidean spaces, we can look for alternative number theories. Some unconventional number theories define an "ultimate class" which has the largest size within that theory.[14] By declaring that strength must be measured in such a theory, we can make sure that Mildred has unsurpassable strength.

Perhaps a similar trick will work for God. Where God needs to be unsurpassable, we define intrinsic maxima, and where we need to make excuses for God, we say no maximum exists. For example, Richard Swinburne does not want to fault God for not making ours a maximally good world, so he claims, "God cannot create the best of all possible worlds, for there can be no such world—any world can be improved by adding more persons to it, and no doubt plenty of other ways as well."[15] But of course, how we tighten or loosen our intrinsic maxima is completely arbitrary. In fact, the whole ontological argument is irredeemably infected with arbitrariness. Our ideas of perfection are too arbitrary to prevent Mildred appearing along with God, even without Mildred we are left with a mess of conflicting metaphysical intuitions, and the concept of an unsurpassable being can only be patched up by arbitrary definitions. In the face of such rampant arbitrariness, the very idea of an argument compelling us to accept God through logic alone becomes a bad joke.

No doubt ontological proofs will continue to tempt philosophers.[16] But, as Kai Nielsen says,

> Unless we happen to love solving puzzles, we are very likely, if we are interested in the philosophy of religion, to sigh with ennui at the appearance of yet another baroque but carefully crafted argument to prove that the denial of God's existence is self-contradictory. There has to be, we think, something wrong somewhere in such an argument: the problem is, can it readily be located?[17]

Our philosophical tradition has long nourished a hope for a shortcut to God. But we cannot bring anything into existence, not even God, by definitional magic. Some real-world facts, however minimal, will have to take the stage.

Perhaps the fact a material universe exists at all is just what we need. Something, metaphysicians think, must have caused the universe; and a God seems to be the kind of power who would go around causing universes. Metaphysically, if not logically, God might be a necessary being, while the universe has no power within itself to come into existence.

Naturalists take the natural universe to be all there is. Ultimately, events have impersonal causes, or they are uncaused—accidental. Now, even if the world were a complete network of natural causes, naturalism cannot explain how it happens that *our* network of causes exists and not another. The universe remains a brute fact, an uncaused reality, an accident. But this is no defect of naturalism. Without venturing back into the swamp of logically necessary beings, there is no avoiding brute facts which happen to be true for no reason whatsoever. As Antony Flew remarks, "The Principle of Sufficient Reason—that there has to be a sufficient reason for anything and everything being as it is, was, and will be—is not, as has often been thought, necessarily true. It is instead demonstrably false."[18] Even theists cannot avoid brute facts, though they can argue that God is a better ultimate fact than the universe. Demanding that everything has a cause does nothing for such a God; we are immediately tempted to ask what caused God, and doctrines about how God is self-caused never get off the ground.[19]

So modern arguments no longer say the universe must have a *cause*, which is God. They say God is the most probable *explanation* for why a universe exists.

Perhaps a God would create a universe—without God, it is less probable that anything would exist. So the fact we are here is evidence supporting a God. Fair enough, except that it is not clear why a perfect, self-sufficient God would create a material universe. Does God get lonely? The Greek Platonists thought direct involvement with the world would threaten their God with imperfection, so they imagined lesser intermediates to be responsible for the universe. Abrahamic religious traditions imagine a more hands-on sort of God, but why a perfect God would cause a universe, directly or indirectly, remains something of a mystery. Still, the existence of a universe might be evidence for a hands-on God.

Let us see how this kind of argument works with a more earthly example. Say we suspect there is an arsonist in town. If we then come upon the charred remains of a house, this evidence will support our suspicion. Some numbers will make the example more concrete. The probability of a fire when there is an arsonist around should be fairly high, so *if* an arsonist is in town (a), there is a 90 percent chance there will be a fire (f); in other words, $P(f|a) = 0.90$. If, before learning about the fire, we consulted some crime statistics and figured there was only a 10 percent chance an arsonist was in town, then we start with a prior probability $P(a) = 0.10$. Now, informally, when a predicted event takes place, our suspicions get stronger. Clearly, unless there is an even greater chance for a fire without arson, our suspicion that an arsonist is around should get stronger after we see a fire; or $P(a|f) > P(a)$. Bayesian statisticians have an equation for this:[20]

$$P(a|f) = \frac{P(f|a)P(a)}{P(f|a)P(a) + P(f|\neg a)[1 - P(a)]}$$

Say there was only a 10 percent chance for a fire with no arsonist in town, so $P(f|\neg a) = 0.10$. Plugging in the numbers, we get $P(a|f) = 0.50$—our 10 percent level suspicion that there might be an arsonist lurking around goes up to 50 percent after a fire.

Now we can do the same, substituting the universe for the fire and

God for the arsonist. If a universe is more likely with a God than otherwise, the fact that there is a universe increases God's probability, just like a fire is evidence for arson. But before we are impressed, let us imagine a priest from the temple of fire showed up on the scene and declared the fire was caused by a fire demon (d). Unless mollified by a temple offering, these invisible spirits always cause fires where they pass, so $P(f|d) = 1$. So observing a fire makes a fire demon more probable, or $P(f|d) > P(d)$. But say that all the probabilities involving the demon were the same as those involving the arsonist, except the prior probability $P(d) = 0.0001$—we do not want to be dogmatic and say fire demons are impossible, but they are a lunatic idea anyway. After the calculation, we end up with $P(d|f) \approx 0.001$—greater than before, but still negligible. The priest walks off in a huff.

Whether the universe (u) is worthwhile evidence for a God (g) depends on three numbers. $P(u|g)$, the likelihood of God creating a universe, should be large; $P(u|\neg g)$, the probability of a universe without God, should be small; and $P(g)$, the probability of God independent of anything we know about universes, should be large. Now, we might bring $P(u|g)$ close to certainty by invoking theological mystery and requiring a hands-on God. But we do not have the faintest idea what $P(u|\neg g)$ or $P(g)$ might be; we have only some metaphysical intuitions worth exactly their sale price. With the fire, previous experience can guide us. Crime statistics can give us a fairly decent idea of how likely it is that an arsonist lives in a community, what percentage of fires can be traced to arson, and so on. Experience also shows us that the fire-demon hypothesis is pretty useless, even though it guarantees fires. But there is no way to select a representative sample of universes and find out what proportion of them were created by a God. We need to find the relevant probabilities without the benefit of *any* previous experience.

One possible solution is to favor simpler hypotheses over complicated scenarios. Scientists like simple theories, and in everyday life, we prefer to explain fires by a single arsonist rather than by a massive conspiracy. We like to assume the least about what we do not know. Richard Swinburne believes "theism is a very simple hypothesis—the simplest hypothesis to provide an ultimate explanation which there could be."[21] If we confine our explanations to the natural world, we find a lot of complexity. Even if everything behaves according to a few elegant laws of physics, these laws remain brute facts. Furthermore, the physical state of the world remains one enormously complicated possibility among countless alternatives. Theism, in contrast, promises to tie everything to a single ultimate cause—a cause so simple it has no limits. Very little information goes into the philosophers' definition of God, just a sense of a perfect being with ultimate creative power.

Unfortunately, simplicity is a complicated mess of a concept. There is no convincing account of exactly what simplicity is and why it should be connected to high probability, so it is doubtful whether there is a logical principle of simplicity fundamental to probabilistic reasoning.[22] Simple representations of our environment use up little memory and give fast results. But this has no obvious connection with truth; our models of the world often sacrifice accuracy to practicality. For example, say we want to avoid getting eaten. A model that takes a month to recognize a predator from visual data is useless, even if it has a near-zero error rate. And if it records the past so perfectly that we need a monstrous brain capable of carrying so much irrelevant detail, we have no way to escape the predator anyway.[23]

But say we figure out how to determine probabilities based strictly on the logical simplicity of hypotheses. Say also that the existence of a perfect necessary being is a nice and simple hypothesis, rather than the vague and messy idea it invariably is. Even then, God is more like a fire demon than an arsonist. An arsonist is a complicated person and arson a complex process, while a fire demon is a pure spirit who simply causes fires. But obviously, the simple hypothesis of a fire demon is so uninformative it provides no improvement over saying a fire was uncaused. In fact, a strictly accidental fire makes more sense—after all, nothing is quite as simple as nothing. God might be a simple hypothesis, guaranteed to cause a universe. A fire demon is a simple hypothesis guaranteed to cause fires. Simplicity and guaranteed results are not enough to recommend a hypothesis.

Unlike Mildred the perfect unicorn, a fire demon might be made into a respectable idea. If offerings to fire demons actually prevented fires, if the priests had a detailed theory about the behavior of fire demons which explained actual fires, we would take fire demons more seriously. Successful hypotheses tie together many real-world facts, showing that what previously might have looked like a haphazard collection of facts fits a simpler pattern. But the simple God of the philosophers is thoroughly useless in explaining the world as we find it. Even if someone came up with a believable argument that the existence of *a* universe made God highly probable, we eventually have to ask if a God is probable given *our* universe. This is quite a different and more relevant question. And if all a theist can do is repeat the same impersonal explanations a naturalist uses, only tacking "God decreed it so" onto them, we might as well not bother with God.

We need a God who is more like an arsonist and less like a fire demon, and not at all like Mildred. For many centuries, we have hoped to prove God from just about nothing but the concept of a perfect necessary being. This is not only a failure, it is a complete waste of time.

THE IMPOSSIBLE GOD

Where does God stand, if the classical proofs fail? This might not be much of a disaster, since demolishing the proofs is not the same as disproving God. We do not believe in unicorns, but not because any metaphysical argument for the existence of unicorns fails. The fact that we have no positive evidence for unicorns is more significant, but even this is not the whole story. Unicorns do not fit our most reliable theories about the world. The magical nature of a unicorn does not sit well with physics, and even an unmagical single-horned horse would be biologically peculiar. We do not have any good reason to modify our theories to allow unicorns.

So it would seem the debate should now shift away from claims of ironclad proof to asking whether a God helps explain our world. But skeptics often try to take a short cut and dispel God with some philosophical magic of their own. For example, J. N. Findlay once produced an "ontological disproof" of God.[24] He agreed with tradition: God must be the greatest, most perfect being. But *if* God must be so great we should not even be able to conceive of a world without God, then metaphysicians have shot themselves in the foot. Given the shoddy track record of ontological arguments, we can no longer speak of logically necessary beings with any confidence. God the unsurpassable being begins to look impossible.

Indeed, metaphysicians have been so spectacularly unsuccessful in proving God, we have to wonder whether God makes sense at all. God's omni-attributes keep flirting with contradictions,[25] we are better off without necessary beings, and philosophers seem determined to make God irrelevant to real life. The traditional arguments do not only fail to prove God, they lead us to suspect that the notion of God is a thoroughly confused one. Until the faithful get their act together and produce a claim which is at least intelligible, there is nothing there even to criticize. In 1864, Charles Bradlaugh wrote:

> The Atheist does not say "There is no God," but he says: "I know not what you mean by God; I am without an idea of God, the word 'God' is to me a sound conveying no clear or distinct affirmation. I do not deny God, because I cannot deny that of which I have no conception, and the conception of which, by its affirmer, is so imperfect that he is unable to define it to me. . . ."[26]

Bradlaugh's words still ring true for the God of the philosophers. But we cannot stop here. Few believe in just a bare metaphysical theism; philosophical proofs are a way of supporting much richer religious traditions. Precisely because God cannot be some sort of ultimate simplicity, theists have breathing room. We can reconsider how we should express religious intuitions about an unimaginably great personal reality upon whom we are all dependent. After all, theologians often try to refine their statements about God; some philosophers criticize our idea of God only in order to redesign it.[27] A God who is not a perfect necessary being can still be incomparably great in relation to all else.

The more skeptical intellectual mood of the Enlightenment prompted many to put the idea of God back on the drawing board. For example, many now adopt a "process theology," with a God vaguely akin to a life force of the universe. Among other attractions, this tries to avoid the peculiar idea of an unchanging God. In Frank T. Miosi's words,

> In the new theology, all reality is continuously undergoing modification and change. God is also part of this, and far from being unrelated to the physical world, God is conditional and is affected by everything that happens in the world. In the new theology, God's "absoluteness" is not the traditional final, unlimited, and unchangeable absoluteness. In this system, God is absolute because the divine experiencing is composed of the totality of all experience, and when these increase, God increases. Since the main quality of existence in this theology is experiencing in continuum, God is not the Unchangeable One but the ever more Becoming, ever more Involved One.[28]

Unfortunately, for all its attempts to connect to the real world, this new God is as immersed in obscure metaphysical declarations as ever. All these doctrines seem to be earnestly making fact claims. However, when a rival metaphysician declares God is not the Becoming One but the One Unalterable in Essence though not in Person, what is an innocent bystander supposed to do? Join the fun with another string of Capitalized Words?

Maybe the problem is with the whole metaphysical approach, applied either to the classical God or to a new-and-improved model. Mary Baker Eddy, the prophet of Christian Science, once insisted her husband had been poisoned. But since there was no evidence, she declared he must have "been killed by metaphysical arsenical poisoning, which leaves no trace."[29] Likewise, traditional metaphysics too often seems to obfuscate rather than illuminate. If someone claims the moon is made of green cheese, we can discuss planetary science, bring up the Apollo landings, maybe even set her right. But if she begins to say the moon is made of perfect cheese which is self-existent in its essential substance, we have to suspect she is not making a real claim at all.

Among modern philosophers, logical positivists are famous for their desire to get rid of metaphysics. They think only those fact claims we have some way of verifying are meaningful. The "God" of the philosophers, since it is a metaphysical term with no possible empirical consequence, is meaningless.[30] In other words, positivists argue that if God leaves no trace on the world, it is no better than Eddy's metaphysical poison.

If the positivist critique of metaphysics is to be useful, however, we have to be clearer about what we mean by verification. Positivists too often put human observation at the center of their philosophy,[31] while physical science routinely claims the existence of objects which are quite unobservable in human terms. For example, we cannot see a proton. Of course, we can see the tracks a proton causes in a bubble-chamber experiment, so we still can translate into human observations what a proton does. So consider quarks. These particles make up protons, and they seem quite impossible to observe individually. The attraction between quarks *increases* with distance, so the energy we put into separating quarks from one another always creates new quarks out of the vacuum. Imagine a rubber band we can never break, because the energy we put into the elastic forms new rubber bands, leaving us with two intact rubber bands, one in each hand. We end up with more bound states of quarks—like protons—before we can isolate any single quark.[32] So there is no sense in which we observe quarks directly. But though an overly zealous positivist might call a quark a "useful fiction," physicists are quite comfortable treating quarks as real.

This sort of thing happens in physics all the time. Consider the neutrino, another elementary particle which was first noticed in some unusual particle decays. Physicists had known for some time that there were quantities such as energy and momentum which were conserved in all interactions. These conservation laws had been repeatedly confirmed by experiment, and had some appealing theoretical explanations. But suddenly there were these "weak" nuclear decays in which it looked like energy, momentum, and a few more usually conserved quantities simply vanished. Either this was an interaction which violated conservation laws, or there was a hitherto unknown particle—the neutrino—which carried off the apparently unconserved quantities. Unfortunately, these rare weak interactions also seemed to be all that neutrinos participated in. So it looked like physicists were inventing a new entity which had no function other than to preserve certain principles they were fond of.

Today, the neutrino is a well-established member of our particle zoo. This is not because we have managed to pick one up and look at it, but because neutrinos have become an integral part of our physical theories.[33] We expect, for example, that weak interactions should produce a burst of neutrinos in a supernova, and that being nearly massless, they should travel to Earth at close to the speed of light. Some of these neutrinos interact with earthly matter, however rarely. And these weak interactions produce decay products in such a way as to provide a signature of passing neutrinos. Looking at signals produced in massive underground tanks of water, we are now able to detect cosmic neutrinos.

Quarks and neutrinos are meaningful, but not because they are recipes to obtain certain direct human observations. To understand what "neutrino" means, we have to become familiar with a whole network of concepts in elementary particle physics. A neutrino makes sense because it has a role in our physical theories. And our successful theories are not metaphysical hand waving; they rely on rigorous reality checks in very precise experiments. By specifying what difference neutrinos make in the world, our physics tells us how to *observe* neutrinos.

This example gives us a way to appreciate the positivists' insight, without demanding that God should fit their narrow sense of verification. God can be elusive, as long as it is elusive like a neutrino and not like a perfect unicorn. Yes, theology produces what positivist A. J. Ayer called "deeply significant nonsense"[34] in abundance. When God makes no contact with reality, we have good reason to be unimpressed by Capitalized Words and ask what difference God makes.[35] Presumably God did not cough up a universe out of metaphysical necessity and then retreat to brood over eternal things. Nonetheless, if we find a theory to explain our world in which God takes on a central role, this will be enough to make sense of God. We need a religious picture of the world, a picture richly connected to our spiritual and everyday experiences—not just some intuitions about ultimate perfection. Even if metaphysical doctrines get us nowhere, if it is possible to weave together the concrete claims of our religious traditions into a theory where everything depends on an awesomely great personal reality, we will see how God is a real possibility.

Driving metaphysicians from the temple means taking actual religions more seriously, including their potentially embarrassing supernatural claims. This is risky. Modern, sophisticated theology avoids possible clashes with secular knowledge, preferring to lock science and God into separate boxes. Plus, it is hard to let go of the certainty a metaphysical approach promises; many feel, like Bishop John Shelby Spong, that "any God that can be killed ought to be killed."[36] But if we have learned anything about the philosophy of religion, we know this way will not work. The best that defenders of God can hope for is a dismal stalemate between skeptics and believers. In fact, the situation is probably worse. Kai Nielsen has long argued that the problems with the God of the philosophers run deeper than flawed proofs and inadequate evidence. Modern metaphysicians talk about God but fail to make claims with real content. We do not know what it means for the unsurpassably great, wholly unanthropomorphic God of sophisticated theism to exist or not. Devout philosophers still say a God exists and cares about us, but such religiously crucial fact claims have become idle words. We cannot even take a leap of faith against the evidence and just accept such a God, for we have no idea what we are supposed to be talking about.[37]

This is not to say the idea of God is hopelessly flawed. If the God of the philosophers is a confused idea, this is a problem only with a rather limited approach to God which never was at the heart of religion anyway. Our religions make plenty of substantive claims about realities beyond nature. If they succeed, we can come back later and discuss how the God we have begun to grasp might be unsurpassably great. But if we can understand our world without a divine power, God will have to go the way of unicorns and fire demons.

A RELIGIOUS THEORY

Religions claim revealed knowledge, a purposeful direction in history, a world shaped by divine creation, miraculous signs from a supernatural realm, even mystic contact with God. Many such beliefs are a disorganized mess, and a lot are plainly superstitious. However, we can work on these intuitions about gods and demons, and about what the world should be like if it were created and ruled by a God. We can start by connecting these ideas to each other, and asking what difference a God would make in the world.

Even metaphysical doctrines of God will be useful for a religious theory, as long as we do not get too ambitious and try to prove God on the cheap. We still want God to be "necessary" in some sense, even though logical or causal necessity leads us into a swamp. Perhaps the notion of necessity tries to capture the intuition that everything *depends* on God.[38] A bed of flowers, for example, may depend on a gardener for its existence. The gardener planted the flower bed for her own purposes, arranged it, and maintains it as long as it exists. There would be no flower bed without the gardener, yet the gardener does not depend on the flowers, however much she may like them. The flowers are not bound to the gardener out of some deep metaphysical necessity, but they happen to, as a matter of fact, depend on her. Our world may have a similar relationship with God: everything might, as a matter of fact, depend on a divine power who created and who sustains the world.

This is still a bit too ethereal, but it is more promising. We can at least picture a cosmic theory which puts a God in a uniquely self-sufficient position while everything else depends on this God. Moreover, there is a traditional proof of God which tells us how this kind of God makes a difference. According to the argument from design, we can infer that there is a God from the intricate functional order we see in the universe. Imagine we stumble upon a flower bed. Observing the orderly way the flowers are laid out, how their colors blend into a harmonious scheme, and how the soil is suited for their growth, we realize there must be a gardener. Without a guiding intelligence, we would expect a disordered mess, not orderly rows and unblemished colors. The same argument, many scientists and philosophers have thought, applies to the universe. We behold a world which sustains life, in which everything from atoms to plants and animals fits together in harmony. These are signs of intelligent design. Our universe could have been an amorphous mess, without life, intelligence, or any of the complex interlocking structures which excite our sense of wonder. Intuitively, many of us think a purposeless, accidental world would look like a featureless primal chaos. The complex order in our world, then, must trace to a guiding intelligence who made the world for a purpose.

If we go no further than that intuition, design becomes yet another metaphysical argument: complex order requires a designer, therefore there is a God. Skeptical philosophers eat this sort of argument alive.[39] If, for example, the design argument is an attempt to explain the universe by drawing an analogy to how humans create complex order, this is the most tenuous of analogies. We only distinguish between artifacts and natural objects by using an immense amount of background knowledge about humans and human creations, the sort of knowledge we have no clue about when it comes to universes. The analogy has other problems as well. Humans are more complex than their creations, so our analogy suggests a complex God. As biologist Richard Dawkins says, "If we want to postulate a deity capable of engineering all the organized complexity in the world, either instantaneously or by guiding evolution, that deity must already have been vastly complex in the first place."[40] So who designed God, an even more complex designer?

If we rein in our desire for certainty, however, we can see design as a possible explanation for our world. We can take a top-down view, where complex functional order comes from intelligent design, and where everything ultimately derives from a creative mind so simple it has no limits. This appeals not only to intuitions that order is a sign of purpose, but also to the common belief that our minds are nonmaterial creative agencies. In contrast, naturalism inspires a bottom-up approach, where we try to find impersonal, material explanations even for intelligence. These are substantially different pictures of the world, and we can fruitfully argue about their strengths and weaknesses.

To flesh out a top-down view, we can start making connections to commonsense convictions and even natural science. In everyday experience, nonliving matter does not spontaneously organize into anything interesting. In fact, left to themselves, things get more disorganized. The books and papers on our desks will, through the haphazard interactions of everyday life, get strewn about to create a mess. But our clutter will never sort itself into neat piles. It seems that mindless, impersonal processes can only maintain order or create chaos; only living, intelligent beings act on dead matter to order it. When we see a neat desk, we infer that an intelligence cleaned it up. We can now enlist science in support of these commonsense observations. After all, the famous second law of thermodynamics tells us physical systems tend toward disorder. In that case, it seems the order we see can only have degenerated from a more perfect past state. As creationist Henry M. Morris puts it,

> The Second Law (the law of decreasing available energy, as the universe heads downward toward an eventual "heat death," with the sun and stars all burned out) tells us that there must have been a primeval creation, or else the universe would already be "dead"! The First Law (law of energy conservation) tells us that no energy is now being created, so the universe could not have created itself. The only scientific conclusion is that *"in the beginning God created the heaven and the earth"* (Gen. 1:1).[41]

Of course, this a crude example. Serious arguments for a top-down view of the world no longer get entangled with creationism; they instead suggest there is a progressive purpose driving biological evolution, or that modern physical cosmology shows us there must be a creator.[42] But Morris's basic approach of presenting a religious picture of the world and trying to support it with science is correct. There is no doubt the creationists' God makes a difference. And even sophisticated design arguments will express similar intuitions. For example, thermodynamics does not prevent refrigerators from existing—they can locally reduce entropy as long as they also increase global disorder. But there is still the puzzle of how, in a world which tends toward disorder, a complex device like a refrigerator can come to exist in the first place. In a presumably closed material universe behaving according to the laws of thermodynamics, it is surprising to find matter clumped up in galaxies, not to mention living, thinking structures. However weak the analogy to the engineers responsible for the refrigerator, an intelligence intervening from outside the natural world is at least a possible explanation. Design arguments will always look for a kind of order where impersonal explanations fall short, and try to account for it by showing it fulfills a specific purpose. Defending a bottom-up view against this approach requires not a search for conceptual flaws in design arguments, but doing the physics to show how order comes out of chaos.[43]

Design is not enough. Even if we find that our world was designed, this is not much to build a religion upon. The Designer might be something quite different than the traditional God. And if design fails, this would be a strangely swift defeat for religion. So a good religious theory might start with design, but it must enrich it with other supernatural claims. After all, if ours is a top-down, mind-first reality, there should exist other effects besides the occasional sign of purposeful order. Perhaps we catch glimpses of a transcendent reality in the extraordinary experiences of saints and mystics. Or perhaps, as Muslim apologists say, we are guided by "instinct, intuitions, inspiration, and prophetic dreams"—even miracles which hint at the power and glory of their source. There is more: "The highest form of guidance has, however, been revealed in a much clearer and more categorical way: through the Word of God revealed to His prophets."[44] The traditional supernatural claims of our religions all testify to a personal reality transcending material existence.

If we can tie together multiple, independent signs of a supernatural reality, we will have a robust religious theory. To the extent we succeed, "God" will have a secure meaning as a central entity in this theory. Consider, once again, the neutrino. This is a solid concept because it helps explain many different experiments, and it is connected to a rich network of concepts in particle physics. Any one of the individual strands of support for a neutrino may be weak, but ripping "neutrino" out of this network would disrupt the whole too badly. We need a similar network, composed of various ideas about design, scriptural revelation, or mystic contact, for a respectable idea of God.

For example, beliefs about angels can form one of the strands of a network of religious concepts. Most religions claim there are spirits who may be higher than humans but are not quite gods. The Catholic cosmos still includes such "spiritual, noncorporeal beings" as angels and devils,[45] and believers from all Abrahamic religious traditions still report encounters with such spirits. Some New Agers make contact with beings who have reached a higher spiritual level and lend a helping hand to humans progressing toward enlightenment. It is often hard to take such beliefs seriously—they have a taint of popular superstition dripping with sentimentality. But *if* there were such beings, God also would become more credible. Angels fit a theistic picture of the world very well, while naturalists do not expect to find spirits of any sort. Such beings could at least establish that there is a spiritual reality distinct from the material world. God, as a pure spirit unlimited by a material body, would become easier to understand if we had examples of immaterial spirits.

Of course, God does not stand or fall with angels or beings from higher spiritual planes. If we discovered disembodied spirits exist, this would be only one strand of evidence supporting God. After all, it is possible that a world where spirits flit around in the night could have no God presiding over it. And if angels are mere superstition, there may yet be a God. There are many different ways to construct a top-down theory of the world, and many ways to fit God into such a picture. In discussing such theories, we have to consider many individually nondecisive arguments, and see how they add up—we cannot rely on philosophically sexy attempts at proof or disproof.

Examples like angels might seem unfair; while popular religion is full of concrete claims, these should not obscure the considerably more developed intellectual traditions which theology draws upon. True enough. But even angels are traditionally more religiously serious beings than the sentimental fluff appearing in stories of guardian angels. Furthermore, as with the design argument, modern defenders of God can refine traditional supernatural claims to produce rather sophisticated ideas. Sociologist Peter Berger, for example, thinks our normal-seeming reality is "haunted by that otherness which lurks behind the fragile structures of everyday life."[46] Through the cracks in the mundane order of the world, we catch glimpses of the spiritual reality behind it all. This is no less a top-down view of our world than popular supernaturalism, and it is also something we can seriously argue about. After all, any religious theory must capture the intuition that there is something beyond and superior to the natural order. There must be something more than an impersonal physical universe and our everyday social life. Reality must be deeply personal; all our explanations of the world must lead back to the immaterial creative mind who is God. Even those philosophers who still favor the traditional proofs do so in order to defend a view of the world in which there is something almost magical about our minds, where we ultimately explain events by purpose and personal choice, and where we are not accidents coughed up by an indifferent world, but souls with a role in a divine plan.[47]

No matter how armchair proofs of God fare, religions paint a different picture of our world than what skeptics claim it to be. They say the harmony of life testifies to a God, the universe is ultimately explained by divine creation, and a salvation history took place as recorded in scriptures. There is magic in the world, sustained by a mysterious but personal transcendent reality. Our moral convictions require a divine source and assurance. These are serious claims; they do not come only in the shape of fundamentalist superstitions. Furthermore, they are vital for theistic religion. For what becomes of God if there is no revelation and the heavens do not declare a creator's glory? Even if we could be persuaded that metaphysical arguments give us a supreme being, how do we then worship this abstraction?

THE GREAT PROGRAMMER

The claims religions craft, like the existence of design in the world or a supernatural realm, should give us a good start on making sense of God.

This does not, however, mean we can just get rid of the God of the philosophers and return to the basics. Religion aims for more than theoretical truth—providing emotional satisfaction, moral vision, and so on—and this means our doctrines of God usually tolerate a good measure of paradox. So while traditional religion makes real claims, it includes a fair share of nonsense; there is a good reason philosophers have worked so hard to clear away some of the conceptual cobwebs from religion. Any sort of theism will face conceptual challenges like reconciling our supposed free will with an omnipotent God who determines everything, or explaining how an all-good God created a world with so much evil. We have to be vigilant about keeping things straight.

It would help if we had some overall framework to help us build a decent religious theory, even if we later discard much of the scaffolding. Our framework should help make sense of divine attributes and provide a stage for divine acts like creation and revelation. At the same time, we should retain a sense that we are struggling to talk about a fundamentally mysterious reality. Traditionally, the way to go about such a task is to describe God metaphorically and analogically.[48] All descriptions of God are partly inadequate, but some are useful nonetheless.

So we need a metaphor, even something like the "toy model" a physicist uses to start attacking a complicated problem—not a full-fledged theory, but rather a tool for exploration. In our technological society, a computer analogy will come in handy. Scriptures always use the language of their time to tell us something of their mysterious and awesome deity; something we are familiar with may work for us. This does not mean replacing the Bible-God with a cyber-God for our time, but using a different set of metaphors to clarify our idea of divinity.

In times past, God used to be likened to the Great Architect, or the Watchmaker responsible for the intricate design in nature. Catching up to the computer age, let us try the Great Programmer. Computer programs can be quite complicated, including simulations of everything from arrays of interacting atoms to living things. Now, thinking more like a science-fiction writer than a metaphysician, we can imagine ourselves in an analogous situation. *We* may be creatures designed by the Programmer. Perhaps the universe is actually something like an enormous computer program, maybe even built to answer some cosmic question.[49]

At first it is hard to take such an analogy too seriously. Stereotypes being what they are, the first image we probably have is God as a deranged hacker. But in fact, the Programmer metaphor is much more powerful than the Architect or Watchmaker, because *computation* is such a mathematically fundamental concept. Our physics, for example, can be described in computational terms, as a "digital mechanics."[50] Especially if design is to figure prominently in our framework, programming will be an excellent metaphor. Indeed, programming has been used as an analogy in design arguments since the very beginning of computers.[51]

Now, as with any analogy, we should not be too literal-minded. God does not type away at a console. The point is that the relationship between God and the universe can be conceived of similarly to that between a programmer and her program. This will help us understand something about the traditional divine attributes, the relation of the Creator to the created, and how mere finite mortals can get to know something about the God behind it all.

First of all, it is very easy to see how the Programmer is completely independent of the program. When we program a computer, *we* determine what happens. Our creation does not constrain us; we respond to what it produces as *we* choose. If our world is analogous to a program, we and everything around us depend upon the Programmer for our existence. God the Programmer, on the other hand, remains independent, creating the world out of free choice.

The program, of course, has rules of operation determining how it runs. These are its "laws of physics," or the instructions on how to transform one internal state to another throughout the course of the program. If we are simulating a cluster of atoms, for example, we may design things so that they behave according to Newtonian mechanics. But there is also the underlying reality of the computer; what we see as atoms bumping into each other on our screen happens because of patterns of electrical activity. In our world, then, the physics we see might only be a secondary reality. The Programmer is in control of the laws of physics, down to the last detail of the program's internal evolution.

When we simulate atoms, "time" is also under our control. Physicists usually solve the equations of motion for small time steps, and iterate the process to see what happens. "Time" for the simulated system is different than the time the program takes to run in the external world of the computer. We may even want to "run time backwards" every now and then, or play with the internal program-time in any way we see fit. So it becomes very natural to think of the Programmer as standing eternally outside our stream of time.[52]

If the Great Programmer is outside our time, we can also understand the act of creation better. Our simulation of atoms may start from a clearly special initial state, or it may be possible to extrapolate internal program-time backward without limit. But in either case, the simulation needs a programmer to bring it into being. This creation does not take place within program-time; there are no previous states meaningful in terms of the program's own operation except perhaps a "chaos" of haphazard bit-patterns. By analogy, the Programmer creates our world out of nothing, acting from outside of time; whether our universe is eternal or bounded in our physical time does not matter.

We program computers with a purpose in mind, and if we go to the trouble of ensuring that complex objects appear within our program, our purpose usually has to do with their behavior. Artificial life researchers, for example, try to create the sort of intricate functional order seen in living things on their computer screens.[53] The Great Programmer, then, must also have had a purpose in creating the world, though we may have some difficulty fathoming it. But this purpose is likely to involve the sentient beings inhabiting the creation. Perhaps the world is a dramatic stage on which they act out their moral choices. Perhaps the religious quests of the creatures are the Programmer's way of reflecting on itself. There are many possibilities. Still, it becomes easy to see that sentient creatures might be involved in the grand design.

So far, the Programmer metaphor is useful in understanding how God is supposed to be so fundamental to the world, and yet so removed and different from it. It can also help us understand how the Programmer has all sorts of omni-attributes *relative to the program*.[54] That is, our Programmer can know everything involving the program, be able to do anything with it, and so forth. If we act like overly ambitious metaphysicians and talk about the most perfect being conceivable, we will run into the usual difficulties—from the old "can God create a stone God cannot lift?" to a host of knottier problems. But with a programming analogy we at least have a God with no limits regarding what it can do with its creation, which should be good enough anyway.

With enough resources, we can answer almost any question regarding the programs we write. Since we can play with program-time as we please, we can make extra runs to answer questions regarding past or future states of the program, and then intervene in the present program-time accordingly. The Great Programmer should also be omniscient, able to answer any question about even the most minute detail of the world.

Omnipotence with respect to the program is also easy to understand. We control our creations, within the limits imposed by our ability and the computer's capacity. The Great Programmer would be able to do most anything with the program.

An all-powerful programmer could easily intervene in the normal course of operations as determined by the program's usual instructions. God may want to provide miracles and revelations in order to disclose important facts to sentient creatures. Furthermore, the Programmer's interaction with the world is by means entirely separate from ordinary physics. When we alter the course of a program, say by directly manipulating the contents of the computer's memory, what happens is entirely unexpected from the point of view of the program's internal operations. Skeptical philosophers often criticize a supernatural God, pointing out that it is difficult to imagine how something entirely beyond nature, not directly interacting with the material world, can still acquire information from the world and act on it. But the Programmer analogy should convince us it can be done.[55]

A stickier question is how to end up with a God who promotes human happiness. Our world seems supplied with a superabundance of nastiness, which causes a great deal of difficulty for theology. God is supposed to be "omnibenevolent," but what this means gets very murky. Even so, we can easily discuss the excuses theologians make for God within our analogy: that we are the creator's creatures to be disposed of any way that pleases the Programmer; that suffering prepares our souls for fellowship with the Programmer in the next life; and so on. In any case, it is plausible to think that the Programmer prefers its creatures to behave in certain ways, and that they can either rebel or conform to the Programmer's will.

The sentient creatures within the program can have souls. The Programmer may choose to preserve the information about their personality and memories after their deaths, to be reactivated in an appropriate heaven or hell—perhaps after the original program is terminated in an apocalyptic fashion. These ensouled creatures may even have been created in the image of the Programmer—sharing in consciousness, personality, ability to make moral choices, and so on—but truly be only an image, not participating in the deeper external reality in the way the Programmer does.

We could continue; almost all traditional ideas about God can be interpreted within the Programmer picture. The Great Programmer is only an analogy, a toy model, but it at least indicates that we can flesh out an intelligible idea of God without resorting to metaphysical obscurities. It gives us a way to think about concepts like omnipotence without getting bogged down in impossibilities. Of course, this is only a beginning—many an attractive proposal is still wildly false. Nothing here absolves us from seeking good reasons to say a God exists. We cannot declare God to be an invisible programmer, a redundant entity which makes no difference in the world. If our analogy blossoms into a cosmic conspiracy theory with a Great Puppeteer undetectably pulling the strings of every atom, we are no better off than with the terminally obscure Gods metaphysicians keep dreaming up.

The Great Programmer would probably be rather elusive, but we can think of ways to look for relevant evidence. We can seek ways in which the natural order is disrupted, indicating a reality beyond the material world. We must need a supreme personality beyond nature in order to explain the world. With a framework on which to hang intuitions about a divine reality, we can now seek specific patterns of divine action, and critically revise our metaphors in the process. The Great Programmer will come into its own if it anchors a supernatural theory of our world. *If* this meets with success, metaphysicians can work on tidying things up and keeping God's superlatives consistent.

Is the Programmer a metaphor with which an orthodox theist might become comfortable? Probably—keeping in mind the limitations of the analogy. If we demand that God be exactly similar to a human programmer, we will begin wondering what sort of superphysical medium our deity works with, or what society of gods the Programmer moves around in. For a satisfying theism, God must stand alone. Perhaps we can stretch our analogy so as to fold God and its "computer" together, picturing a universe which is part of a God who includes but is larger than its creation. But even though there is room to play around, every analogy has its breaking point. If God is to remain somewhat mysterious without being terminally obscure, this is only to be expected. The perfect metaphor does not exist; the Great Programmer is helpful, but that is all.

Our toy model breaks when we stretch it, and it is too tidy and simplistic. The tale of a personal God acting in human history will be fuzzier. But the model does its work, showing us that we can do better than hand waving when talking about a religious theory. If we cannot at least gather the typical supernatural claims religions make together under a programmer-like framework, it is hard to see how a God makes sense at all. Certainly, God cannot be a cosmic hacker; certainly, God must be larger than this metaphor if it exists. But we must have a handle on divine reality somehow, otherwise "God" is no more than a word invested with feelings of deep significance.

IS GOD A PHILOSOPHICAL PROBLEM?

God makes sense when woven into a network of supernatural claims—a religious theory describing a fundamentally spiritual world. Such a God is not a conceptual necessity conjured up by philosophers, but a possibility to be examined in the light of the best of our knowledge.

If this is correct, the way we typically argue about God must change. We usually think of the existence of a God as a philosophical problem; indeed, as one of *the* problems of philosophy. From everyday armchair theorizing to the kind of detailed analysis occupying the pages of academic journals, our debates over God focus on conceptual puzzles and grand metaphysical schemes. But after millennia of philosophy, neither the questions nor the basic answers regarding divinity have changed much. We now have a clearer understanding of the issues, and we have undoubtedly learned a good deal from the peripheral questions raised during the quest for God. Yet there is also an air of sterility to the whole debate. We too often seem to be spinning our philosophical wheels. The problem is not God alone; philosophy often has difficulty producing cognitive progress unless it helps launch a new science. Perhaps, as Henry Kyburg says, one reason for this lack of progress is that philosophy deals with "the (remaining) *hard* problems."[56] But continual floundering is also a sign of stagnation; it is at least reason to explore other approaches.

So we need a different style of debate. Religious believers claim God is a fact, and with modern science, we have learned a few things about how to sort out the likely from the improbable among fact claims. To judge whether there is a God, we should be asking concrete questions inspired by the intuitions about a supernatural realm expressed by our religions. We should be discussing analogies like the Great Programmer, not wrangling over lists of infinite attributes.

Contemporary philosophers themselves often express skepticism about metaphysical thought, even about the mainstream philosophical tradition as a whole. Some even suspect that the deep significance we attach to traditional philosophy is mainly a long-standing prejudice of our intellectual culture. Taken too far, this suspicion threatens to reduce philosophy to a conceptual therapy of sorts, making us lose sight of some real and interesting questions. And suspicion of philosophy can too easily turn into a resistance to all criticism. There are already too many theologies with no reality checks, too many academics in religious studies who insist religion is a unique phenomenon which can be approached only on its own terms.[57] Even so, philosophy has become too confined to its academic niche, isolated from science, obsessed with language. There is room to think that other disciplines do better muddling through with questions of fact, that philosophy is in serious need of reconstruction.[58]

Of course, such skepticism can also work against science. If we quit talking about metaphysical certainties, what deep reason is there to believe our sciences? Why should we trust modern physics and say the earth is billions of years old, and ignore stories that it was created in six days?

There is *some* merit to such worries. After all, what we consider worthwhile evidence is not entirely independent of the theories we already trust. It is tempting to give up and say we just have to take God or a godless world as a fundamental metaphysical commitment. However, not all theories are self-contained, and we do not walk about inhabiting hermetically sealed worldviews. We do have reality checks, and while we cannot do without some kind of theoretical framework as a *starting point*, we are also able to consider rival claims. In our history, a six-day creation was long our starting point. Modern science has been an excellent way to break out of our cultural assumptions and learn more. We have made progress on many problems which used to produce philosophical stalemates.

With religious questions as well, the best way to make progress is to involve our science, our history—the best of our secular knowledge.

Serious religious belief and unbelief are both handicapped by their over-reliance on conceptual arguments. Too many of us profess to believe science and religion are separate domains; too often we shudder to think they may have something to say about each other. Liberal theists say the fact of biological evolution has no significance for religion. Atheists say "Science can never lead us to God. It can't even try,"[59] and they concentrate on refuting the classical proofs. Neither are taking their fact claims seriously enough.

There are good practical reasons behind the way we shy away from letting religious and secular knowledge claims touch one another. We now live in a religiously diverse social world, where no *one* doctrine reigns supreme. Our religions can no longer be sustained as common social beliefs;[60] but we can stuff our convictions in separate compartments and put up protective barriers of faith around God. However practical this arrangement may be, it does not help us take God seriously. As Charles P. Henderson points out while arguing that modern science reveals the glory of God,

> To defend religion by carving up reality so as to leave certain realms to science and certain realms to religion is a well intended maneuver that in one stroke eliminates the possibility that God can be what God must be; namely, the organizing principle of reality itself.[61]

Theists want to affirm that principle; atheists deny it. We have enough of an idea of how to decide between rival fact claims to at least make a good beginning.

Our debates over God should rely on our natural and human sciences—without apologies. This is not to deny that a debate over religious fact claims will have its own distinct character. God is not a problem in physics or history—the questions we need to address cut broadly across all our knowledge. They call for a wide perspective as well as specialized expertise. Theology was once called Queen of the Sciences, providing a unifying framework for all intellectual disciplines. It can no longer serve the same way—its subject matter is either too dubious or too specialized on particular scriptures or traditions. Philosophers, however, can still take on such a broadly integrative role. They just need to resist the temptation to get something for nothing.

NOTES

1. Gerald Bray, *The Doctrine of God* (Downers Grove, Ill.: InterVarsity, 1993), p. 82. See also pp. 213–14.
2. Ibid., p. 95.

3. *Catechism of the Catholic Church* (English translation, Boston: Libreria Editrice Vaticana, 1994), ¶31.

4. See Henry Margenau and Roy Abraham Varghese, eds., *Cosmos, Bios, Theos: Scientists Reflect on Science, God, and the Origins of the Universe, Life, and* Homo Sapiens (La Salle, Ill.: Open Court, 1992). This does not, however, mean devout scientists depend on such arguments to fit science in their religious life; see John Marks Templeton and Kenneth Seeman Giniger, eds., *Spiritual Evolution: Scientists Discuss Their Beliefs* (Philadelphia: Templeton/Giniger, 1998).

5. James Thrower, *Western Atheism: A Short History* (Amherst, N.Y.: Prometheus Books, 2000).

6. Bray, *The Doctrine of God*, p. 110.

7. See Anselm's *Proslogion*, translated in *The Existence of God*, ed. John Hick (New York: Macmillan, 1964), pp. 25–27. Anselm makes some dubious metaphysical assumptions. Also see Peter A. Angeles, *The Problem of God: A Short Introduction* (Amherst, N.Y.: Prometheus Books, 1980), pp. 11–13, but the argument can be restated without these.

8. J. L. Mackie, *The Miracle of Theism: Arguments For and Against the Existence of God* (Oxford: Oxford University Press, 1982), pp. 41–49.

9. Alvin Plantinga, *God, Freedom, and Evil* (New York: Harper & Row, 1974), pp. 104ff; Charles Hartshorne, *The Logic of Perfection* (La Salle, Ill.: Open Court, 1962), p. 55.

10. Michael Martin, *Atheism: A Philosophical Justification* (Philadelphia: Temple University Press, 1990), pp. 89–90.

11. Richard M. Gale, *On the Nature and Existence of God* (Cambridge: Cambridge University Press, 1991), pp. 224–37. (I1) and (I2) are paraphrased from Gale's propositions 76_3 and 87.

12. Plantinga, *God, Freedom, and Evil*, pp. 90–91.

13. Martin, *Atheism*, chap. 12.

14. Willard Van Orman Quine, *Set Theory and Its Logic*, rev. ed. (Cambridge: Harvard University Press, 1969). A maximal number arises because the ultimate class is defined so as to prevent its being a member of any set, and have the same cardinality as its power set.

15. Richard Swinburne, *Is There a God?* (Oxford: Oxford University Press, 1996), p. 17.

16. E.g., Clement Dore, *On the Existence and Relevance of God* (New York: St. Martin's Press, 1996), chap. 1.

17. Kai Nielsen, "Against Ethical Rationalism," in *Gewirth's Ethical Rationalism: Critical Essays with a Reply by Alan Gewirth*, ed. Edward Regis Jr. (Chicago: University of Chicago Press, 1984), p. 59.

18. Antony Flew, *God, Freedom, and Immortality: A Critical Analysis* (Amherst, N.Y.: Prometheus Books, 1984), p. 52.

19. Martin, *Atheism*, pp. 96–106, 118–24.

20. Vic Barnett, *Comparative Statistical Inference*, 2d ed. (New York: John Wiley & Sons, 1982), p. 192; Colin Howson and Peter Urbach, *Scientific Reasoning: The Bayesian Approach* (La Salle, Ill.: Open Court, 1989).

21. Swinburne, *Is There a God?* p. 43. My Bayesian description of a cosmological argument is based on Swinburne's *The Existence of God* (Oxford:

Clarendon, 1979), chap. 7. He argues nontheistic hypotheses have a very low prior probability, pp. 287–88. For critiques, see Martin, *Atheism*, pp. 106–18; Mackie, *The Miracle of Theism,* pp. 95–101; Keith M. Parsons, *God and the Burden of Proof* (Amherst, N.Y.: Prometheus Books, 1989), chap. 2.

22. Elliott Sober, "Let's Razor Ockham's Razor," in *From a Biological Point of View: Essays in Evolutionary Philosophy* (Cambridge: Cambridge University Press, 1994). In particular, "objective" Bayesian statistical inference, which determines prior probabilities based on simplicity or maximizing entropy, does not work without a background theory to fix our choice of statistical model. I see no way of avoiding arbitrarinesss except by using real-world information—*sampling* the world.

23. Human brains compromise in this way too. Our memories are not static tape recordings extractable by hypnosis or other techniques, but active reconstructions. Our reasoning is a grab bag of heuristic procedures, often violating the rules of probabilistic inference. See Thomas Gilovich, *How We Know What Isn't So: The Fallibility of Human Reason in Everyday Life* (New York: Free Press, 1991). Even if a Bayesian account of inference is correct, this is only in an ideal sense; Howson and Urbach, *Scientific Reasoning*, pp. 292–95.

24. J. N. Findlay, "Can God's Existence Be Disproved?" in *New Essays in Philosophical Theology*, ed. Antony Flew and Alasdair MacIntyre (London: Macmillan, 1955). Against such views, see Richard H. Jones, "The Religious Irrelevance of the Ontological Argument," reprinted in Richard H. Jones, *Mysticism Examined: Philosophical Inquiries into Mysticism* (Albany: State University of New York Press, 1993).

25. Anthony Kenny, *The God of the Philosophers* (Oxford: Clarendon Press, 1979), chap. 7.

26. Charles Bradlaugh, "A Plea for Atheism," reprinted in Gordon Stein, ed., *An Anthology of Atheism and Rationalism* (Amherst, N.Y.: Prometheus Books, 1980), p. 10.

27. E.g., Introduction to Gale, *On the Nature and Existence of God*.

28. Frank T. Miosi, in Paul Kurtz, L. Fragell, and R. Tielman, eds., *Building a World Community: Humanism in the Twenty-first Century* (Amherst, N.Y.: Prometheus Books, 1989), p. 266.

29. Julius Silberger Jr., *Mary Baker Eddy: An Interpretive Biography of the Founder of Christian Science* (Boston: Little, Brown, 1980), p. 143.

30. A. J. Ayer, *Language, Truth and Logic*, 2d ed. (New York: Dover, 1952), pp. 114–20.

31. This is particularly clear in Machian positivism; see Gerald Holton, *Science and Anti-science* (Cambridge: Harvard University Press, 1993), chaps. 1–2.

32. Elliot Leader and Enrico Predazzi, *An Introduction to Gauge Theories and Modern Particle Physics* (New York: Cambridge University Press, 1996), vol. 2, chap. 21.

33. Ibid., vol. 1.

34. A. J. Ayer, "The Claims of Philosophy," reprinted in A. J. Ayer, *The Meaning of Life* (New York: Charles Scribner's Sons, 1990).

35. Kai Nielsen, *God, Scepticism and Modernity* (Ottawa: University of Ottawa Press, 1989); Flew, *God, Freedom, and Immortality,* chap. 6; Kenneth H.

Klein, *Positivism and Christianity: A Study of Theism and Verifiability* (The Hague: Martinus Nijhoff, 1974).

36. John Shelby Spong, *Resurrection: Myth or Reality?* (San Francisco: Harper, 1994), p. 22. Such sentiments, often dressed in verbiage inspired by Paul Tillich, seem to attract liberal churchmen concerned the modern world is passing their religion by. See John Shelby Spong, *Why Christianity Must Change or Die: A Bishop Speaks to Believers in Exile* (San Francisco: Harper, 1998); John A. T. Robinson, *Honest to God* (Philadelphia: Westminister, 1963). If any God deserves oblivion, it is this one.

37. Kai Nielsen, *Philosophy and Atheism: In Defense of Atheism* (Amherst, N.Y.: Prometheus Books, 1985), chap. 5.

38. Nielsen, *God, Scepticism and Modernity,* pp. 69–77, argues this sense of "necessity" does not produce a real fact claim, but I think it is useful as a starting point.

39. Martin, *Atheism*, chap. 5; the difficulties are not limited to analogical design arguments.

40. Richard Dawkins, *The Blind Watchmaker: Why the Evidence of Evolution Reveals a Universe Without Design* (New York: W. W. Norton, 1987), p. 316.

41. Henry M. Morris, "The World and the Word," *Back to Genesis* 93 (1996).

42. In practice, these arguments tend to have serious conceptual defects; see "The Pseudo-Problem of Creation in Physical Cosmology," *Philosophy of Science* 56 (1989): 373; Adolf Grunbaum, "Pseudo-Creation of the Big Bang," *Nature* 344 (1990): 821.

43. Taner Edis, "Taking Creationism Seriously," *Skeptic* 6, no. 2 (1998): 56.

44. Thomas Ballantine Irving, Khurshid Ahmad, and Muhammad Manazir Ahsan, *The Qur'an: Basic Teachings* (London: Islamic Foundation, 1979), p. 28.

45. *Catechism of the Catholic Church*, ¶¶328–36; Peter Kreeft, *Angels and Demons: What Do We Really Know About Them?* (San Francisco: Ignatius, 1995). Many theologians remain interested in supernatural entities other than God; Anthony N. S. Lane, ed., *The Unseen World: Christian Reflections on Angels, Demons, and the Heavenly Realm* (Grand Rapids, Mich.: Baker, 1996).

46. Peter L. Berger, *A Far Glory: The Quest for Faith in an Age of Credulity* (New York: Free Press, 1992), p. 145.

47. E.g., Richard Swinburne's discussion in *Is There a God?* is pervaded by assumptions about special qualities of mind. However, this does *not* mean theism requires a stark supernaturalism exemplified in a crude mind-body dualism. Even conservative believers can be more sophisticated; e.g., William W. Stevens, *Doctrines of the Christian Religion* (Nashville, Tenn.: Broadman, 1967), pp. 125–40. On the more occult side, see David Ray Griffin, *Parapsychology, Philosophy, and Spirituality: A Postmodern Exploration* (Albany: State University of New York Press, 1997), who defends a "naturalistic theism" which smoothly connects supernatural and material reality. Calling this "naturalistic" is highly misleading, however, as it is very much a top-down picture of the world.

48. Without metaphor and analogy, revelation of a fundamentally mysterious God is impossible. See Bernard Ramm, *Special Revelation and the Word of God* (Grand Rapids, Mich.: William B. Eerdmans, 1961), pp. 40–43.

49. E.g., Stanislaw Lem, *Memoirs of a Space Traveler* (San Diego: Harcourt Brace Jovanovich, 1983), pp. 35–51. Computer scientist Edward Fredkin speculates on this theme in Robert Wright, *Three Scientists and Their Gods* (New York: Times Books, 1985). Douglas Adams fans, of course, know that the answer is "42."

50. Edward Fredkin, "Digital Mechanics," *Physica D* 45 (1990): 254.

51. Even Charles Babbage did so, much before a working computer became possible. See Geoff Simons, *Is God a Programmer? Religion in the Computer Age* (Brighton, UK: Harvester Press, 1988), pp. 74–78. Simons develops the Programmer analogy further in his book.

52. In other words, an Augustinian conception of God and time. Bertrand Russell, *History of Western Philosophy* (London: George Allen & Unwin, 1961), pp. 351–53.

53. Claus Emmeche, *The Garden in the Machine: The Emerging Science of Artificial Life* (Princeton: Princeton University Press, 1994). Of course, artificial life is a process of blundering our way through, precisely because the end results are too complicated for a conventional engineering design approach. So the artificial life analogy would undermine the notion of intelligent design if carried too far.

54. Omni-attributes are sometimes called "relative attributes," in the sense of "the attributes of God in relation to . . . creation." Stevens, *Doctrines of the Christian Religion*, p. 47.

55. Paul Edwards, *Reincarnation: A Critical Examination* (Amherst, N.Y.: Prometheus Books, 1996), pp. 301–303, poses the "modus operandi" problem: how does a "pure mind" interfere? The Great Programmer analogy suggests God need not be a pure mind (whatever that means), while remaining above the natural order.

56. Henry E. Kyburg Jr., *Science & Reason* (New York: Oxford University Press, 1990), p. 13. This is a somewhat peculiar statement, considering that there is no lack of obnoxiously difficult problems in our sciences.

57. As a discipline, religious studies is too often concerned with liberal apologetics rather than explanation, guided by an attempt to solve the problem of the plurality of religions by melting them all into a mystical transcendent irreducible to material and social realities. Donald Wiebe, *The Politics of Religious Studies* (New York: St. Martin's Press, 1999); Timothy Fitzgerald, *The Ideology of Religious Studies* (New York: Oxford University Press, 2000); Russell T. McCutcheon, *Manufacturing Religion: The Discourse on Sui Generis Religion and the Politics of Nostalgia* (New York: Oxford University Press, 1997).

58. Mario Bunge, *Philosophy in Crisis: The Need for Reconstruction* (Amherst, N.Y.: Prometheus Books, 2001); Kai Nielsen, *After the Demise of the Tradition: Rorty, Critical Theory, and the Fate of Philosophy* (Boulder, Colo.: Westview, 1991).

59. Gordon Stein, "Why Scientists Should Study Philosophy," *Free Inquiry* 15, no. 1 (1994): 61.

60. Peter L. Berger, *The Heretical Imperative* (New York: Anchor, 1980).

61. Charles P. Henderson Jr., *God and Science: The Death and Rebirth of Theism* (Atlanta: John Knox, 1986), p. 16.

Chapter Two
Let There Be Life

What a book a devil's chaplain might write on the clumsy, wasteful, blundering, low, and horribly cruel works of nature!
—Charles Darwin, in a letter to Joseph D. Hooker (1856)

Religious conservatives suspect the theory of evolution chases God out of our world, reduces us to material beings, and turns us into accidents. They are correct. If life was specially designed, or if evolution were about progress, we could imagine a God responsible for it all. But with evolutionary biology, the creativity in nature is no longer the product of a supernatural mind but the result of blind, purposeless trial and error. Evolution accounts for life as we see it, warts and all—without a designer.

SPECIALLY CREATED

Traditionally, God ruled over an earth-centered universe devised with humans in mind. The glory of the heavens and the wondrous harmony of nature declared the majesty of their creator. Everything, from the stars above to the species populating the earth, was specially created to serve a divine purpose.

Nowadays we live on an obscure planet in a mediocre galaxy, and even the history of life appears to be a product of natural processes. A world where the signs of divine creation are everywhere has become difficult to imagine. Of course, we might run into scriptural literalists who insist the earth and its life-forms were made but a few thousand years ago, and in the United States, such creationism remains a major headache for science educators. Our intellectual culture also harbors some smoldering antievolution sentiment; every year a scholar or two can be counted on to attack Darwin or modern evolutionary theory. Even so, the whole dispute seems rather old; a revisiting of long-settled matters.

Now that creationism is relegated to the intellectual fringe, and that liberal religious people accept evolution, it would seem the Darwin wars are not relevant to today's questions about science and religion.

This impression, however, can be misleading. Special creation was once an intellectually serious idea. Darwinian evolution was crucial in undermining the design argument, and it is still vital for today's debate.

In 1691, one of the founders of natural history outlined a view of nature which would dominate science and natural theology for nearly two centuries.[1] In *The Wisdom of God Manifested in the Works of the Creation*, John Ray explained how the fixity of species fit a divine purpose. His was a static view of nature, in contrast to the ever-changing complicated mess we perceive today. As Ray saw it, the divine plan had to be free of the imperfections implied by large-scale or irregular change. But Ray's picture of nature did more than satisfy theological needs—it was a theoretical framework which explained the structure of living things by asking what God created them *for*. A bird is meant to fly, so its wings must be well designed for that purpose. Nature must be a harmonious web of purposes, so the bird must exist in a delicate balance with its environment. Furthermore, the designs of life show a certain unity. We can classify creatures in neat hierarchical groups; we see no abominations like mammals with feathers or with eight legs. Unity of design and harmony of life is just what we would expect from a single designer with a definite overall purpose.

Though we now think Ray was mistaken, he successfully identified some common themes in nature. Living things exhibit good functional forms; a bird's wing does have an aerodynamic shape. This fits very well with the traditional attribution of complex functional order to intelligent design. And species are fairly stable entities. Seeing them as fixed—with variations merely deviations from Platonic forms—was perfectly reasonable, especially when we did not yet know the Earth was so old.[2]

So creationism used to make sense of life. There was some philosophical discontent, such as David Hume pointing out that, as a metaphysical doctrine, the design argument was not as ironclad as theologians would like. But his criticism had little effect. Design, for thinkers like Ray, was part of a robust religious view of the world, where God was not a metaphysical extra who never appears on stage but the ultimate source of the pervasive purposes in life.

When creationism collapsed, this was not because of philosophical difficulties or even because of a direct contradiction by data. There were always things which did not fit smoothly; parasites, for example, strained the picture of benign harmony, even if imperfections could be expected in a fallen world. In time, more uncomfortable facts like extinct species and an old earth accumulated. As evolutionary alternatives developed, they explained the imperfections and transformed how scientists saw the successes of special creation. We ended up thinking that species change, give rise to other species, and go extinct. Populations adapt to their environments and exhibit good "design"[3] because genes promoting reproductive success have a better chance to make it to the next generation. We find a nested hierarchical order when classifying species because they have a common ancestry. Modifying creationist scenarios could keep them consistent with the evidence, but compared to evolution they became a tangled mess of excuses.

Thus the argument from design in life died the death of a thousand excuses, killed off by Darwin rather than Hume.[4] And with its passing, more trouble followed. It was bad enough that the handiwork of God became the product of blind nature, but other vital religious doctrines came into question. Without a historic Creation and Fall, today's Christians still have to find ways to reinterpret Original Sin and the Atonement. Muslims have to confront the fact that the Quran, the direct word of God, says humans were specially created. These are not trivial matters since, outside of a network of specific religious claims, we only have the impossibly obscure God of the philosophers.

And so evolution has always been accompanied by theological debates, conversions of old doctrines into vague metaphors, and projects of reimagining theistic metaphysics. The effort, though, is one-sided. Whether in reinterpretation or rejection, it is always the theologians who react to science, while biologists have long stopped being influenced by theology in their work.

The creation-evolution dispute, then, seems a classic battle between science and religion. After all, special creation was an important religious claim, anchoring a once-compelling view of God and our place in the divine order. Discarding this for an account which makes no reference to anything transcending nature is about as clear a retreat for religion as one could ask for, if not an outright defeat.

On the other hand, many of our religions have adapted to evolution. Though creationism retains substantial support among conventionally religious people,[5] liberal believers mostly agree that interpreting scriptural creation-stories literally is a mistake, and that God and evolution are compatible.[6] Perhaps evolution, within its proper sphere, does no harm to religion—it may even help strip away some nonessential beliefs and reveal divine creativity more fully.

Historically, the relationship between science and religion seems similarly ambiguous. Though they do not complement one another in happy harmony, neither do they clash in heroic conflict.[7] Communities making serious fact claims occasionally disagree, but they also find ways to contain the sources of friction. So asking what evolution means for God is still important. Darwinian evolution has been famously corrosive of religion; if theology can change and legitimately deflect the challenge, critics of God have some serious rethinking to do. If not, our doubts must grow stronger.

RESISTING EVOLUTION

Creationists seem fixated on the Genesis story. Rooted in a populist Protestant culture, they expect both nature and Scripture to be accessible to commonsense interpretation[8]—the plain reading of God's word cannot be sacrificed to counterintuitive theories like evolution. Indeed, such uncompromising opposition to evolution as to concoct a bizarre "creation science" as an alternative is largely an American, evangelical Protestant peculiarity. And as such, it is easy to dismiss as a narrowly sectarian position which has little to do with the Gods of other traditions.

This picture needs modification, however, considering the strong opposition to evolution in the Muslim world. For example, recently a brand of Turkish creationism has arisen which borrows wholesale from the Protestant version, and which has enjoyed considerable success.[9] They copy the Americans selectively, omitting signature doctrines like flood geology because the Quran is more ambiguous about such matters than the Bible. So it is all the more striking how much of creation science they adopt intact. Though Islam is quite different in culture and history, and Muslims are not obsessed with Genesis, they find common concerns addressed by Protestant creationists. In particular, they want to defend an ideal of created order which infuses nature with moral guidance deriving from specially designed roles for each living thing. Much of American creationism turns out to be free of Protestant doctrinal idiosyncrasies; it can be pressed into service by most any Abrahamic fundamentalism.

However, though not as sectarian as often thought, creationism still appeals to too narrow a constituency to explain why conservative theists resist evolution. This resistance is not mere fundamentalism. Our religions envision a top-down, spirit-first reality. Even in the realm of what we see, there must be unbridgeable gaps separating order from chaos, animate from inanimate, life from nonlife, humans from other animals, and consciousness from nonconscious matter.[10] The lower stages in this hierarchy of being should depend on the higher stages, not the other way around. And so, explaining the development of life without reference to a higher order of reality is unthinkable. Seyyed Hossein Nasr, a respected scholar of religion, denounces evolution and observes that

> From the purely religious point of view also, the evidence against evolution is universal even in traditions such as Hinduism, Jainism, and

> Buddhism where cosmic history is envisaged on grand scales and where there has been perfect awareness among those who read their sacred Scriptures that the world has been around much longer than six thousand years, that other creatures have preceded man on earth, and that the geological configuration of the world has changed. . . . In all sacred Scriptures and traditional sources whether they speak of creation in six days or of cosmic cycles lasting over vast expanses of time, there is not one indication that higher life forms evolved from lower ones. In all sacred books man descends from a celestial archetype but does not ascend from the ape or some other creature.[11]

A world a few thousand years old would have made better sense for a universe centered on a personal God and the salvation history of its chosen people. But this is hardly essential. A static world would fit a hierarchically structured reality better, but if creation changes more than our traditions once allowed, so be it. Perhaps God creates by modifying earlier forms. The problem is, Darwinian evolution goes much further than forcing some painful changes on theology. It radically undermines the whole top-down universe, situating creativity squarely in the material world. So the pious approach evolution with suspicion, while infidels put "Darwin fish" symbols on their cars. For good reason.

Evolution changes everything. To understand life, and particularly ourselves, it suggests we look to earthly history rather than to celestial archetypes. We begin to see ourselves as self-important apes, though we still prefer to think we are incomplete angels.

Evolution bolsters naturalism, however, not just by expanding our understanding of nature and anchoring our species securely within. Even more compelling is the Darwinian *mechanism* for generating our complex histories. Impersonal natural causes are never completely at home in the monotheist universe, and Darwinian evolution relies not only on this alternative to God, but also on something worse: chance. It allows us to think that complex functional order can emerge from disorder, without external help.

The basic idea is simple yet powerful. We ratchet up complexity step by step, by retaining novelties thrown up by historical accidents. For this, we first need some structures, however pedestrian, which replicate themselves and thereby pass information down to further generations. Then we need a world operating by Murphy's Law—ensuring things will occasionally go wrong while replicating. Those few accidental errors which change our replicators in a way which promotes better replication will be retained in successive generations. And so, little by little, we bridge the gaps theists have seen in nature.

Imagine a planet covered with many streams and rivers which is inhabited by some dim-witted humanoid natives. They need to cross their local streams very often, jumping from one stone to another, using the occasional fallen tree, and so on. But these natural "bridges" have keepers who take tolls from users. Since this is a decent living, there is competition for bridge keeping. Every now and then, a rival keeper sets up a new bridge close by, copying a bridge which seems to be earning well in a neighboring village, or even copying the bridge she hopes to supplant. She tries to replicate the design as closely as possible; cutting stones from the quarry to have the same sort of shape and felling trees that look similar. She cannot, however, make a perfect copy. Her stones will be just a bit different in size, have a protrusion where the original had a concavity, and so on.

Now, bridges have to support their keepers. Unless a bridge keeps up with the local competition, it will be abandoned and decay. More importantly, others will not copy an unsuccessful bridge. But though copying errors will usually result in a worse bridge, since the errors are random, there will also be rare accidental improvements in bridge design. These will tend to survive in the next generation, displacing earlier versions. For example, a natural bridge might contain a log wedged between stones, partially under water, and impeding the flow of water by gathering floating debris. In some copies of this design the log will be lifted further up out of the water, in some sunk further below. Some copies will have the log wedged slightly more securely, some less. Bridges which are slightly more stable and less flow-impeding will also be slightly more convenient for the customer and need less maintenance, and so support a more comfortable keeper. Aspiring keepers will be more likely to copy her bridge. Eventually there will be bridges containing segments where a log is wedged tightly between large stones and lifted clear from the water. In time there will be bridges consisting entirely of such segments. Older designs will be replaced, but some will survive in certain environments. On streams which flood violently and frequently, the log structure will be swept away too often to be viable. Bridges there will be big boulders set at intervals people can step across.

The bridges will keep improving. For example, structures where a protrusion on a rock happens to meet a depression in a piece of wood will be more stable. This device will, if things go well, evolve into interlocking structures allowing more complex bridges, raised higher off the level of the water. Some of these, originating over milder waters, will migrate to regions with flooding waters. Having overcome the deficiencies of the wedged-log bridges, they will replace the old boulder bridges. Eventually, many bridge species will be quite sophisticated. They will be adapted to their local conditions, have an efficient distribution of loads, and be well designed against wind and flood. Different designs will be successful in different environments: there will be some suspension bridges, and some arched stone bridges with no wood at all. In fact, the planet will produce some very good engineering designs, though the project will be way over budget and millions of years past deadline. But there will be no designer. The dim-witted humanoids will know no more about engineering than a million years before; all the keepers ever do is copy other bridges and collect tolls.

It is just such a process which is central to the history of life. Blind genetic variation filtered by natural selection is not the full story; modern biology is also concerned with processes like genetic drift, and different accidents, like mass extinctions, which radically affect the history of life. Nevertheless, the creativity in biology is entirely due to the same Darwinian process as in our bridge-building tale.

The picture we end up with is unambiguously bottom-up; naturalistic rather than mind-first. No external guidance, no overall purpose enters into the Darwinian mechanism. Its thrust is, in fact, just the opposite. Random variation-and-selection is not restricted to biology; it can erect bridges in science-fiction stories as well as populate our planet with a multitude of species. It is generalizable to provide naturalistic explanations for any complex functional order, even our own intelligence. As Daniel C. Dennett describes it, Darwinism acts like a "universal acid" eating away at any idea of top-down design.[12]

Not only does evolution make the Designer superfluous, it is hard to imagine even one receded into the metaphysical background. In biology, evolution is driven by reproduction and the competition of the next generation for the scarce resources which allow them to reproduce in turn. Untold numbers die, and some survive to continue the cycle. If a population finds itself in comfortable circumstances, it will soon expand to exploit its resources more fully and end up back at the edge of starvation. Where nature is delicately balanced, it is a balance of suffering, not a happy harmony. This is not the work of an all-powerful designer choosing the best means to achieve any religiously intelligible end.[13] A creationist textbook states the problem fairly accurately:

- Evolution is inconsistent with God's omniscience. The history of evolution, as interpreted by evolutionary geologists from the fossil record, is filled with extinctions, misfits, evolutionary cul-de-sacs, and other like evidences of very poor planning. The very essence of evolution, in fact, is random mutation, not scientific progress.
- Evolution is inconsistent with God's nature of love. The supposed fact of evolution is best evidenced by the fossils, which eloquently speak of a harsh world, filled with storm and upheaval, disease and famine, struggle

> for existence and violent death. The accepted mechanism for inducing evolution is overpopulation and a natural selection through extermination of the weak and unfit. A loving God would surely have been more considerate of His creatures than this.[14]

In fact, the difficulty for God here runs deeper than the considerable nastiness in nature. Darwinian evolution is such a relentlessly naturalistic process that speaking of benevolent purpose, efficiency, and other anthropomorphic attributes is entirely out of place. It is hard to imagine any personal force behind evolution, even one in a nasty mood. At bottom we have variation and selection, full stop.

If this picture is accurate, theists face a formidable challenge. Some maintain that it is not accurate, that we should not prematurely dismiss evolution as God's way of creating. Theological conservatives, though, have too much at stake to keep diluting and reinterpreting their doctrines, so they have more reason to resist evolution.

The question then is: can this be done without fundamentalism, without becoming ridiculous? Recently, a number of "intelligent design" (ID) theorists have taken the field, who have some claim to make such a case. Though conservative Christians, they downplay sectarian commitments, opposing evolution with the basic theistic notion that material reality cannot be autonomous—that creativity must reside in the independent mind who infuses order into an otherwise meaningless world.[15]

In their effort to make a serious case, ID proponents have scored a few philosophical points. In countries like the United States, ensuring that only evolution is taught in the science classroom depends on the support of liberal religious people who think of evolution as a divinely guided process. So defenders of evolution tend to rule out creationism by declaring that science can only consider naturalistic explanations—no miracles, no ghosts, no creators. There might be a deeper supernatural reality behind the phenomena, but science, they say, can tell us nothing about such metaphysical claims which go beyond nature.[16]

This position is politically convenient but intellectually suspect. Nineteenth-century biologists did not come to think special creation was a hypothesis they were not allowed to entertain. They rejected it, deciding evolution explained life better. And intelligent design is still, on the face of it, a straightforward fact claim. Indeed, that is how we explain human artifacts, and it would seem legitimate to ask whether an intelligent agent is directly responsible for complex structures in nature. Such an explanation would be similar to historical accounts, where a historian narrating the Roman conquest of Gaul uses personal causes like Caesar's political ambitions to make sense of the physical and documentary evidence. Though beyond nature, the traditional God is a personal agent, and divine acts *might* be an integral part of the best explanation. These would not be arbitrary interventions, but acts which make sense in the context of a religious theory telling us about divine intentions.

Disallowing personal agents from scientific explanations is arbitrary. Perhaps evolution cannot account for life, and we *do* need intelligent design. The business of science, surely, is to help us figure out whether this is the case; if it cannot even address the question, there must be something wrong with our understanding of science.

ID advocates, then, cannot be defeated by the wave of a philosophical wand. They correctly argue that science cannot be restricted to a predefined set of naturalistic possibilities, and they are justifiably disturbed by how God becomes optional in a culture which affords science the authority to describe our world, but insists that science and religion belong in separate spheres. Any conservative theism which demands that God have an explicit hand in ordering the world must resist theories which remove God from the action; as long as there are religious people who take their traditions seriously, there will be opposition to evolution. And if ID can be shaped into a theoretically sophisticated, empirically well-anchored proposal, it will have a legitimate place in scientific debates, regardless of attempts to define it out of bounds.

BRINGING BACK THE DESIGNER

So, can Darwin's mechanism do the job? First of all, there is the old question of how to develop new complex structures when intermediate forms have to be functional. To get from fallen logs to the Golden Gate, we need a path of gradual improvement. At each step of the way, we must have a better bridge. What if no such path exists? In the real world, what good is, say, half a wing? As creationist Henry M. Morris puts it, "A new structural or organic feature which would confer a real advantage in the struggle for existence—say a wing, for a previously earth-bound animal, or an eye, for a hitherto sightless animal—would be useless or even harmful until fully developed."[17]

In ID circles, this argument has been resurrected by biochemist Michael Behe, who argues that biological structures exhibit "irreducible complexity" at the molecular level.[18] *All* components of such structures must be present for them to work, and so no path of small improvements exists. Assembling them requires massive coordinated change, which is extremely improbable by accident.

Recall the frequently flooding streams in our bridge story. There, log bridges were less successful than sturdy boulders, and so there was no intermediate between primitive bridges and structures raised high above the waters. But we did not need a sequence of bridges all gradually better at crossing *flooding* streams. Elevated bridges could evolve elsewhere and spread. The bridges need not always get better at a single function in a single environment. With life, the idea of gradual improvement in a single direction is even more misleading. An organism has no fixed function; replicating genes is all that matters. Under some circumstances, an ability to fly will be selected for; under others, losing flight will be advantageous. Being able to succeed in many ways dramatically increases the likelihood of functional intermediates. A species' evolution can be roughly thought of as a population moving on a "fitness landscape."[19] This is like a mountainous landscape with a few low fertile valleys in which species can flourish. Now, imagine a blind robot in a rugged countryside, who wants to move to low ground as much as possible. There is a lower valley just west of the one it lives in, but there is a hill in between. Our robot tests the ground in its immediate vicinity, and keeps moving to lower spots. If constrained to move between east and west, the robot will never find the valley. But if there is another dimension it can explore—moving north to south as well—it can go *around* the hill, eventually ending up in the western valley. The more dimensions, the better the chance of there being a path.[20] And life is nothing if not multidimensional.

With insects, for example, there are at least two proposed uses for half a wing. Proto-wings may have been useful at first only for heat regulation, and later became gliding implements, then flying wings. A more recent theory suggests rudimentary wings were useful at first to skim across water surfaces, as a form of locomotion intermediate between swimming and flying still seen in some primitive insects.[21] Or consider the path between prokaryotes and eukaryotes. Prokaryotes are primitive cells, not much more than genetic material and some chemical soup wrapped in a membrane. Eukaryotes are highly complex, containing many specialized organelles, some even with their own genetic material. Popular pictures of evolution lead us to expect gradually increasing complexity within a lineage of originally prokaryotic cells. But the path life actually took may have been more interesting. A eukaryote is probably a result of symbiosis—a partnership between different prokaryotes which evolved into a union.[22]

The half-a-wing argument is thus a nonstarter. Antievolutionists will, on occasion, come up with tough cases; the very multidimensionality of the fitness landscape means that discovering the path evolution took will often be very difficult. Without clues to history like organelles with their own genetic code in eukaryotes, ideas like symbiotic evolution might not have surfaced even as speculation. All biologists can say is that in cases where we have detailed evidence, we have success tracing the path of evolution; otherwise, though we may have gaps in our knowledge, these are not gaps we need the Designer to bridge.

However, there is more to ID than bad biology. After all, the fundamental claim of ID does not concern intermediate forms, nor is it even specifically about biology. It is that true creative novelty, or meaningful *information*, must be a product of intelligence. The classical argument from design is too often only a weak analogy, ready to be abandoned for the first naturalistic alternative. To avoid this weakness, mathematician William Dembski uses information theory to detect design. We distinguish design from accident, he says, by seeing if our data exhibits contingency, complexity, and specification.[23]

Contingency means that an information-conveying system must allow many possible arrangements. Not all order is evidence of purpose. Objects we drop fall rather than drifting off in random directions, but this only manifests a simple physical law. In contrast, it is physically as easy to type "`uTq.gkwf(fFerj...`" as to type "`a real argument`." It is precisely this lack of physical constraint which allows a variety of messages.

To rule out pure chance taking over in the absence of simple constraints, Dembski demands *complexity*. A world of physical laws and random events will occasionally produce something which makes sense, like a monkey at a typewriter banging out "`hello world`." But the longer and more complex the message, the more unlikely this is.

Finally, *specification* is crucial for telling what sort of data is meaningful. Finding π encoded in a radio signal from space would suggest an intelligent source, while any particular random string, though just as improbable, is merely noise. We must be able to specify meaningful patterns before the fact; otherwise, given thousands of crank-hours at work, we can find messages in anything, like a plan of history in the Great Pyramid.

Dembski proposes a rigorous form of these criteria,[24] meant to identify information-bearing data even when we cannot decipher its meaning. With a rigorous procedure to identify a particular sort of order indicating intelligent origin, the design argument would be much improved.

Consider astronomer John Playfair's argument, two hundred years ago, that the solar system must have a divine origin because of its stability and because all planets orbited in the same direction and plane.[25] After all, the laws of physics also allow planets cavorting in unstable orbits, rotating any which way in orbital planes with no relation to one another. Blind chance, in fact, would almost certainly have produced such an arrangement, and very soon we would have had no solar system at all. A stable, orderly system is special; our welfare and very existence depend on it. This puzzle, Playfair said, would be solved if the solar system was designed with us in mind.

Playfair, however, did not reveal any real pattern to his improbable order; he only postulated an intelligent cause. And any half-decent theory of planet formation shows us how planets form in a single plane and orbit in the same direction, explaining this order through physical constraints. In Dembski's terms, the contingency aspect of Playfair's case was too weak.

Life, however, exhibits just Dembski's sort of specified complexity. A strand of DNA, for example, carries information. It is not constrained; in isolation, a string of biochemical gibberish is no different in chemical stability than one which codes for a vital protein. Functional DNA is very unlikely to come about by random combination of base pairs, and the order it exhibits is not an after-the-fact invention. So, if Dembski's design detector reliably sorts out artifacts from the haphazardly cobbled together, even when we know little about the functions of the artifacts, it seems we should infer that organisms are also, at some level, products of intelligent design.

Of course, it may be that Dembski's filter simply fails to distinguish between explicit design and evolution—both may generate information.[26] So Dembski tries to show that a Darwinian mechanism *cannot* create specified information, because this is conserved. Unintelligent processes can never add new content. Consider a message string, "`3:45pm`." A computer program can change this to "`15:45`"; no information is gained or lost thereby. Or it might degrade the message by rounding to the nearest hour, leaving "`4:00pm`." If the string was input to a program which e-mails meeting times, it could be converted to "`Next department meeting 3:45pm`," but the additional comment, though useful, is not really new. Such a program could only be used to transmit meeting times; this information is built into its initial design.

Random processes do no better. A noisy channel might, with a lot of luck, produce "`Christmas party 3:45pm`," but there is no reason to trust it. Variation-and-selection can add no meaningful novelty to a message because all it does is reveal information in preprogrammed selection criteria. So the creativity producing information-rich structures like living beings cannot be captured by blind naturalistic processes.

We can apply these ideas to the physical world. Newtonian physics, for example, conserves information at a microscopic level; a complete description of particle positions and velocities at any time also determines all past and future states. Following Dembski, we might suspect that if complex structures appear at some point, this is not a genuine novelty, since these were implicit in previous states.

This suggests a clockwork deism, where the information provided through an initial design unfolds in time, manifesting in complex macroscopic structures. However, if life exhibits specified complexity, this is at a much more coarse-grained level of description than the particles which make up an animal. And there, even a Newtonian universe does not conserve information. Not being able to track individual particles, physicists use macroscopic variables like temperature and pressure. If we bring objects at different temperatures into contact and let them reach equilibrium, they will end up the same. No measurement can recover their original temperatures, and they will not spontaneously acquire different temperatures again.

But then, Dembski allows that naturalistic processes can degrade macroscopic information. In fact, his version of ID revives the creationist suspicion that evolution is incompatible with the second law of thermodynamics—how can complexity originate and progressively increase in a world which tends to disorder? Local entropy decreases are perfectly fine, provided that the global entropy ends up increasing, but there is more to this question. Systems which cause a local decrease of entropy, like refrigerators, are themselves complex systems. While exporting entropy to their environment, they do not create information other than that specified in their design. Information, it seems, is required to organize mere matter; the question is where this information comes from, not just local entropy change.

Dembski expresses important theological ideas, and raises interesting questions. His greatest weakness is his discomfort with even a minimal sense of evolution: that life-forms descend with modification from older species. There is no serious doubt about this fact. So his ideas could work better not as a sophisticated creationism, but together with more liberal thoughts about the compatibility of God and evolution. After all, the main challenge to purposeful design is the Darwinian mechanism, not bare evolution. And since Dembski's ID is broad enough to include deism, it is worth exploring how design and descent-with-modification might go together.

To restore purpose to evolution, and align it with the hierarchical view of reality proposed by religions, liberal theologians portray it as a progressive process.[27] God infuses life with a progressive tendency seeking ever higher complexity and consciousness. John F. Haught portrays this as a metaphysical necessity, so God is not only compatible with evolution but actively demands progressive development:

> if God is infinite love giving itself to the cosmos, then the finite world cannot possibly receive this limitless, gracious abundance in any single instant. In response to the outpouring of God's boundless love the universe would be invited to undergo a process of self-transformation. In order to "adapt" to the divine infinity, the finite cosmos would likely have to intensify, in a continuously more expansive way, its own capacity to receive such an overflowing love. In other words, it might endure what we know scientifically as an arduous and dramatic evolu-

> tion towards increasing complexity, life, and consciousness. In the final analysis, it is as a consequence of the infusion of God's self-giving love that the universe is excited onto a path of self-transcendence, that is to say, evolution.[28]

Though steeped in metaphysical hand waving and a rather strange notion of love, Haught's emphasis on progress brings him close to Dembski. Both consider creative novelty to be beyond the ability of materialism to explain. Haught also speaks of God as "the ultimate source of the novel informational patterns available to evolution," and argues that information structures reality hierarchically, trying to affirm Seyyed Hossein Nasr's views without denying evolution.[29]

More scientifically aware liberal arguments also fit ID well. Uncomfortable with the remote God of clockwork deism, John Polkinghorne and Arthur Peacocke look to the breakdown of Newtonian physics to find space for God to act on the world. Quantum indeterminism and dynamical chaos, they say, introduce an "openness" to the universe, giving us a glimmer of a top-down sort of causality connected to "active information." It even allows us to suspect that evolution was supernaturally tweaked to ensure our otherwise improbable emergence on the scene.[30] In the context of ID, this means God can infuse new information into the world through evolution, manifested in its progressive direction.

Dembski's ID, then, bridges the gap between conservative and liberal theology. For conservatives, it provides a way to escape anti-intellectual fundamentalism. For liberals, it promises to put some backbone into their speculations about how God might direct evolution. Theistic thinkers of all stripes have to challenge the Darwinian vision of life, bringing back the Designer. And in all cases, they pose the question of explaining creative, progressive novelty in a bottom-up fashion. This needs an answer.

ORDER FROM CHAOS

In the popular imagination, evolution moves up a ladder ascending toward humanity.[31] A creative force permeates reality, gradually coaxing order out of chaos, spirituality out of brute material existence. This was perhaps best expressed by the paleontologist and Jesuit priest Pierre Teilhard de Chardin. His God is the source of the progressive evolutionary processes in the universe, *and* its final goal, its "Omega Point." Not content with a mere metaphysical gloss, he says that "evolution has a precise *orientation* and a privileged *axis*. . . . I believe I can see a direction and a line of progress for life, a line and a direction which are in fact so well marked that I am convinced their reality will be universally admitted by the science of tomorrow."[32]

Unfortunately, this line of progress has not become visible to scientists. In evolution, new species branch out every which way, rather than following a mainstream of development with occasional lesser deviations. Cockroaches and bacteria are as much the current end point of evolution as a bunch of self-important primates. Far from being universally admitted, Teilhard's ideas are irrelevant to modern science. Biology today is thoroughly Darwinian, explaining life without invoking grand purposes.

Curiously, the Darwinian revolution did not take place in Darwin's lifetime. He convinced biologists that life had evolved, but natural selection and a purely naturalistic approach did not take hold. In the decades after Darwin, many different evolutionary theories took the stage, and almost all made room for a purposeful direction in the history of life. Many biologists took a developmental view, believing evolution was an unfolding of potentialities intrinsic to the structures of living things. Life became more complex by a progression which was in the nature of being—not through accidents and blind trial and error. A God could well be the force guiding this progress, though this would be a deity more like that of a German philosopher than a Hebrew prophet. So Darwin was eclipsed, and biologists of the time treated selection as a minor, conservative factor in evolution.[33] Nature could still satisfy moral hopes and supply religious meaning, and philosophers found much in biology to inspire their deeply purposive metaphysics.[34]

Between the 1930s and 1950s an "evolutionary constriction" took place, reducing the number and kind of theories in play. Modern genetics had much to do with this, along with other maturing biological disciplines. Biologists did not reach a tidy consensus; we still have open questions on matters like the significance of genetic drift, the details of speciation, and how much of large-scale change is concentrated in speciation processes. But after the constriction, these were the important issues. The remaining debates concerned only blind, bottom-up mechanisms. Purposive theories disappeared, and evolutionary biologists no longer talked of guiding aims or progressive forces manifesting themselves in biological structures.[35]

Still, even without immediate purposes, the cumulative changes which emerge from evolution need an explanation. The history of life is not dominated by any single direction, of course—the superabundant diversity among insects is as much a part of the story of evolution as improved cognitive capabilities. We are more interested in the latter because it is *our* story. While increasing complexity and brainpower is a biological trend, it is not central to evolution like Teilhard imagined it to be.[36] Even so, we would like to know where the trend comes from.

The question is important because many trends are made up of irreversible changes. A famous textbook illustration tells how a moth population during the Industrial Revolution developed a darker color, blending better with the newly soot-covered bark of trees. The problem is, after the air got cleaner, light color was selected for and the moths got lighter again. We never had a new species of moth; the color change was *reversible*. But speciation is irreversible. A daughter species may go extinct, but once it branches out, it will not evolve back into the ancestral form. The mechanisms of evolution, however, are reversible—as with the moths. How, then, do speciation and trends occur?

This problem is familiar to physicists. Our dynamical laws are time-reversible: no one can tell if a movie of a collision between billiard balls is being run in reverse, just by watching it. Technically, our laws are fully invariant only under a "CPT" switch, which means we have to replace everything with its antiparticle and go to a mirror-image world, along with reversing time. But this is still a simple symmetry. Our world is a very different place. A dropped glass shatters. Watching a movie of a thousand shards gathering to make an intact glass and rising to someone's hand, we know it is run in reverse. Even in a mirror-image antimatter world, the glass would not come together. Macroscopic physics, particularly thermodynamics, is irreversible. Yet at the microscopic level, we have only time-reversible laws.

It turns out irreversibility is due to asymmetric boundary conditions and the fact that the information available to us degrades in time. We usually associate physics with predictability—if we know how a cannonball emerges from the cannon, we can calculate where it lands. Of course, we never know the exact initial state of the cannonball, but this only translates into a small error in our knowledge of where it will land. But complex physical systems are often chaotic, which means uncertainties in our initial knowledge grow exponentially in time. To track a chaotic cannonball, we need a constant flow of new information updating our knowledge of its current state. Worse, physical systems can exhibit randomness. Unlike with dynamical chaos, there is a limit beyond which no amount of information can improve our predictive ability. In the long run, random and chaotic systems haphazardly wander through their possible states. We can still find equations for their overall statistical properties, but these macroscopic descriptions will now exhibit irreversibility.[37]

Imagine a game with ten players seated in a circle. One player starts with one hundred chips, others have none. Every turn, they roll a die for each chip they possess. If they roll a 6, they pass the chip to their left; a 1, to their right; anything else, they keep it. Now, the game run backward in a mirror-image world has the same rules. But events progress irreversibly. Though we cannot predict the exact course of the game, we can tell things will even out. The person with all one hundred chips will lose them, until the game settles down to a point where each player has about ten, with some small fluctuations around that number. The game does not have a built-in chip-equalizing tendency; the rules say nothing about a fair distribution. But if we *start out* from a state where one player has everything, we know where the game will go. At each turn, the game can attain many more equitable states than the one where chips move back to the original owner.

Evolution proceeds similarly. A daughter species might itself give rise to a new species which exactly resembles its own parent. But among all the possible new branches, one identical to the original parent is a remote possibility. Speciation is irreversible, and extinction really is forever.

As in our game, irreversible trends in evolution also reflect an asymmetric starting point. Stephen Jay Gould gives the example of a lineage of marine species originating in shallow water. On average, more and more species will inhabit deeper waters over time. But this is because new varieties will face less competition and will be more likely to survive if they inhabit the deep. Deep water need not be an intrinsically better place to live—it is just where newcomers can eke out a living. Likewise, we find trends toward complexity and brainpower because—and only because—life started out simple and stupid.[38] Perhaps the bacteria already have the good spots, so latecomers must get by through developing brains and whatnot. And even if ours is a good spot, we have stumbled on it accidentally, not because of a cosmic force impelling life toward improvement.

There is still something peculiar about this explanation. After all, according to the second law of thermodynamics, the world tends to disorder and decay. Complexity, even if it seems to emerge accidentally, goes in the opposite direction.

Again, this is a problem physicists have had to face. Not only does life evolve, but stars form, galaxies cluster, and all this does not crumble into the featureless mess of thermodynamic equilibrium. So evolution can only take place within a system driven away from equilibrium.

An untended desk gets messier, until it cannot get any worse. Say we suddenly double the size of our desk. Now, relatively speaking, the desk is orderly: the old mess covers only half the desk. The mess will creep and conquer the new area, reaching equilibrium again. But now imagine our desk continually expands, at a rate faster than the mess can creep. Since the maximum possible messiness grows faster than the mess, the desk moves away from equilibrium. We have an increasingly sharp idea where the mess is concentrated on the overall area of the desk.

This is a crude example, but the principle is important. When the possible states a system can occupy grow in time, *both* entropy and order can increase. Since starting points set irreversible trends, let us look at the beginning. The big bang starts at maximum entropy, or complete disorder. As the universe expands, however, we find a situation much like our inflating desk. The universe's entropy increases, but its maximum possible entropy grows even faster. The difference in between is space for order to form spontaneously.[39] The nature of our microscopic forces makes this possibility an actuality. Gravity is a purely attractive force, so at cosmic distances, thermodynamic equilibrium is unstable. Even the tiniest fluctuations in density will tend to grow. In an expanding universe shaped by gravity, matter will clump up in black holes and galaxies.

The argument applies to life as well. In fact, increasing both entropy and order works best in systems with few physical constraints, which can carry information. Evolution is like a process which not only varies the letters printed in a message, but can increase the number of letters, even come up with new alphabets. In the course of evolution, the possibilities open to life grow much faster than the diversity which is actually realized. We again have something like the inflating desk.[40]

Moving away from equilibrium does not guarantee any interesting order forms. Among the locally entropy-reducing structures which will form naturally, we now need something which *replicates* itself, using the energy available in its locally open system. In a world where entropy increases, where things often break down, there will be random imperfections in this replication. Most variants will, of course, show no improvement or do worse. But some will replicate better in some local environments. Our replicators will compete with one another to commandeer energy, or food, to reproduce. Each generation will include new variants, a few which may do quite well somewhere. Since we start with crude replicators, some variants will be functionally more sophisticated. Under conditions where they replicate well, we will see more of them. In this way, Darwinian processes generate biological information.[41]

Evolution, then, is not a peculiar exception to universal decay but a consequence of the same sort of physics which gives rise to the second law of thermodynamics. Strange as it may seem, order comes from disorder. In systems driven away from equilibrium, complex structures may spontaneously form, using random noise. The right conditions are not always easy to attain; too much noise and we get the sort of boiling inferno where no order lasts long, too little and we end up with a crystalline order which freezes everything in place. Nevertheless, in a large, varied universe there will be environments poised between order and chaos. And in such places it would be surprising if complex structures did not form and flourish.

With hindsight, the Darwinian process has an air of inevitability about it. Still, the universality of Darwinian processes needs more emphasis. We often focus on the fundamental laws of nature and their consequences, so at times it can seem the science of complexity is but a nuisance foisted upon us because we cannot keep track of many-body systems; a mere artifact of imperfect knowledge. Actually, much of what we have learned about complexity is valid under a wide range of dynamical laws and initial conditions: concepts like irreversibility, self-organization, and Darwinian variation-and-selection are not very sensitive to the underlying microscopic physics. So studying complexity requires more than traditional physics, calling on fields such as biology and computer science.[42] The exact history realized in a universe does, of course, depend on microscopic details. But just obtaining local pockets of specified complexity is not too difficult. They will come about, under random initial conditions, with a variety of dynamical laws.

All of this is rather counterintuitive. It is common sense to look for an intelligence responsible for complex order, and to see purpose behind progress. But biologists do not see progress toward any predefined goal. They may debate the limits of adaptationist reasoning, but few doubt that accidents are the source of novelty in life. They discuss controversial new ideas, for example, about occasional mass extinctions which decimate life so badly that surviving such events is as much a matter of luck as of fitness. From large mutations to large geologic catastrophes, our increasing awareness of accidents at different levels and scales is changing how we think accidents shape life.[43] But nowhere do we see a privileged axis of progress, or any evidence of purposeful guidance like beneficial mutations not occurring haphazardly.[44]

Physics also weighs in, providing a world of accidents, of seething, mindless dynamism, and showing us how to get order from chaos. This further leads us to think randomness is the source of novelty in our world, which is then shaped by selection. So evolution does not progress in any purposive sense: it blindly blunders its way through the possibilities of life. More complex life exists now only because there was none starting out—not because of any progressive force, cosmic goal, or underlying design.

It could have been otherwise. Our religious traditions give us every reason to expect we should discover the hand of God in life. Yet it is not there.

DARWIN IN MIND

When explaining life, physics and biology do not invoke anything beyond nature. But then, there may just be more to evolution than is captured by natural science. Both ID and liberal hand waving try to reach beyond a physical perspective, either by identifying something extra which must be added to the picture, or by accepting the science and insisting theology can still deepen our understanding of what has happened.

In particular, information-based arguments for design remain appealing, regardless of physics. This is because they draw upon deep-seated intuitions that creative intelligence is something more than the operation of a mere mechanism. Imagining a combination of laws and chance, however elaborate, to be genuinely *creative* is quite a stretch. And Dembski's criteria to detect design may well single out a kind of order somehow related to creative intelligence. So without some account of the place of intelligence within nature, we might still suspect that naturalistic explanations of complex order overreach.

Many a science-fiction tale tells how a hero defeats a computer by posing a problem it was not programmed to deal with. It then starts saying "does not compute!" in a synthetic yet anxious voice, and finally goes up in smoke. Unlike the rule-bound machine, however, we think human intelligence at its best is flexible and innovative. We confront situations beyond what we have prepared for, and if we do not always succeed, we still often come up with novel approaches to the problem.

As made clear by Dembski's argument that information is conserved, it is difficult to see how new content can be generated mechanically. Artificial intelligence researchers ask us to imagine machines that perform a variety of complicated tasks, learning about and responding to their environment in sophisticated ways. But if these machines remain within the bounds of their programming, it is natural to attribute intelligence not to them but to their designers. ID voices this suspicion: that no preprogrammed device can be truly intelligent, that intelligence cannot be explained by natural processes.

Outside ID circles, similar intuitions underlie many respectable, long-standing criticisms of the prospects for machine intelligence; particularly those based on Gödel's incompleteness theorem.[45] *Any* rule-bound system has blind spots because it is unable to step outside of a predefined framework, and it would seem a genuinely creative intelligence would not suffer from such rigidity.

It turns out, however, that all that is needed to add the required flexibility to a machine is to let it make use of randomness. A random function can be used to break out of any predefined framework; it serves as a novelty-generator.

This needs some explanation. Flipping a perfect coin, we get a sequence of heads and tails like `thhththhhthhttht` . . . , continuing indefinitely. Mathematically, what makes this random is its complete lack of pattern. With a sequence alternating `thththththththth` . . . , we can identify a pattern and predict what the next toss will show, but with the random sequence, any guess is as good as another. No recipe—no algorithm—can correctly specify more than a finite number of the heads or tails in a random sequence.[46] It is patternless, completely free of all predefined rules.

In that case, a machine which can flip a coin now and then can avoid being permanently stuck in a preprogrammed rut since it can make decisions not determined by any rule. It can explore options unforeseen in its initial design. This is still far short of what we want from intelligence. But we can now prove a "completeness theorem" showing that all possible infinite sequences of heads and tails can be expressed as a finite combination of rules and randomness, including rules arrived at randomly. Equivalently, all functions—all possible behavior sequences—can be treated as a mixture of rule-determined and random behavior. So if all we claim is that intelligent beings are flexible in a way not captured by rules, randomness alone does the trick. There is no other option.[47]

Randomness supplies the novelty; now we need to use this for actual creativity. And we already know an excellent mechanism for putting bare novelty to work: natural selection. Dembski's claim that randomness does not help create content is incorrect; a Darwinian process is different from altering a message through fixed selection criteria. *Everything* is subject to random modification—there are no predetermined criteria, nothing but mindless replication and retention of successful variants.

There is a long way from randomness and selection to something as convoluted as human creativity. But it is almost certain that Darwinian processes are fundamental to creative intelligence. So our current sciences of the mind are full of Darwinian ideas: Gerald Edelman's "neural Darwinism"; the "Darwin machines" and cerebral selection mechanisms of William Calvin; a second replicator in the realm of culture, "memes," spread by our ability to imitate; Daniel Dennett's multiple levels of Darwinian mechanisms depending on competing processes to assemble our stream of consciousness; and more.[48] Workers in machine intelligence are increasingly relying on Darwinian mechanisms to produce flexible learning and creativity in computers—even to begin exploring art and original engineering designs.[49] Artificial intelligence is no longer a matter of canned, preprogrammed output, but open-ended, evolutionary behavior. Of course, many of these ideas are speculative, a lot are bound to be dead ends. Still, variation-and-selection is beginning to be vital for theories of mind as well as biology.

This leaves hopes to get to an intelligent designer through information in a peculiar position. It appears religious thinkers through the centuries have been right to notice how designed artifacts and living things are similar. And Dembski's ID is a good stab at pinning down what they share; his criteria of contingency, complexity, and specification reveal a special kind of order they have in common. But ironically, these criteria, if they work at all, only detect Darwinian processes at work. The complexity of life is directly produced through evolution, but an artifact also is an indirect product of the variation-and-selection processes that must be a part of creative intelligence.

So today, the Darwinian challenge to the design argument looms larger than ever. Before evolution, it was reasonable to think that complex order required an intelligence separate from mindless physical processes. And Darwinian evolution was at first just an alternative to design, though a vastly superior explanation of life. Now Darwin is taking over our minds, our own designs. We can no longer even contrast evolution to the work of a nonmaterial intelligence. The idea of supernatural design relied on an analogy to our own actions in the world; now this analogy cannot even get started.

Theologians do not give up easily. Some acknowledge that evolution is not goal-directed, and interpret variation-and-selection as God inspiring self-creativity in the world. God acts *through* the randomness which generates novelty. If so, Denis Edwards points out, theology can respect both the randomness of nature and the purposiveness of divine action. After all,

> Does not the doctrine of divine transcendence suggest that God might achieve purposes in a way that radically transcends all human notions of achieving purposes? Once this transcendence of "divine action" is acknowledged, then it is reasonable to think of divine purposes as being achieved through a process, random mutation, which in scientific terms is understood as chance-filled and purposeless.[50]

This attempt to have it both ways draws upon a venerable tradition of God acting through "secondary causes." Metaphysicians have always worried about reconciling divine sovereignty with free will and the order of nature, so they have said God is not a cause like those in everyday life, but is responsible for bringing these causes together.[51] Consider a classical design argument, like John Playfair inferring divine purpose from the improbably stable orbits of planets. The planets behaved according to the impersonal secondary causes of physics; there was no explicit divine intervention. But the pattern in which these physical causes came together indicated a deeper, purposeful cause behind it all.

Of course, chance is an absence of cause. But then, our metaphysical traditions also insist everything has a cause, hence "chance" merely represents our ignorance of the real causes of an event. A very nonordinary cause, like a transcendent God, may be just the thing behind what looks like chance from a scientific perspective.

Even in science, the concept of chance began as a placeholder for ignorance. Statistical reasoning emerged in social science, as a device to find predictability in varied, complicated individual actions when considered in the aggregate.[52] Physicists then took ideas of randomness on board in order to predict the behavior of large systems where there was no hope of tracking each individual particle. The underlying dynamics remained deterministic, and "chance" did, in fact, seem to stand for ignorance.

Of course, what classical statistical physics did was to hide randomness in the initial conditions. And as modern physics developed, we found a world full of random events. With general relativity, randomness appears at the boundaries of spacetime. And quantum mechanics is notorious for its pervasive dynamical randomness. No one can say, for example, when a radioactive atom will decay. We can predict the statistical behavior of large numbers of atoms, but individual decays are random; we can infer no cause behind them.[53] Neither does this lack of rhyme or reason stay decently confined to a weird subatomic realm. Chaotic dynamical systems amplify small differences, even quantum fluctuations. The physical world, it seems, is in large part a game of dice.

Even good old-fashioned thermodynamics is a sign of true randomness in the world. An often-used measure of missing information is Shannon's "information entropy," which is mathematically almost identical to the expression for physical entropy in statistical physics. So it can be tempting to adopt a "subjective" interpretation of entropy, saying it expresses our lack of knowledge about the world. However, entropy is also an objective thermodynamic quantity. These perspectives can be reconciled when there is real randomness in our dynamics.[54] This randomness sets an objective limit beyond which it is not possible to acquire information to improve our predictive ability.

There is a common theme here. In physics, randomness appears at the boundaries, in the fundamental theories, the points where we have reason to believe that no amount of effort will produce further information. And in fact, this is just what we should expect. Recall our alternating sequence, `ththth`. . . . We can begin to think about a mechanism behind this regular alternation. In fact, we can specify the whole sequence, no matter how long, without wasting a lot of ink. A random sequence, however, does not have a short description. To specify the sequence, we have to list the actual heads and tails in it, one by one. Randomness is where explanation ends: it is what is left when we can do no better than describe how things are and leave it at that. It is what we have to take as given and say no more.

So naturally, our fundamental theories at any moment will exhibit randomness, in their boundary conditions and in their dynamical laws which stand as brute facts. This is not a matter of physicists throwing up their hands and admitting ignorance about what lies beyond. Instead, it means our best efforts cannot find any further pattern to explain, in all likelihood because there is none.

Of course, no one can stop a metaphysician from declaring a hidden cause determines everything in a random sequence. But we do not learn anything by attaching causes to randomness in this way. Without a pattern allowing us to legitimately infer a causal structure behind the sequence, at the end of the day, we still can do no better than list how each coin toss came out. Saying random events are determined by the will of God does not tell us any more than saying they are caused by the whims of hyperdimensional dancing rabbits. It does not make sense for everything to be caused; if we are interested in explaining the world rather than making excuses for God, it is better to refrain from inventing causes where we cannot infer them.

What, then, of the transcendent God who assembles secondary causes? Such a deity will remain popular, but there is no reason to take it seriously. Design arguments have always tried to identify patterns in the world which indicated intelligent design, and tied these to concepts of purpose which were recognizably anthropomorphic enough to be religiously meaningful. But in a world of Darwinian creativity, design arguments do not get off the ground. Then, looking for God behind randomness takes us into the territory of the classical cosmological argument. Now, in the patternless noise of randomness, purpose finds no foothold. We might speculate, like John Polkinghorne, about how post-Newtonian physics allows for God to invisibly act in the world. But this is empty talk, unless some of what we now think of as random is not so—that it actually shows a pattern made sense of by a religious theory. This is a very strong claim, which needs real evidence to back it up. And it is no good to try and have it both ways like Denis Edwards; when divine actions become indistinguishable from chance, theology becomes content-free verbal fog.

Our world is full of random, uncaused noise. Through Darwinian processes, this noise gives rise to creative novelty, including, very likely, the creativity of our own brains. Yet our religions insist that we are fearfully and wonderfully made, that creativity demands something beyond the material world. It is hard to imagine how much more thoroughly this vision can be undermined.

GOD AFTER DARWIN

The small but influential Institute for Creation Research occupies a modest campus in southern California, staffed by earnest Christian laypeople and creation-scientists with Ph.D.s from secular institutions, all convinced that "true science" shows evolution is complete nonsense. They also have a very well done Museum of Creation and Earth History on site. About thirty thousand visitors each year walk through exhibits describing the six days of creation, Noah's Flood, and the dire effects of the false religion called evolution.

To someone immersed in modern science, the museum is bizarre yet oddly compelling. One wall is devoted to a drawing which impresses viewers with the overwhelming chemical complexity of just one eukaryotic cell. Nearby is a display of decaying objects, illustrating the second law of thermodynamics. Creationists ask if we can seriously believe that cell came to be without intelligent design, and it is hard not to sympathize. Biologists from nearby universities occasionally bring students through, to get them thinking about how evolution might answer such questions. These answers, though, do not have the commonsense appeal of intelligent design.

The randomness which fuels evolution is the first stumbling block. To survive in a complicated world, we continually seek patterns in our experience. It is better to mistake order in sensory chaos than risk missing a vital pattern—better to see a hungry lion in a trick of the light than to mistake a real lion for a trick. Indeed, we are notoriously unable to deal with random data, habitually interpreting it to conform to transient or nonexistent patterns.[55] The most important part of our environment, and the part which takes most intelligence to cope with, is other humans. Hence our brains are adapted for perceiving design and purpose behind haphazard information, since other humans produce a wide variety of ambiguous signals.[56] There are good Darwinian reasons for our aversion to accidents.

Randomness is just the beginning. At the origin of life, we look for simple replicators, capable of some degree of autocatalysis, as precursors to present-day biochemistry. For example, one candidate to explain how genetic information came to be DNA-based is that it evolved out of a simpler "RNA world."[57] There is little doubt *something* like this can work, but creationists demand a precise reconstruction of how life arose, replicated in the laboratory. This is extraordinarily hard to do. Randomness and selection produce an unpredictable history, since what survives by sheer luck or just priority can constrain the future. These are "frozen accidents," the way an archaic programming language like FORTRAN hangs on in science not because it is better, but because it is so impractical to rewrite millions of lines of data analysis code. There is no telling where evolution will lead. For prebiotic history, we have nothing like a fossil record; though chemical evolution is an exciting field making real progress, we might never find out exactly what happened.

How can this collection of counterintuitive principles and studied uncertainties compete with a display of the complexities of a cell, followed by a comfortably anthropomorphic explanation? Creation-scientists love public debates with evolutionary biologists; it is a lot easier to emphasize how crazy evolution sounds than to explain why it works. So creationism holds on tenaciously, well protected by religious populism and by everyday common sense. Evolution is a solid fact in mainstream science, and creation a gross falsity. But in the wider world, the various species of creationism will find their niches and survive, maybe even flourish.

Liberals are not so comfortable confronting science; they will continue to avoid trouble by keeping Darwin safely confined to biology. After all, plenty of biologists consider evolution to be perfectly compatible with faith, and argue forcefully against old-fashioned creationists and ID proponents alike who suggest otherwise.[58] Theology contributes nothing to their understanding of biology, but their work does not get in the way of their piety either.

Philosophers also do their bit for the harmony of science and religion. Michael Ruse, for example, criticizes biologists Richard Dawkins and E. O. Wilson, who have based their skepticism about religion on evolutionary biology.[59] Indeed, *if* Darwinian evolution was about nothing more than the history of life on earth, Dawkins and Wilson would be claiming too much. And since Ruse is also impressed by the many excuses theistic philosophers find to keep saying God guides the process, he finds little left to worry about.

So liberal hopes that Darwin poses no danger will also survive, protected not by creationist common sense but by a newer conventional wisdom which limits science and religion to their separate spheres, and requires no reality checks on theology. This is, to be sure, a relatively harmless view which does not get in the way of serious science. But it results in compatibility without coherence, religion without consequence. Darwinism naturalizes creativity across the board, not just in biology; and disconnected from creativity, God is crippled. If this universal Darwinism holds, then historian of biology William B. Provine appears to be correct: "There are no purposive principles whatsoever in nature. There are no gods and no designing forces that are rationally detectable. The frequently made assertion that modern biology and the assumptions of the Judaeo-Christian tradition are fully compatible is false."[60]

This is not to say that a Darwinian theist is being wildly inconsistent, or that evolution immediately banishes God to the outer darkness of useless ideas. While Darwinism makes a divine reality very doubtful, creative novelty and complex order are not supposed to be all God is responsible for. We might still find weighty reasons to think there is a God, and *then* come back to see God was behind the apparent randomness in our world. However, we cannot start with a straightforward design argument. The precise arrangement of stars in the sky is extremely improbable, and while astronomers explain some overall patterns of clustering, they are happy to leave a lot to accidents. An astrologer might look for a yet unknown astrological pattern behind the night sky, hoping to find something astronomers have overlooked. But the prospects for this are dim, and the same goes for finding a God behind Darwinian accidents.

Instead, we might encounter signs of God elsewhere. Perhaps within, if we have specially created souls, as Catholics say. Or, looking for God in nature, we might yet find a divinity who works through secondary causes. Digging deeper in physics, we might conclude God determined the laws of physics in order to allow our existence. If so, we could still see intricate order as a manifestation of a deep creative force underlying the world. The accidents of evolution would then be part of a larger pattern involving God.

In 1691, John Ray put forth his vision of a majestic, specially created world. Even at the beginnings of modern science, such views had rivals. Ray was especially concerned to avoid the new ideas of "mechanick theism" which made God set the stage, and then disappear behind the scenes.[61] This diminished God, limiting divine action to a remote act of creation. And though the divine creation was grand, it still was like a spiritless machine blindly going through its rigid motions. Ray would have none of this, and many today agree. But today, evolution has so undermined the traditional God that, if there is a God to be revealed by natural science, "mechanick theism" is the only starting point left.

NOTES

1. John C. Greene, *The Death of Adam: Evolution and its Impact on Western Thought* (Ames: Iowa State University Press, 1959), p. 5.

2. Niles Eldredge, *Time Frames: The Rethinking of Darwinian Evolution and the Theory of Punctuated Equilibria* (New York: Simon and Schuster, 1985), pp. 32–34, 185–88.

3. Design, when used without implying underlying purpose, remains a useful biological metaphor. Elliott Sober, ed., *Conceptual Issues in Evolutionary Biology* (Cambridge: MIT Press, 1984), pt. 5.

4. Hume stops short of disagreeing with design as an explanation of the world; David Hume, *Dialogues Concerning Natural Religion*, Part XII: reprinted in David Hume, *Writings on Religion*, ed. Antony Flew (La Salle, Ill.: Open Court, 1992). On how Darwinism leads to a critique of the design argument superior to Hume's, see James Rachels, *Created From Animals: The Moral Implications of Darwinism* (Oxford: Oxford University Press, 1991), pp. 118–20.

5. According to a series of Gallup polls in the United States, belief in a recent creation of humans has remained at about 45 to 50 percent since 1982; *National Center for Science Education Reports* 13, no. 3 (1993): 9.

6. Roland Mushat Frye, ed., *Is God a Creationist? The Religious Case Against Creation-Science* (New York: Charles Scribner's Sons, 1983).

7. John Hedley Brooke, *Science and Religion: Some Historical Perspectives* (Cambridge: Cambridge University Press, 1991).

8. James Gilbert, *Redeeming Culture: American Religion in an Age of Science* (Chicago: University of Chicago Press, 1997).

9. Taner Edis, "Cloning Creationism in Turkey," *Reports of the National Center for Science Education* 19, no. 6 (1999): 30; Taner Edis, "Islamic Creationism in Turkey," *Creation/Evolution* 14, no. 1 (1994): 1.

10. Arthur C. Custance, *Science and Faith* (Grand Rapids, Mich.: Zondervan, 1978), pp. 134–49. Some religious philosophers still defend the notion of an "animal soul" or life force irreducible to physics; e.g., Gary R. Habermas and J. P. Moreland, *Immortality: The Other Side of Death* (Nashville, Tenn.: Thomas Nelson, 1992), p. 51.

11. Seyyed Hossein Nasr, *Knowledge and the Sacred* (Albany: State University of New York Press, 1989), p. 236. When arguing that evolution is intellectually unacceptable, however, Nasr echoes the Protestant creationists with whom he does not want to be associated.

12. Daniel C. Dennett, *Darwin's Dangerous Idea: Evolution and the Meanings of Life* (New York: Simon and Schuster, 1995).

13. Evolution also undermines traditional theodicies, such as attributing evil to the freely chosen disobedience to God of moral agents. Furthermore, natural selection goes a long way toward explaining the nastier aspects of life, while theists must make excuses. Rachels, *Created from Animals*, pp. 103–107.

14. Henry M. Morris, ed., *Scientific Creationism*, 2d ed. (El Cajon, Calif.: Master Books, 1985), p. 219.

15. J. P. Moreland, ed., *The Creation Hypothesis: Scientific Evidence for an Intelligent Designer* (Downers Grove, Ill.: InterVarsity, 1994); William A. Dembski, ed., *Mere Creation: Science, Faith & Intelligent Design* (Downers Grove, Ill.: InterVarsity, 1998).

16. Robert T. Pennock, "Naturalism, Evidence, and Creationism: The Case of Phillip Johnson," *Biology and Philosophy* 11, no. 4 (1996): 543. Trying to rule creation out of scientific consideration on philosophical grounds has been attractive to defenders of evolution; for a critique, see Taner Edis, "Taking Creationism Seriously," *Skeptic* 6, no. 2 (1998): 56.

17. Morris, *Scientific Creationism*, p. 53.

18. Michael J. Behe, *Darwin's Black Box: The Biochemical Challenge to Evolution* (New York: Free Press, 1996); Karen Bartelt, "A Scientist Responds to Behe's 'Black Box,'" *Reports of the National Center for Science Education* 19, no. 5 (1999): 34; Niall Shanks and Karl H. Joplin, "Redundant Complexity: A Critical Analysis of Intelligent Design in Biochemistry," *Philosophy of Science* 66 (1999): 268.

19. Mikhail V. Volkenstein, *Physical Approaches to Biological Evolution*, trans. A. Beknazarov (Berlin: Springer-Verlag, 1994), chaps. 8, 9.

20. This is analogous to a task of finding minima of a complicated energy surface, with only local information available at any step. Increased dimensionality makes it easier to find trajectories which go around barriers, so that "relaxing the dimensionality" is sometimes an effective optimization technique. G. M. Crippen, "Conformational Analysis by Energy Embedding," *Journal of Computational Chemistry* 3 (1982): 471.

21. Stephen Jay Gould, *Bully for Brontosaurus: Reflections in Natural History* (New York: W. W. Norton, 1991), chap. 9; James H. Marden and Melissa G. Kramer, "Locomotor Performance of Insects with Rudimentary Wings: Sailing on Water versus Gliding in Air," *Nature* 377 (1995): 332; M. Averof and S. M. Cohen, "Evolutionary Origin of Insect Wings from Ancestral Gills," *Nature* 385 (1997): 627.

22. Lynn Margulis, *Symbiosis in Cell Evolution: Life and Its Environment on the Early Earth* (San Francisco: W. H. Freeman, 1981).

23. William A. Dembski, *Intelligent Design: The Bridge Between Science and Theology* (Downers Grove, Ill.: InterVarsity, 1999). For a critique, see Taner Edis, "Darwin in Mind: 'Intelligent Design' Meets Artificial Intelligence," *Skeptical Inquirer* 25, no. 2 (2001): 35.

24. William A. Dembski, *The Design Inference: Eliminating Chance through Small Probabilities* (New York: Cambridge University Press, 1998). For criticism, see Brandon Fitelson, Christopher Stephens, and Elliott Sober, "How Not to Detect Design—Critical Notice: William A. Dembski, *The Design Inference*," *Philosophy of Science* 66, no. 3 (1999): 472. However, their objections do not seem fatal.

25. Greene, *The Death of Adam*, p. 83.

26. Wesley R. Elsberry, review of *The Design Inference*, by William A. Dembski, *Reports of the National Center for Science Education* 19, no. 2 (1999): 32.

27. Even theological conservatives can respond to evolution positively if they interpret it as progress. See David N. Livingstone, *Darwin's Forgotten Defenders: The Encounter Between Evangelical Theology and Evolutionary Thought* (Grand Rapids, Mich.: William B. Eerdmans and Edinburgh: Scottish Academic Press, 1987).

28. John F. Haught, *Science and Religion: From Conflict to Conversation* (New York: Paulist Press, 1995), p. 71.

29. John F. Haught, *God After Darwin: A Theology of Evolution* (Boulder, Colo.: Westview, 2000), p. 73.

30. John Polkinghorne, *Belief in God in an Age of Science* (New Haven, Conn.: Yale University Press, 1998), chap. 3; Robert John Russell, Nancey Murphy, and Arthur R. Peacocke, eds., *Chaos and Complexity: Scientific Perspectives on Divine Action*, 2d ed. (Vatican: Vatican Observatory Publications, 1997). Polkinghorne also expresses some vague sympathy with ID; John Polkinghorne, *Faith, Science & Understanding* (New Haven: Yale University Press, 2000), chap. 5.

31. Stephen Jay Gould, *Wonderful Life: The Burgess Shale and the Nature of History* (New York: W. W. Norton, 1989), pp. 27–45.

32. Pierre Teilhard de Chardin, *The Phenomenon of Man*, trans. Bernard Wall (New York: Harper & Row, 1959), p. 142.

33. Peter J. Bowler, *The Non-Darwinian Revolution: Reinterpreting a Historical Myth* (Baltimore: Johns Hopkins University Press, 1988).

34. H. James Birx, *Interpreting Evolution: Darwin and Teilhard de Chardin* (Amherst, N.Y.: Prometheus Books, 1991), chap. 2.

35. William B. Provine, "Progress in Evolution and Meaning in Life," in Matthew H. Nitecki, ed., *Evolutionary Progress* (Chicago: University of Chicago Press, 1988). Developmental ideas are not completely dead; e.g., Mae-Wan Ho and Peter T. Saunders, eds., *Beyond Neo-Darwinism: An Introduction to the New Evolutionary Paradigm* (London: Academic Press, 1984), owes a lot to the developmental tradition. Ho and Saunders's "new paradigm" has not much affected biology, however, let alone redirected it fundamentally.

36. Francisco J. Ayala, "Can 'Progress' Be Defined as a Biological Concept?" in Nitecki, *Evolutionary Progress*.

37. Pierre Gaspard, "Diffusion, Effusion, and Chaotic Scattering: An Exactly Solvable Liouvillian Dynamics," *Journal of Statistical Physics* 68 (1992): 673; Joel L. Lebowitz, "Boltzmann's Entropy and Time's Arrow," *Physics Today* 46, no. 9 (1993): 32; Huw Price, *Time's Arrow and Archimedes' Point: New Directions for the Physics of Time* (New York: Oxford University Press, 1996). Price points out that chaos is not sufficient, and that directions in time crucially depend on cosmological boundary conditions. However, these views are complementary. An increase in entropy depends on the order of information transfer, and not temporal order; see E. T. Jaynes, "Information Theory and Statistical Mechanics II," *Physical Review* 108 (1957): 171. Price's objection, that the physics should be the same for times $t + 1$ and $t - 1$ (p. 44) does not hold, because $t - 1$ is closer to $t = 0$ where we obtained our initial information about the system. Symmetry holds for times t and $-t$, and *this* requires, as Price notes, a cosmological solution.

38. Stephen Jay Gould, "On Replacing the Idea of Progress with an Operational Notion of Directionality," in Nitecki, *Evolutionary Progress*. Gould further argues that it is a mistake to focus on trends, since complexity appears in the tail end of a distribution still characterized by bacterial simplicity; Stephen Jay Gould, *Full House: The Spread of Excellence from Plato to Darwin* (New York: Harmony, 1996).

39. Victor J. Stenger, *The Unconscious Quantum: Metaphysics in Modern Physics and Cosmology* (Amherst, N.Y.: Prometheus Books, 1995), pp. 227–30.

40. Daniel R. Brooks and E. O. Wiley, *Evolution as Entropy: Toward a Unified Theory of Biology*, 2d ed. (Chicago: University of Chicago Press, 1988). See also Stuart Kaufmann, *Investigations* (Oxford: Oxford University Press, 2000), who argues we cannot finitely prestate the configuration space of a biosphere, so novelty is always possible.

41. Thomas D. Schneider, "Evolution of Biological Information," *Nucleic Acids Research* 28, no. 14 (2000): 2794.

42. Remo Badii and Antonio Politi, *Complexity: Hierarchical Structures and Scaling in Physics* (Cambridge: Cambridge University Press, 1997).

43. David M. Raup, *Extinction: Bad Genes or Bad Luck?* (New York: W. W. Norton, 1991); Trevor Palmer, *Controversy: Catastrophism and Evolution: The Ongoing Debate* (New York: Kluwer Academic/Plenum, 1999); Virginia Morell, "Size Matters: The Genes Behind Adaptation," *Science* 284 (1999): 2106; Sid Deutsch, "The Case for Large-Size Mutations," *IEEE Transactions on Biomedical Engineering* 48, no. 1 (2001): 124.

44. Research indicating possible Lamarckian evolution or directed mutation can in fact be explained in a Darwinian manner. Donald G. MacPhee, "Directed Evolution Reconsidered," *American Scientist* 81, no. 6 (1993): 554. For an example of a theistic context for directed development, see Custance, *Science and Faith*, pt. 4.

45. Roger Penrose, *The Emperor's New Mind* (Oxford: Oxford University Press, 1989); Roger Penrose, *Shadows of the Mind* (Oxford: Oxford University Press, 1994).

46. Gregory J. Chaitin, *Algorithmic Information Theory* (Cambridge: Cambridge University Press, 1987), chap. 7.

47. Taner Edis, "How Gödel's Theorem Supports the Possibility of Machine Intelligence," *Minds and Machines* 8, no. 2 (1998): 251.

48. Gerald M. Edelman, *Bright Air, Brilliant Fire: On the Matter of the Mind* (New York: Basic, 1992); William H. Calvin, *The Cerebral Code: Thinking a Thought in the Mosaics of the Mind* (Cambridge: MIT Press, 1996); Susan Blackmore, *The Meme Machine* (Oxford: Oxford University Press, 1999); Dennett, *Darwin's Dangerous Idea*.

49. David B. Fogel, *Evolutionary Computation: Toward a New Philosophy of Machine Intelligence*, 2d ed. (New York: IEEE Press, 2000); Peter Bentley, ed., *Evolutionary Design by Computers* (San Francisco: Morgan Kaufmann, 1999).

50. Denis Edwards, *The God of Evolution: A Trinitarian Theology* (New York: Paulist Press, 1999), p. 53. See also Holmes Rolston III, *Genes, Genesis, and God: Values and Their Origins in Natural and Human History* (New York: Cambridge University Press, 1999); Polkinghorne, *Belief in God in an Age of Science*, chap. 3.

51. Majid Fakhry, *A History of Islamic Philosophy*, 2d ed. (New York: Columbia University Press, 1983), pp. 46–48; William Lane Craig, *Divine Foreknowledge and Human Freedom* (Leiden, The Netherlands: E. J. Brill, 1990), chap. 13. Darwin himself used the idea in some of his early thoughts about evolution; Greene, *The Death of Adam*, pp. 265, 271.

52. Ian Hacking, *The Taming of Chance* (New York: Cambridge University Press, 1990). Interestingly, prescientific uses of "chance" also referred to human affairs, including as an expression of God's will, though there was no coherent idea of randomness behind it. See Keith Thomas, *Religion and the Decline of Magic: Studies in Popular Beliefs in Sixteenth and Seventeenth Century England* (New York: Oxford University Press, 1971), pp. 118–24.

53. Some physicists, particularly David Bohm, postulate causes behind quantum randomness. This takes the form of hidden variables, or permanently inaccessible information which is moreover propagated superluminally! This approach may occasionally be of value in addressing conceptual problems in quantum physics (see contributions in Franco Selleri, ed., *Quantum Mechanics versus Local Realism* [New York: Plenum, 1988]). However, quantum mechanics remains formally the same under Bohmian interpretations, and hence still harbors true randomness. There are no causes we can *infer* behind quantum randomness, we can only declare them to exist if such is our metaphysical taste.

54. Rèmy Lestienne, *The Creative Power of Chance*, trans. E. C. Neher (Urbana: University of Illinois Press, 1998), chaps. 10–13.

55. A. Lazcano and S. L. Miller, "The Origin and Early Evolution of Life: Prebiotic Chemistry, the Pre-RNA World, and Time," *Cell* 85, no. 6 (1996): 793. Also, Daniel Segre, Dafna Ben-Eli, and Doron Lancet, "Compositional Genomes: Prebiotic Information Transfer in Mutually Catalytic Noncovalent Assemblies," *Proceedings of the National Academy of Sciences* 97, no. 8 (2000): 4112, proposes that lipids were instrumental in getting the RNA-world started.

56. Thomas Gilovich, *How We Know What Isn't So: The Fallibility of Human Reason in Everyday Life* (New York: Free Press, 1991), chap. 2.

57. Stewart Elliott Guthrie, *Faces in the Clouds: A New Theory of Religion* (Oxford: Oxford University Press, 1993), chaps. 3, 4.

58. E.g., Kenneth R. Miller, *Finding Darwin's God: A Scientist's Search for Common Ground Between God and Evolution* (New York: Cliff Street, 1999).

59. Michael Ruse, *Can a Darwinian Be a Christian? The Relationship Between Science and Religion* (New York: Oxford University Press, 2001).

60. Provine, in Nitecki, *Evolutionary Progress*, p. 65.

61. Greene, *The Death of Adam*, pp. 8–13.

Chapter Three
The Gods of Modern Physics

There are, who disavow all Providence
And think the world is only steered by chance;
Make God at best an idle looker-on,
A lazy monarch lolling on his throne.

—John Oldham (1682)

Physics is on its way to a Theory of Everything. Maybe. Along the way, though, many Gods claim support from modern physics. There is the Designer who set all the physical constants just right for us, and the Creator who brought our universe into being. And, of course, there is the cosmic mind we are all a part of, maybe even the mathematical genius revealed in the harmonies of natural law. Reality is not as spiritually uplifting as these hopes. Modern physics describes an impersonal universe of purposeless material processes. Deep down, ours is an accidental world.

PHYSICS IN THE NEW AGE

Physics is supposed to be the fundamental science, probing the nature of space, time, matter, and motion. Cosmologists and high energy physicists study the most basic subatomic particles and forces, and the early history of our universe. They talk about the very beginning of time, when the basic forces were unified, and the cosmos was in a state of sublime symmetry captured by just a few equations. To be sure, there was plenty of complexity hiding in this simplicity. As the universe cooled down, the fundamental symmetries were broken. With many particles, simple interactions combined to produce maddening complexities and surprising collective behaviors like superconductivity. But these are secondary manifestations of an underlying simplicity. The truly fundamental facts about the world are those which particle physicists and cosmologists play with.

Of course, this is a caricature; few physicists really think the rest of science is stamp collecting. Still, many of us imagine theoretical physicists, personified by a rumpled image of Einstein, to be a modern priesthood making arcane pronouncements about the true nature of reality. In that case, perhaps religion lurks beneath the surface of physics. Though we chased God out of biology, maybe a deity would not be directly involved with such secondary aspects of creation anyway. We may discover a designing mind, not in the complicated mess of mundane reality but in the equations of fundamental physics.

Historically, our ideas of God and our conceptions of material reality have always been related.[1] For example, ancient physics and cosmologies were quite congenial to a deity. Matter was like an inert lump of substance, which meant motion had to be an independent principle. Since motion was not the natural state of matter, a prime mover had to begin, and perhaps sustain, a changing world. Objects moved according to their natural purposes. The cosmos was earth-centered, underlining our importance in a grand design which was a moral as well as a physical order.

The scientific revolution brought us mechanical explanation, changing our picture of material reality and our ideas of God along with it. A universe infused by spirit and purpose at every level was replaced by bodies in motion described by unbending laws of nature. This was troublesome for old-time religion; however, the mechanical universe meant opportunity as well as difficulty.[2] Institutionalized monotheism always has individualist, magical, and occult rivals, and a *limited* mechanical philosophy proved very helpful in combating freelance supernaturalism. If the world is mainly a machine, it is not a web of deep meanings and occult correspondences one consults an astrologer to disentangle. In such a world, the mundane does not gradually shade into the magical; only procedures authorized by orthodox institutions put us in touch with divinity. The supernatural becomes the ghost in the machine. Mechanist philosopher René Descartes declared that animals were mere automatons; but human consciousness, it seemed, required a nonmaterial soul. After mechanical philosophy and the triumph of Newtonian physics, the supernatural became more sharply defined. Human personality was not an extension of an animistic physics where everything was permeated by spirit; our free will, moral capacity, and rational soul put us wholly beyond physics. A miracle was now not just a wondrous demonstration of power, but a violation of natural law, a clear sign of something transcending nature.

Unfortunately, the "mechanick theists" were not able to limit the demon they let loose. Modern science went on a rampage, seeking ways to explain everything in sight as a consequence of natural, physical processes. This was so wildly successful, it became increasingly difficult to set anything in the world apart from material reality. Instead of achieving security in a domain protected from occult exploitation, the supernatural began to fade away altogether. The laws of nature took center stage; explanations referred to laws, experiments sought lawlike regularities.

Many intellectuals responded to Newtonian physics by becoming deists. God created the world, set its laws, and then disappeared from the scene. However, with the advancement of science, the classical metaphysical reasoning in support of this sort of God—that the universe requires a first cause, that laws require a lawgiver—began to look rather sterile. An absentee God was religiously useless anyway. Taking a different approach, devout scientists claimed that the laws of nature they found pointed to a creator. Newton thought there were instabilities in planetary orbits which required divine interference to correct. Later physicists, like today's creationists, argued that since the second law of thermodynamics meant the universe was becoming more disorderly, something must have ordered the universe at the beginning.[3] But such signs of God had a habit of evaporating as we learned more. Even before Darwin, physical science had prepared the ground for God becoming an unnecessary hypothesis.

Today's physics goes further in discovering a universe far from the traditional religious imagination—indeed, far from common sense. Newtonian physics was at least decent enough to keep space and time, matter and energy separate; its particles were solid, its waves knew their proper place. Worse, while religion denies that deep down anything is an accident, modern physics inflicts randomness on us. The microscopic world is a state of seething, mindless dynamism; a realm where everything in our commonsense physics, from the solidity of objects to events requiring causes, has a habit of breaking down.

So how is a religion supposed to cope? Though physics does not inspire the resistance evolution does, some religious people, such as young-earth creationists, have little choice but to reject much of today's physics. And again, just as with evolution, many popular apologists adopt an easier alternative, proclaiming an amazing correspondence between modern physics and old-time religion instead. A few practitioners of creative interpretation even announce that physical cosmology confirms the Genesis narrative in detail.[4]

Of course, serious religious thinkers do not choose such disreputable means to harmonize physics and religion. Philosophical theists are much more likely to claim that modern physics supports the classical arguments for God. They say modern cosmology shows us a universe fine-tuned to produce intelligence, thus reviving the argument from design. And they argue that the big bang which started our universe must be an event of supernatural creation from nothing. Where science reaches its limits, theology comes to the rescue with a reassuring smile and a more complete explanation.

Some make an even more ambitious argument. Naturalism, the story goes, depended on a Newtonian clockwork universe in which we could explain everything by mechanical processes. Now, however, the spooky world of quantum mechanics is supposed to banish the machine and return spirit to the center of the universe. Physics was but a small part of Newton's intellectual life, which was largely devoted to studying biblical prophecy and the occult science of alchemy. Post-Newtonian physics, many now believe, opens the door to a mystical, consciousness-infused universe which is more in line with the fuller vision of Newton.

This enthusiasm about modern physics peaks in our alternative, individualist religious tradition. New Agers tell us that physicists, the supernerds, have at long last come upon the same mystic truths as gurus and hermits, thereby redeeming themselves:

> Theorists speculate that once upon a time the superforce created space, created matter, nudged the forces we see today into being, and thereby shaped our universe. (Sounds suspiciously like God, doesn't it?) . . . this all seems as though science has come full circle and is now willing to admit that there is a unifying force in the universe, an overriding presence that can explain the workings of the cosmos for the simple reason that He/She/It created the cosmos. . . .
>
> Increasingly, mathematicians talk like mystics, and scientific journals read like holy writ. Physics is becoming indistinguishable from metaphysics. Scientists trained in the rigorous scientific methods, graduates of the school of naturalism, have pushed that naturalism to its furthest extreme—to the extent that the most unlikely people have become *super*naturalists. New Age travelers should be aware of this quiet revolution so that they do not become unduly confused when accosted by physicists wearing clerical robes, chanting formulaic mantras.[5]

Such beliefs about the kinship of mysticism and modern physics are, of course, based on the most superficial of parallels.[6] In our popular culture, physics is all too easily misused as an authority to back ideas we are already predisposed to favor. However, this is only part of the story. Some physicists and philosophers have made serious claims to undermine naturalism, based on modern physics. After all, any God worthy of the name must in some way be responsible for the universe and its basic physical order. We would expect some signs of God in physics, whether this deity is a designer, creator, or cosmic mind. Newtonian physics let us think the world is only natural, nothing more. We must now see what today's physics has to say.

EGOCENTRIC COSMOLOGY

A century ago, cosmology was not a science. We had more metaphysical prejudices than theoretical tools, and little observational evidence. Physicists could speculate, but the best that could be said for their efforts was that they were an improvement over myths of gods making the world.

After Einstein's theory of general relativity, physicists began to talk about all of spacetime as one geometrical entity, and ask what its shape might be. Astronomers discovered that galaxies were moving away from one another, hence the universe was expanding. In that case, by general relativity, a "singularity" had to lurk back in time—starting out the universe with a huge explosion from a single point. We then observed the background radiation which was the cooled-off remnant of the big bang. Cosmology had become a science, with real theories connected to sophisticated observational tests: Not a science as solid as the physics behind transistor radios, but respectable nonetheless.

God usually does not fare well when origin questions become the province of science rather than mythology. However, a serious cosmology also gives religious theories a chance to acquire some real teeth. The design argument, for example, can get a new lease on life.

A physical theory has roughly three elements we can play with: dynamical laws, parameters, and boundary conditions. General relativity tells us how energy curves spacetime, and quantum mechanics explains how elementary particle states evolve; such broad theories are the laws. Parameters are physical constants like the speed of light. Boundary conditions are constraints such as the initial state of a system. A falling rock is described by the laws of gravity, parameters like the strength of gravity, and boundary conditions like the initial height we drop it from.

Figuring out what kind of world we would get if we had different dynamical laws is difficult. Varying parameters and boundary conditions, however, is fairly straightforward, and physicists do this all the time. We do not see quantum effects in everyday life, because Planck's constant $\hbar$ is very small in our universe. An apple does not vanish and reappear outside of a shopping bag, though electrons can tunnel through energy barriers. But we can imagine a world with a much larger $\hbar$, in which carrying fruit home from the market would be a very peculiar task indeed.

When we fiddle with the parameters and initial conditions of cosmological theories, we seem to get many "anthropic coincidences"—very tight restrictions on what the parameters must be if the universe is to support human life. For example, our universe is very large and flat. To the inhabitant of a small planet, say one mile in diameter, it would be obvious she lived on a curved surface, while for us in everyday life, the earth might as well be flat. As far as astronomers can tell—up to intergalactic distance scales and more—our universe is very nearly flat. But from general relativity, we might have expected the universe to be extremely curved, wrapped up into a four-dimensional spacetime ball with a radius of about 10^{-35} meters. It seems the big bang spewed forth lots of matter, at an expansion rate ensuring a large and nearly flat universe. If this initial expansion rate had been less by one part in 10^{17}, the universe would collapse back on itself long before life had any chance to form. If it were larger by the same miniscule fraction, the matter in the universe would quickly be dispersed far and wide, leaving a lifeless, mostly empty space.[7]

Manipulating physical parameters in tiny ways often produces possible worlds inhospitable to any interesting organization of matter. This has led some physicists to suspect, in Freeman Dyson's words, that "the universe in some sense must have known we were coming"[8]—the fundamental physics of the universe seems fine-tuned in order to produce us. Some even propose an "anthropic cosmological principle," suggesting that the existence of humans explains how the universe is.[9] Perhaps we *are* the center of the universe.

With cosmologists glorifying the human ego, theologians cannot be far behind. The fine-tuning observation is a classic prelude to a design argument. In past centuries, physicists could see God in the improbable order and stability of planetary orbits, now it is in the improbably fine-tuned expansion rate. As an added bonus, this version of design involves the grand vistas of cosmology. Theists with a mystical or metaphysical bent find anthropic cosmology quite attractive.[10] A cosmic designer is nicely removed from the sordid affairs of the world, not like a God who has to go around specially creating cockroaches and cabbages.

Not all physicists are impressed, however. Experience shows that intelligent design is a fragile explanation, never accounting for much. It is too easy to convince ourselves the cosmos is centered on the human ego. Anthropic principles produce very little in the way of real knowledge or experimental tests; they look like, as Heinz Pagels put it, "needless clutter in the conceptual repertoire of science."[11] The fine-tuning is, of course, an interesting problem to solve, but to most physicists it suggests deeper physical principles yet to be discovered.

The first thing to look for is an underlying physical constraint. Fine-tuning comes about when we have arbitrary physical parameters we can tweak as we like. But if a better theory reduces the number of free parameters, the fine-tuning often disappears. Consider, for example, mass. Inertial mass is a measure of how difficult it is to accelerate an object. There is also gravitational mass, determining how strongly gravity acts on something. In Newtonian physics these two masses could well be different for an object, but we *always* observe inertial and gravitational masses to be the same. This seems like incredible fine-tuning, though since it does not involve the human ego, it does not excite the theologians. In any case, we already have a better theory. General relativity ties gravity to the curvature of spacetime, so the two kinds of mass have to be the same.

More unified theories of fundamental physics will, similarly, reduce the number of free parameters which anthropic arguments rely on. Perhaps the universe is so large because it is constrained to be large, and the fact that a large universe is necessary for our sort of life is purely incidental.

An example of narrowing the possibilities for fine-tuning is James Hartle and Stephen Hawking's "no-boundary" proposal.[12] General relativity causes trouble because of its arbitrary boundary conditions. Anything can come out of the initial singularity, so naturally universes like ours look unlikely. Hartle and Hawking's quantum cosmology solves the problem by getting rid of the boundary. Then, if we include a rather simple form of matter called a massive scalar field, the early universe quickly enters a phase of almost exponential expansion. On coming out of this phase, we find a very large universe still expanding at almost the critical rate.

This rapid expansion of the early universe has an even more central role in "inflationary" cosmologies. A small universe is curved very tightly, which means it contains a large amount of gravitational energy, even without matter. With many theories of particle physics, this leads to an exponential stretching of spacetime after the universe starts out very small. The presence of scalar fields makes some parts of the universe expand, and however proportionally small these parts may have started out, they inflate to make up almost all the volume of our universe.

Inflationary theories do more than give us a large, flat, and smooth universe. They also hint that whole universes can inflate into being starting from nothing but a quantum fluctuation.[13] This sounds strange—where, then, does all the energy in the universe come from? As it happens, present estimates of the total energy of the universe—all the energy of matter and radiation plus the negative potential energy due to gravitation—are consistent with zero. A universe does not need energy to inflate into being.

Inflation also naturally leads to the idea of multiple universes, since in that case, our universe is one of many bubbles in a "false vacuum." Moreover, those regions of the universe which inflate sprout new inflationary bubbles. Cosmologist Andrei Linde describes just such a "self-reproducing inflationary universe."[14] In each universe-bubble, the low-energy, postinflation physics is slightly different in random ways. And if we have reason to believe there are many different universes with varying physics, fine-tuning arguments lose their force. If the cosmos is continually generating different arrangements, something like our bubble is likely to come about at some point.

Multiple universes lead to another speculative but interesting idea. If the physics in a "baby universe" derives from its parent, but can be slightly different, this would be like passing information from generation to generation with some chance for this information to mutate. The stage is set for Darwinian evolution. This idea is not that bizarre. Life is a peculiar sort of order, falling between the regimentation of a crystal lattice and the boiling chaos of a gas. Complex, self-organizing systems drive themselves to a narrow critical region on the edge of order and chaos. This looks like fine-tuning of parameters to critical values, which is one of the reasons biology has been the traditional inspiration for the design argument. So it is quite possible that the anthropic coincidences in our universe are due not just to physical constraints, but also to quasi-Darwinian processes.

Lee Smolin has, in fact, explored such a Darwinian cosmology.[15] Smolin takes black holes, singularities in spacetime from which not even light can escape, to be birthplaces of baby universes. Creating a black hole also inflates a new and separate spacetime, with physics slightly different from ours. In this case, universes creating the most black holes will reproduce at higher rates, and so will dominate among all the universes in existence. Most universes will be large and near flat because these are efficient black-hole producers.

Of course, all of the above proposals are uncertain. It is hard to say which, if any, will survive in another decade or two. Even now, our views of the flatness of the universe, inflation, and baby universes are in flux, changing with new observational evidence and with new ideas about achieving a full-blown theory of quantum gravity.[16] Plus, with possible other universes, and ever-more extreme conditions and energies, gathering even indirect evidence for such theories becomes a challenge. However, ideas like multiple universes, inflation, or a boundary-free universe arise naturally from our current physics. They generate interesting questions and stimulate real research. Looking for intelligent design in fine-tuning, in contrast, leads to a dead end; it does nothing but promise us we are special.

Still, the design argument refuses to die. For whenever the latest fuss about the incredible improbability of complex order or human life is put to rest, someone will always step back and try again. Physicist Paul Davies is impressed by anthropic cosmology, and is unconcerned if a theory like Smolin's succeeds in linking black holes to the existence of life. If so, "we can still wonder why the laws of nature are such that this linkage occurs. How fortunate that the requirements of life match those of the baby universes so well."[17]

Fortunate, indeed. And there might be some reason for this. But at some point we have to stop; *no* theory can so completely explain the world and our presence in it that no more questions are possible. After a while, hinting at significant coincidences and improbabilities at the limits of our knowledge is not even useful as a spur for research. Without some form of a better theory, and without a way of sampling the world to anchor our probability judgments in reality, there simply is nowhere to go. Following Davies, we would end up like philosophers who demand a complete Sufficient Reason for everything, and try to sculpt a God out of the resulting metaphysical fog.

Still, Davies has a point: we *can* ask questions about the laws of physics. Stephen Hawking criticizes inflationary cosmology because it still leaves the initial conditions at the big bang free—there is no guarantee that our sort of universe will emerge.[18] But if that is a defect for a theory, then Hawking's scalar fields, or even his physical laws, are just as arbitrary. Laws are as much part of the information that goes into a theory as are parameters and boundary conditions—just as to a computer, a program and its input are both strings of bits, and both together make up the information determining its output. A boundary-free universe is an attractive physical idea, but it has no automatic conceptual advantage over a theory that takes certain initial conditions as given. Someone like Davies can always ask Hawking why his laws happen to be such that humans might exist. The best reply is that those laws seem to fit the world we observe, and if Davies or anyone else can find a deeper pattern in what we see, then we will adopt their theory. If there is no real alternative, Hawking's laws can stand as given, with no deeper explanation and no apologies.

In fact, varying everything, including laws, illustrates just how bad an explanation intelligent design is. If humans are central to the world, there is *no need* for such a vast universe, or so many billions of years passing before life emerges. With our kind of physics, these are in fact necessary for life to form. But since we can question everything, we should also wonder why a kind of physics friendlier to God is not true. Why, for example, do we not have an earth-centered cosmos, in which physical bodies move to fulfill intrinsic purposes? Or take the worlds described in the great poems of Dante or Milton. Their fictional universes are designed as a setting for a cosmic religious drama, and are an integral part of their stories. We can easily imagine all sorts of worlds in which we would not need tortured, after-the-fact interpretations to find religious significance in nature. Then why *our* cosmos, among the infinity of possibilities which allow intelligent life?

Clearly the cosmic design hypothesis has a massive improbability problem of its own. There is too much "unwarranted design" in our world—too many features which are unnecessary, about which divine purpose provides no clue whatsoever.[19] Ideally, the rationality and benevolence of God should explain why the divine creation is as it is and not like something Dante could have imagined. But theologians are too busy trying to excuse the obviously nasty, inefficient, and purposeless aspects of our world to try and explain anything. At this point, some theists may be satisfied to say God's ways are incomprehensible to our finite minds. However, asking physicists to be impressed by such a cosmological hypothesis is a bit much.

Retooling intelligent design as an explanation for cosmic order seems, at first, promising. Solid naturalistic theories like evolution are the bane of the design argument, so we move to the limits of our current knowledge, where theories are much more uncertain. But curiously, cosmic design is a weaker proposal than intelligent design in life. While cosmological fine-tuning arguments rely on similar intuitions of improbable order, their empirical base is smaller, and even the half-baked ideas of ID have more of theoretical substance to discuss. The egocentric universe began to die once we adopted mechanistic explanations, and discovered that we inhabited an insignificant corner of a large universe. Today, it is sustained only by our still-insecure egos, and the impossible demand for complete explanations. The heavens declare no gods; religious theories must look elsewhere to find substance.

THE BIG BANGER

The anthropic universe is a mirage. But many theists believe modern cosmology makes a strong case for God even without anthropic fine-tuning. The big bang, after all, indicates that time had a beginning. The galaxies we observe now were hurled apart from a primeval state, where under extreme physical conditions everything was stripped of organization, all the forces were unified, and spacetime itself was crumpled up into a singularity. All this sounds much like the Abrahamic God calling the world into being from nothing. The big bang requires a big banger.

Historically, infidels have usually thought the universe must always have existed. But believing only God could be eternal, many theists in Abrahamic traditions have insisted otherwise. Naturalism does not immediately lead us to expect the universe has a beginning; maybe it was no accident that the physicist who came up with the "primeval atom" theory, a precursor of the big bang, was also a Catholic priest.[20] Perhaps we must conclude, with astronomer Robert Jastrow, that

> [The scientist] has scaled the mountains of ignorance; he is about to conquer the highest peak; as he pulls himself over the final rock, he is greeted by a band of theologians who have been sitting there for centuries.[21]

Of course, Jastrow's comment is exaggerated at best; theologians hardly predicted the big bang. If our universe had turned out to be closed, hence with an end, this would not have meant that apocalyptic visions of the end of the world were on target. And if one version of theism successfully predicted a beginning for the universe, this is still not that impressive. After all, even a stopped clock is right twice a day. The big bang would strongly support a God only if such a beginning requires a cause outside of nature. Christian philosopher William Lane Craig's "kalam cosmological argument" claims exactly this. In outline, he argues:

(1) Everything that *begins to exist* has a cause of its existence.
(2) The universe *began to exist*.
(3) Therefore the Universe has a cause of its existence.[22]

Craig then says this cause must be the God of classical theism.

This is obviously a close relative of the argument that everything must have a cause, and the universe is caused by God. But the familiar cosmological argument runs into trouble because, among other reasons, metaphysicians have to arbitrarily exempt God from needing a cause. Craig's version gets around that difficulty by demanding only that beginnings have a cause. An eternal God needs no cause, and perhaps an eternal universe might have been uncaused as well. But cosmology says our spacetime did, in fact, have a beginning. Hence it has a cause, which might as well be some sort of God.

Craig is not content with mere fallible science, and so claims an eternal universe is *impossible*. In fact, any "actual infinity" is supposed to be impossible, though there is, as usual, a special exemption for godly infinities. The reason behind this impossibility is that Craig cannot imagine how infinities could exist. In a library with an infinite number of books, for example, all integers would be used up in labeling the books, making it absurd to think of adding a new book.[23]

Now, there is nothing formally wrong with infinities, though transfinite mathematics is strange compared to garden variety arithmetic. The librarian in an infinite library adds a new book by labeling it "1," and renumbering each old book n as $n + 1$. Though strange, infinities are not physically impossible;[24] if they play a legitimate role in a physical model, a metaphysician has no business scolding physicists. Many have been uncomfortable with relativity because of the supposed absurdity of curved spacetime, or have thought quantum mechanics defective because it included randomness or did not fit their requirements for reality. Physicists usually manage to accept the strangeness, and eventually discover the sky is not falling after all.

Still, where metaphysics fails, it seems physics does the job. The big bang looks like a point where the universe started, before which there was literally nothing. And though there remain open questions, such as the nature of the large quantities of "dark matter" required by standard cosmologies, the big bang does not have any serious alternative.[25]

The big bang is popularly imagined as a great explosion; at one time there was nothing, then matter erupted into previously empty space. However, the standard big bang is the beginning of spacetime itself, not an event in time. Since we are not very good at picturing curved four-dimensional geometries, imagine a universe where space is one-dimensional. Also say this space is curved into a circle, so that it remains finite, but has no boundaries. The inhabitants of this universe can only travel back and forth along the circle, getting back where they started if they go far enough. Then, as in our universe, let us say space is expanding, so the radius of the circle starts from zero and increases over time. To keep life simple, let us also have our space eventually contract. Now we can picture all of spacetime: it looks like the surface of a sphere. The time dimension is an axis of the sphere, while space is a circular cross-section of the sphere perpendicular to that axis.

At first glance, it seems like this spacetime has a moment of creation. As we go back in time, the radius of space gets smaller and smaller, and finally collapses onto a point. Before that, nothing at all. Unfortunately, this is an illusion. On a sphere, there is nothing special about any point, including the point where the radius of space becomes zero. We can choose a different axis for the sphere, and cross-sections along that axis will vanish at a different point. It does not help to say the time axis is special, since relativity is all about performing rotations which mix time and space. Asking about a time before the beginning of our spherical spacetime is like asking what lies north of the North Pole. There is no such thing. For a true beginning, we need a real feature of the geometry, not an arbitrary point singled out by our choice of coordinates.

General relativity gives us exactly what we need: The initial singularity from which everything emerges is a *boundary* of spacetime.[26] And a boundary point is very special. Though the time before the big bang is still like a place north of the North Pole, at least our pole is no longer arbitrary. Matter does, in fact, begin at the singularity. While we cannot speak of a cause which precedes the universe in time, there may be a cause which is geometrically prior to the universe, or even the Great Programmer acting from outside of physical time.

Unfortunately, the state arising from such a singularity is fundamentally random[27]—*nothing* can be inferred from it. At the singularity, there is no physical law, no causation; the singularity itself stands uncaused. The big bang does not absolve a religious theory of finding patterns explained by a God; and sheer patternlessness is a bad place to look. It is possible that physicists are mistaken about this randomness, that a God chose what emerged from the big bang.[28] But we cannot legitimately say this without finding a purposeful pattern in this outcome; in other words, we would need a successful design argument.

Attaching a creator to the boundary is metaphysical skullduggery. In any case, even the existence of a boundary point is far from a secure fact. General relativity must only be an approximate theory, since it is oblivious to quantum physics as it stands. We do not have a complete theory of quantum gravity yet, despite enthusiasm among physicists for approaches such as string theory and loop quantum gravity.[29] However, we expect quantum effects to smear out the singularity, so that it is no longer a single geometrical point where physics breaks down. In fact, theories like an inflationary universe and a boundary-free universe all incorporate quantum effects, and in all of them, the big bang is no longer a special boundary point. In multiple universe theories, what seem to be singularities in general relativity link different universes to one another; in Andrei Linde's scenario, we have an eternal self-reproducing universe of connected bubbles. Some string models suggest a single universe with an infinite period of development preceding the big bang. And with the Hartle-Hawking no-boundary proposal, we have a single, finite universe, but one like the boundaryless sphere with no special point to attach a creation event. As Hawking remarks, "if the universe [has] no boundary or edge, it would have neither beginning nor end: it would simply be. What place, then, for a creator?"[30]

Cosmology does not give us a creation event. But whether quantum gravity will erase the singularity or not, it is clear our world cannot be eternal in any commonsense way either. It is still natural to think everything must have a cause, so even something as peculiar as a finite but beginningless spacetime might need a supernatural cause. Though they fail upon close examination, cosmological arguments still have an intuitive appeal. For example, Darwin realized the traditional designer-God was incompatible with evolution, but he seems to have remained undecided about whether the universe requires a first cause.[31] And Craig tells us that the fact that a beginning requires a "creating cause" is "so intuitively obvious, especially when applied to the universe, that probably no one in his right mind *really* believes it to be false."[32] Perhaps. But modern physics is no respecter of intuition. In fact, uncaused events are the rule in our universe.

Deeply ingrained intuitions are useful as a rough guide to our everyday world, but no more. Even Newtonian physics strains common sense; it is natural to think that throwing a ball imparts it some impulse, which is exhausted in flight, returning it to its natural state of motionlessness. It was not easy to come up with the idea that bodies move, and keep moving, unless some force acts to change their motion. With relativity, we rotate space and time into one another, a very strange idea. Quantum mechanics is so counterintuitive that physicists have never been able to come up with a comfortable picture of how it works. So we cannot demand that physics satisfy our common sense. In particular, two aspects of our fundamental theories force us to rethink causality: randomness and time reversibility.

We cannot infer a cause for random events, and the physics in cosmology is full of randomness. The state coming out of the standard big bang is random. Black holes destroy all information about the matter that falls in them. They also create random states by capturing particles out of the quantum vacuum and giving their partners the energy to exist independently.[33] And of course, quantum physics is all about uncaused events. Even the quantum vacuum is not an inert void, but is boiling with quantum fluctuations. In our macroscopic world, we are used to energy conservation, but in the quantum realm this holds only on average. Energy fluctuations out of nothing bring short-lived, particle-antiparticle pairs into being, which is why the vacuum is not emptiness but a sea of transient particles. An uncaused beginning, even out of nothing, for our spacetime is no great leap of the imagination; with quantum cosmologies, this becomes a respectable idea.[34]

Our notion of causality singles out a direction in time: causes precede effects. It is then no surprise that since our basic physics is full of uncaused events, it is also time-reversible. In the microscopic realm, we can think of an antiparticle as a particle moving backward in time and we can have connections between events extending into the past as well as the future, all without the time-travel paradoxes generated when we worry about someone going back in time and killing their grandfather. As physicist Victor J. Stenger points out, "In the quantum case, since the same result as backward causality can be obtained without the signal from the future, no information flows into the past and no causal paradox ensues."[35] We live in a world of macroscopic causality and irreversible changes. But our everyday intuitions about time and causality can be a hindrance in thinking about the microscopic realm, which is often more naturally understood when we do not single out a direction in time. Macroscopic causality emerges, like irreversibility in general, from a physical substrate of uncaused events and reversible interactions.

Today causality is a sophisticated mathematical concept, analyzed in artificial intelligence as well as philosophy. And making causal connections may well be more fundamental to being able to model the world than probabilistic reasoning.[36] But it is dangerous to follow metaphysical tradition and assume that what is basic to our reason is also fundamental in physics. Modern physics gives us a microscopic world without causality as we know it. And this means we should be skeptical about philosophical arguments which appeal to our intuitions about causes. The principle that beginnings require a cause *sounds* very compelling, but it is plain wrong. We have learned a lot about causality; there is no excuse today for treating it like it comes to us out of the blue in terms of metaphysical intuitions.

When confronted with a demand that the universe have a cause, skeptics have usually pointed out that God was not much of an explanation. This is true enough, but not really a positive argument. After mechanistic explanation became popular, infidels liked to restrict causality to the chain of causes in an eternal material universe, pointing out that no supernatural cause was then necessary. This was plausible enough, but still a rather defensive argument. Today's skeptic can do better. In all likelihood, the universe is uncaused. It is random. *It just is*.

A QUANTUM SPIRIT

Modern physics is altogether too self-sufficient to support a God who stands apart from the universe. The God of more mystical traditions, however, is both radically Other and manifested *in* the world. So for some religious thinkers, the "new physics" is an opportunity to reenchant our world, uniting matter and spirit once again.

According to "quantum mysticism," physics reveals a universe shaped by consciousness.[37] And in popular and New Age culture, the word "quantum" seems to mean magic. Quantum physics is supposed to grant us psychic powers, show us we create our own realities, and open the door to a wonderful life of healing and wholeness.[38] The world, in this view, is a network of occult correspondences. For example, to an astrologer, the microcosm of human life corresponds to the macrocosm of the heavens. It does not matter that there is little overt interaction between humans and planets; unlike mere Newtonian forces, quantum consciousness forges a connection of *meaning*. If we associate Mars with war, this is enough to link it to aggression in our spirits. Quantum physics, apparently, gives us a world in which consciousness is central, meaning is a palpable reality, and everything is an undivided occult whole.

In such a universe, God is a cosmic mind rather than an external creator and judge. We discover divinity by looking within. As New Age philosopher Ken Wilber puts it, divine consciousness is the highest level in a hierarchy of being extending from mere matter to the purest spirit:

1. *Physical*—nonliving matter/energy
2. *Biological*—living, pranic, sentient matter/energy
3. *Mental*—ego, logic, thinking
4. *Subtle*—archetypal, transindividual, intuitive
5. *Causal*—formless radiance, perfect transcendence
6. *Ultimate*—consciousness as such, the source and nature of all other levels[39]

Though it seems based on new physics, quantum mysticism paints a familiar picture of a world infused with spirit. It is motivated not so much by physics as by deep-seated convictions about the nature of our minds. Our folk psychology describes an immaterial mind, a magical consciousness with a unity which resists analysis. Modern science, in contrast, has a stubbornly materialist bent: the brain is an extremely complex piece of biological circuitry, and mind is what the brain does. Curiously enough, looking to explain minds without going beyond a physical brain means that basic physics is not directly relevant in studying minds. Physics just allows systems capable of complex information processing; after that, neurobiologists and cognitive scientists investigate minds in their own ways and find that the supposed undifferentiated wholeness of the mind gives way to piece-by-piece examination.[40]

Against such materialism, defenders of spirit can dig in their heels and embrace some sort of immaterial soul. Unfortunately, this dualism is too obviously obscurantist, and doomed to retreat as science advances. A more attractive option is to notice that neuroscientists and researchers in machine intelligence rely on the old physics—macroscopic, classical mechanics—to set the stage for their work. But this physics is supposed to be overthrown, or at least limited. Perhaps, then, the mind is not physical in crassly mechanistic Newtonian terms, but is still intimately connected with a new physics where an irreducible consciousness comes built into the basic fabric. Some prominent physicists have thought quantum mechanics is just such a theory. Eugene Wigner, for example, believed "it was not possible to formulate the laws of quantum mechanics in a fully consistent way without reference to the consciousness";[41] and went on to speculate that a nonmaterial consciousness was the ultimate reality. So at the edges of science, quantum psychology flourishes, from analogies between quantum wave functions and the unity of consciousness, to harmonies of physics and the occult notions of Jungian psychoanalysis.[42]

Such ideas find support from some of the genuine puzzles in quantum physics. Consider a system which can be in either of two states, like a light switch on a wall. In everyday life, and in classical Newtonian physics, the switch is either up, $\uparrow$, or down, $\downarrow$. A quantum state, however, can be a mixture in which these states are superposed on one another:

$$|\psi\rangle = a\,|\uparrow\rangle + b\,|\downarrow\rangle$$

The state $|\psi\rangle$ describes all we know about the switch, including the fact that when we measure it we will find $\uparrow$ or $\downarrow$, with a probability determined by the coefficients a and b.[43] Say we get $\uparrow$. Then the new state will no longer be a mixture, but $|\psi\rangle = |\uparrow\rangle$. Once we find $\uparrow$, further measurements will confirm that knowledge.

This behavior is certainly very strange. Still, nothing so far suggests a role for consciousness. The possibilities we might see depend on the experimental context, but since the results are random, we do not *choose* what we observe.

Now comes the interesting part. One thing in quantum mechanics stands out as a definite, classical state: the result of the measurement. Indeed, everything in everyday life behaves like classical objects; no one observes light switches in a limbo, half up and half down. We—macroscopic, classical beings—use macroscopic, classical instruments to probe quantum systems, and get classical, definite results. Yet quantum mechanics is supposed to describe everything, not abruptly stop when objects become large. Our instruments and ourselves must be described by some enormously complicated quantum state. So how does our world of definite measurements come about?

Let us say we have an instrument, a switchalyzer, which looks at microscopic quantum switches. It displays a $\Uparrow$ when it measures $\uparrow$, $\Downarrow$ when $\downarrow$. But since the switchalyzer is part of the quantum world, it is reasonable to suppose it ends up in a mixed state like $a\,|\Uparrow\rangle + b\,|\Downarrow\rangle$. If this is so, we are in trouble. To measure the macroscopic mixed state of the switchalyzer, we can build a switchalyzer-alyzer, but that will have the same problem. No chain of mere devices will give us a definite measurement, until a conscious observer comes on the scene. *We* never see mixed states. So we might suspect that consciousness is what picks out a definite classical world from among the quantum mixtures.

Before we accept magical shortcuts, however, we should look more closely at the measurement process. Historically, quantum mechanics has been defined by assuming a classical measurement procedure was available. For most ordinary calculations, this works perfectly. However, a side effect of sweeping the quantum-classical crossover problem under the rug is that in quantum mechanics, measurement looks like an elementary operation not subject to further analysis. The mixed state $|\psi\rangle$ stands isolated from the world, we stick in a classical switchalyzer, and out pops a classical result $\Uparrow$. In real laboratories, measurement is a complicated physical process, switchalyzers are not simple macroscopic two-state systems, and everything interacts with a very complex external environment.

Complex, macroscopic systems mean irreversible behavior. We have no way of keeping track of all quantum interactions which make up a classical measurement, so we constantly lose information to haphazard interactions with the environment. As with thermodynamics, however, we can look for appropriate macroscopic variables to describe overall statistical properties of the system, even though these variables will behave very differently than the microscopic physics. In the process of measurement, the mixed quantum state "decoheres" through its interaction with the switchalyzer and its environment, and the whole system evolves toward a set of states with classical properties. Long before any conscious observer glances at its readout, the switchalyzer converges onto a state much like a classical ⇑.

Decoherence, then, brings us very close to a classical world. In that case, as physicist Wojciech H. Zurek observes, there is no reason to suspect that consciousness has a role in quantum physics:

> Relevant observables of individual neurons, including chemical concentrations and electrical potentials, are macroscopic. They obey classical, dissipative equations of motion. Thus any quantum superposition of the states of neurons will be destroyed far too quickly for us to become conscious of quantum goings-on: Decoherence applies to our own "state of mind."
>
> . . . Conscious observers have lost their monopoly on acquiring and storing information. The environment can also monitor a system, and the only difference from a man-made apparatus is that the records maintained by the environment are nearly impossible to decipher. Nevertheless, such monitoring causes decoherence, which allows the familiar approximation known as classical objective reality—a perception of a selected subset of all conceivable quantum states evolving in a largely predictable manner—to emerge from the quantum substrate.[44]

This is not to say the quantum-classical crossover is a completely solved problem. Though many physicists now think decoherence is fundamentally important in preventing macroscopic mixed states, there are still open questions. However, it appears decoherence is a large part of the solution, and even approaches which downplay macroscopic decoherence formulate quantum mechanics without conscious observers.[45] Quantum measurement used to be something of a theoretical blind spot, the domain of philosophers as well as physicists. But though still a complicated technical problem, quantum measurement is increasingly an experimental matter, with significant practical applications.[46] We no longer need burden physics with speculations about consciousness in measurement.

Consciousness does not seem to have a direct role in basic physics. However, quantum mysticism may be able to do without a mind which magically determines the observed world. Classical physics pictures a world of particles zooming around, affecting each other by various forces. These forces always weaken with distance; furthermore, if relativity is correct, no information can travel faster than the speed of light. Now, what if at the quantum level we find a connection between particles which is more like occult correspondence than ordinary interaction?

Many physicists have been troubled by the way quantum mechanics does not respect our classical prejudices. Einstein, most notably, thought there had to be a deeper, better-behaved theory, to which quantum mechanics was only an approximation. David Bohm came up with a solution. He rewrote the equations of quantum mechanics to look like Newtonian equations, except that there was an additional force responsible for quantum effects. Everything now had a definite classical reality from the start. The additional force, however, was unusual. It did not fall off with distance, and it was not calculable from any sources analogous to charges or masses. Particles could be described by definite classical variables, but some of these variables were hidden—there was no possible way of learning their values. And the relationships between these hidden variables were maintained by signals moving faster than light. Bohm would later make a virtue of these properties, and say that reality is an unfolding of an "implicate order" linking everything together in a grandiose mystical unity.[47] An omnipresent "force" responsible for instantaneous effects could well lead to a universe of occult correspondences.

Physicists tend not to be overly impressed with such ideas.[48] The implicate order has no observable physical consequence; all calculations end up identical to conventional quantum mechanics. The notion of undetectable signals flying around the universe faster than light is not very helpful. Even as a different perspective on quantum physics, Bohm's view is misleading. For example, it seems like Bohm's universe would satisfy Einstein by restoring determinism. This, however, is an illusion. Consider a casino where the dice are fair. We can always say that what the dice shows is determined by invisible demons who look up the result of the next roll in a random number table located in hyperspace. So there is no such thing as chance, as the results are all predetermined. Perhaps. But the dice rolls are *still* random, because they are patternless. The demon theory relocates the source of randomness in hyperspace, but still has to take the contents of the random number table as given. The story is the same with hidden variables—quantum mechanics remains just as random as before. Bohm's implicate order is completely irrelevant to the real-world implications of quantum mechanics.

If quantum mechanics is to support any universal wholeness, we need some real, observable physical effects. A demonstration that occult action at a distance is possible would be a good start. Quantum mystics claim that influences of our conscious choices can be transmitted arbitrarily far across the universe, undegraded with distance, faster than the speed of light.

A common illustration of this claim is the famous Einstein-Podolsky-Rosen (EPR) paradox. Let us prepare a pair of our quantum switches, making sure they are set in opposite directions; if one is ↑, the other is ↓. Our quantum state is then $|\psi\rangle = a|\uparrow\downarrow\rangle + b|\downarrow\uparrow\rangle$. Now, let us separate the switches from one another, taking care not to disturb how they were set. At some point, we measure one of the switches, and find ↑. We then instantly know the other switch is described by $|\downarrow\rangle$, no matter how far away it is.

This is not as impressive as it first looks. Say we had two classical switches instead, set them to ↑↓, and sealed them in different boxes. Then we send one of the boxes to Mars. When we open the box here on Earth, and see it is ↑, we instantly know the one in the Martian box is ↓, without waiting out the minutes it would take for a radio signal to travel back and forth. But it is obvious that there is no faster-than-light signal communicating to the Martian switch that it must be ↓. We gain no new information when we realize the Martian switch is ↓; we just work it out from the way the switches were first prepared.

Quantum switches are, of course, trickier. In a classical switch, "up" and "down" are fixed directions. When we measure a switch, we find it up or down according to how it was originally set—we do not see it pointing east or west. But for quantum switches, we choose the direction along which we measure ↑ and ↓ *at the time of our measurement*, not when we prepare the switch. When we observe a quantum switch—in real life, the "spin" of a particle—on Earth, we find ↑ along a direction of our choice: upward, toward the North Star, whatever. The switch on Mars will then be described by $|\downarrow\rangle$ along the same direction. Whether our switches end up ↑ or ↓ is random; we have no control over it. But the alignment of the Martian switch's state was determined on Earth, *independent* of the constraints built into our initial preparation of the switches. At first glance, it seems like our act of choosing a direction has an instantaneous physical effect on Mars.

Before jumping to conclusions, however, we should remember that quantum strangeness is not news; the material question is whether information about our choice of direction was instantly sent to Mars. If so, we could realize a science-fiction writer's dream and build a faster-than-light communicator. But this does not work. The colonists on Mars have to choose a direction to observe the switch, but they do not know anything about the direction chosen on Earth. Though they will measure an ↑ or ↓ along their chosen direction, that will tell them absolutely nothing about what Space Control on Earth decided. Transmitting information faster than light by exploiting quantum mechanics is quite impossible.[49] As with the classical case, all the information we have comes from our knowledge of how the switches were first prepared, and our observation here on Earth. There is nothing paranormal or mystical about EPR, only a quantum world which does not behave according to our classical intuitions.[50]

Proclamations of a quantum spirit are likely to remain popular; if nothing else, the strangeness of quantum mechanics provides plenty of opportunity for obfuscation. But though quantum and classical physics differ, the basic problem Newtonian physics presented for religion remains. Ours seems to be a very material world. It is a dynamic place; matter is not an inert lump of substance. Even the quantum vacuum seethes with activity. But for all its energy, this is a godless, soulless, directionless world. We, including our minds, seem all too much a part of this world. Where, among the equations and laws and randomness, can a God fit in?

LIFE, THE UNIVERSE, AND EVERYTHING

Though modern physics describes a thoroughly material world, physics still has a vaguely spiritual reputation. Historically, physics had the task of reading the mind of God, and the culture of physics still carries traces of this past. There are always physicists ready to interpret physics mystically,[51] and physicists, perhaps more often than other scientists, use "God" as a metaphor—especially in popular writing, but sometimes in the technical literature as well.[52] Of course, this signifies a rational order underlying the confusion of everyday reality, rather than a full-blown theistic God. Nevertheless, the metaphor can take on a life of its own. Albert Einstein was puzzled at the orderliness and mathematical beauty he saw in the universe, and speculated that a superior intelligence was revealed in "the harmony of natural law."[53] The culture of physics emphasizes elegant mathematical order and fundamental symmetries, not lowly accidents, and occasionally this inspires a kind of Platonic mysticism about the laws of nature.

If we find beautiful symmetry principles behind the chaos of everyday life, it is natural to think this indicates a deeper reality of great significance. It might even be something to be religious about. This sense of overwhelming importance is even clearer at the fringes of physics. Practically no one dreams up crank theories about the Secrets Of Reality And A Better Sex Life by mutilating entomology. But physics is *fundamental*, so revolutionary changes solving all our problems must be grounded in the "new paradigms" of physics. Quantum psychology will restore wholeness and meaning to life, dispelling our alienation.[54] Nonlinear mechanics will integrate our vision, helping us construct an environmentally sound approach to the world. So naturally, someone will always stamp "God" on any important physical theory. Psychoanalyst Edmund D. Cohen tells us:

> The major implication of relativity physics for depth psychology is that the unconscious part of the psyche exists outside space and time. It touches the inanimate universe, other psyches, and perhaps even a creator supreme being in ways that we are, by nature, unable to observe directly, and can reach—if at all—only by inferences from indirect observations.[55]

Anyone with a passing acquaintance with relativity will recognize this as nonsense. But such fringe ideas rely on a very mainstream sentiment: physics is central. Physics is supposed to be the key to life, the universe, and everything.

If we need a cosmic perspective to give direction to our lives, looking to physics makes sense. But in that case, we are out of luck. Physics has endless intellectual fascination and considerable technological use, but it has discovered no cosmic purpose, indeed, precious little of direct human significance. A good novel will do a lot more to give us a perspective on life than any number of books on black holes or superconductors. Physicists like Einstein may admire the symmetries of physical law, but it is an *impersonal* order which excites them. Theorists like to describe as much as possible with the least number of principles; their holy grail is a set of fundamental equations which can fit on a T-shirt. A deity would only be conceptual clutter disturbing the harmony.

More importantly, overemphasizing the symmetries of basic physics produces a distorted picture of science. A compact set of equations encompassing all of physics would be a remarkable achievement. However, those equations would also explain very little in our world. The more simple and elegant our basic physics, the more important accidents are. This sounds paradoxical, since we think of elegant mathematical laws as the very opposite of a distastefully arbitrary collection of accidents. But in fact, they are inseparable, the way the randomness of a coin flip is intimately tied to the symmetry between heads and tails.

Symmetry is one of the most powerful concepts in modern physics. Conservation laws, for example, arise from the symmetries of basic physical laws; energy is conserved because our physics does not vary with time. Moreover, the fundamental interactions are all derived from requiring that basic equations be invariant under certain symmetry transformations.[56] Theorizing about basic physics is largely a search for the right symmetries formulated on a sufficiently exotic abstract geometry. These symmetries are most obvious at high energies. Consider a collection of small, atomic-size magnets. The equations describing magnetism are rotationally symmetric; they do not single out any direction in space. At high temperatures, the energy of individual magnets will swamp their mutual interactions. So our collection of magnets will point in haphazard directions, making the overall magnetization of the collection average out to zero. Indeed, if there were a total magnetization, this would indicate a preferred direction in space, going against the symmetry of our equations. When things cool down, something interesting happens. We can no longer neglect the interactions between our atomic magnets; and these interactions favor the magnets lining up in the same direction. Once we get below a critical temperature, the most stable configuration of a collection of magnets will be when most of them line up with one another. This means an overall magnetization, in an arbitrary direction in space, even though the underlying equations are all rotationally symmetric.

Magnets are a simple example of symmetry breaking, in which the low-energy state of a physical system is less symmetric than the equations describing the system. The direction a collection of atomic magnets ends up pointing toward is completely arbitrary—all directions are still equivalent. The magnet settles on a direction determined by environmental noise, or even something as mean as a quantum fluctuation; the state frozen out at low temperatures is entirely accidental.

Much the same kind of symmetry breaking takes place within the more sophisticated equations of elementary particle physics. For example, electromagnetism and the weak force responsible for nuclear decays are unified at high energies. As a result of symmetry breaking, they are very different in our low-energy physics. So we can look at highly symmetric, elegant fundamental theories as a description of *potentialities*. The equations on a T-shirt do not determine our world; that is mainly the result of a series of accidents freezing out the low-energy physics we live by. Highly simple, symmetric theories contain very little information compared to the chaotic richness of life. What they tell us is what sort of coins we flip to get our world. And of course, the more compact and elegant the fundamental theory, the more scope for accidents.

Once we realize symmetries are frameworks for accidents and not echoes of a mystical harmony, we can shift our focus to the particular accidents which form our world. Fortunately, there are plenty of deep theories besides those of basic physics to help us understand an accidental world. These theories tend to be universal, in the sense that they are not very sensitive to details of the underlying physics. Natural selection works independently of the physical sources of accidental variation; hence it is a widely applicable principle. Time asymmetries in our macroscopic world, like thermodynamic irreversibility, exist independently of fundamental physics; it does not even matter whether this physics is quantum or classical.[57] The mathematical theory of computation is mostly insensitive to the details of physical computers. Concepts like information, accidents, and complexity cut across all sciences; and powerful theories about them, like Darwinian evolution, are as fundamental as any physical law.

Giving accidents their due is especially important in a time when research into particle physics faces a real danger of stagnation. We study elementary particles and forces at extremely high energies, where the symmetries are unbroken. Physicists do this with the aid of particle accelerators, among the largest, most expensive, and most complicated scientific instruments ever built. They boost subatomic particles to the brink of light speed in vast underground tunnels tens of miles in circumference, and slam them into one another. But some of the present generation of theories require accelerators the size of solar systems to be tested. We may be approaching a point where it is increasingly difficult to uncover the next level in the deep structure of our world.[58] If so, treating physics as the quest for the fundamental harmonies is doubly misleading. Physics has succeeded not because of intuitions of mathematical elegance, but through a close interaction of theory and experiment. In particle theorist T. D. Lee's words,

- *First Law of Physicists:* Without experimentalists, theorists tend to drift.
- *Second Law of Physicists:* Without theorists, experimentalists tend to falter.[59]

Without theory, science becomes stamp collecting; even separating facts from mistakes depends on our theories. And without experiment, there is nothing to guide theorists but their prejudices.

But even if we reach a point where theoretical particle physics drifts, this means little for the rest of science. We already know accidents are fundamental, whether they appear in the symmetry breaking which froze our low-energy physics in place, or in the mutations which are the raw material for evolution. They are not mere accretions on the true reality of mathematical harmony. Furthermore, an accidental world is not a formless chaos to be shaped by the gods. It is a deeply *historical* world. The evolution of the universe is constrained by the frozen accidents of the past, but novelties also keep arising from, again, accidents. Ours is not a world to be summed up in a few equations; so far, there always seems to be more to find out, new patterns to be noticed, surprises within what we thought we already knew. A historical world is often inelegant, sometimes cruel, but it is always interesting. Whether the search for more basic physical laws stagnates or not, the end of fundamentally important science is nowhere yet in sight.[60]

As far as we can tell, we need nothing outside of physics to explain our world. Elementary particles and their interactions probably make up all we see. However, this does not, as is too often thought, imply that particle physics is *the* basic science and everything else is working out its implications. Setting fundamental laws above all else is itself an accident of our intellectual history. We have tended to believe that theology and philosophy addressed the really deep, fundamental questions. Physics was less pure, limited to describing the material world. Other sciences were even less dignified, requiring thinkers to further dirty their hands with the fallen world. Today, we are more skeptical about deep metaphysics, more impressed with the successes of natural science. We should not, however, dethrone theology merely to substitute physics as the Queen of the Sciences. The sciences are an interdependent web, not a hierarchy. To do justice to an accidental world, physics must lose some of its aesthetic luster and its claims to read the mind of God. But then, coming down to earth from the harmonies of pure thought never hurt any science.

AN UNNECESSARY HYPOTHESIS

In 1799, the great French mathematician Pierre-Simon Laplace published his *Celestial Mechanics*. In this triumph of Newtonian physics, Laplace also confirmed the long-term stability of the solar system, doing away with the occasional divine adjustments Newton thought necessary. The story is that Napoleon asked Laplace what part God played in his system, and Laplace answered, "I have no need of that hypothesis."

Newtonian physics started causing trouble by its very independence from theology. A science where God was not built-in from the foundations was not necessarily a problem, as the idea of God would only be strengthened by the abundant independent confirmation which would surely be forthcoming. Events turned out otherwise; classical physics continually shrank the space for a God active in the world. Physical law began to stand on its own, rather than being the result of divine legislation.

Modern physics radically changed how we understand the basic physical structure of the universe. But it has only exacerbated the difficulties Newtonian science caused for God. In today's physics, God is even more obviously an unnecessary hypothesis. More importantly, modern physics brings accidents to the forefront as well as physical law. Even our firmest intuitions about causation apply only to a macroscopic world emerging from a random substrate, where events just happen. And even the most ambitious fundamental laws are but frameworks for the kinds of accidents which make our world.

This spells disaster for religious theories. We do not expect to be able to see the hand of God in everything. If we cannot fathom the divine reasons behind why it rained last Sunday, or why the sky is blue, this is hardly a great challenge to religion. But when God vanishes from physics, indeed, from all natural science, it begins to look like there is no God after all. If there were a cosmic mind, a creative power, a divine purpose behind everything, we should see traces of this God in our world. We do not.

As usual, this is not a strict disproof. Theists still hope God is somehow responsible for what we see in science, if we just look below surface appearances. Physicist and theologian John Polkinghorne speaks of a God conceived not as a causal agent, but a shadowy figure behind everything. Not only do gaps in our knowledge point to God, but God is generally a "sustainer . . . of 'gaps' and regularities alike."[61] Unfortunately, this God makes as much sense as a Santa Claus who is useless in explaining anything about Christmas, yet still is somehow the moving spirit behind the holiday season. The problem remains: we do not see God in the creation of the heavens, in the beneficent design of earthly creatures, in a consciousness which encompasses the universe. And without some such pattern, the reality of God is very doubtful.

So is the argument about God over? Not yet. Natural science is not all there is to know. Historian Michael J. Buckley observes that science has corroded faith ever since Newtonian physics became a source of truth independent of God. Religious experience throughout history, though, may do the trick:

> In turning to some other discipline to give basic substance to its claims that God exists, religion . . . is admitting an inner cognitive emptiness. If religion does not possess the principles and experiences within itself to disclose the existence of God, if there is nothing of cogency in the phenomenology of religious experience, the witness of the personal histories of holiness and religious commitment, . . . or the life and meaning of Jesus of Nazareth, then it is ultimately counterproductive to look outside the religious to another discipline or science or art to establish that there is a "friend behind the phenomena." . . . To attempt something else either as a foundation or as a substitute, as did the *Newtonian Settlement*, is to move into a progress of internal contradiction of which the ultimate resolution is atheism.[62]

This makes sense, but it is also too obviously a defensive retreat. The idea of God has always included claims about which sciences like physics or biology have something to say. Otherwise, God would only be a peculiar entity allegedly explaining religious experience, and not a unifying principle for all of reality. Those scientists and theologians who looked to nature to find God had good reasons to do so. It was always possible that natural science would confirm the glory of God. It just did not happen. There is no averting disaster by retroactively calling the whole enterprise a religious mistake.

Still, religious history and experience is the next place to look. If something like a God discloses itself in the course of human history or in the special revelations granted to prophets and seers, we *might* have to reconsider whether the accidents of natural science are only seeming accidents. This will, however, be a very difficult thing to do. Our most reliable knowledge comes from natural science, with its solid results and powerful theories. This knowledge is not comforting. We seem to live in a strictly natural, impersonal world, indifferent to our hopes and fears. Some of us find this world liberating, some of us see it as an unbearable emptiness. In either case, the universe does not care.

NOTES

1. Richard Sorabji, *Matter, Space and Motion: Theories in Antiquity and Their Sequel* (London: Duckworth, 1988).

2. John Hedley Brooke, *Science and Religion: Some Historical Perspectives* (Cambridge: Cambridge University Press, 1991), chap. 4.

3. This was fairly common until recently. E.g., Arthur Eddington, *The Nature of the Physical World* (Cambridge: Cambridge University Press, 1930), pp. 84–85; Paul Davies, *God and the New Physics* (New York: Simon & Schuster, 1983), pp. 10–11.

4. E.g., Gerald L. Schroeder, *The Science of God: The Convergence of Scientific and Biblical Wisdom* (New York: Free Press, 1997); Hugh Ross, *The Genesis Question: Scientific Advances and the Accuracy of Genesis* (Colorado Springs, Colo.: Navpress, 1998).

5. Alice and Stephen Lawhead, *Pilgrim's Guide to the New Age* (Tring: Lion, 1986), p. 9. A classic New Age view is Fritjof Capra, *The Tao of Physics: An Exploration of the Parallels Between Modern Physics and Eastern Mysticism* (New York: Bantam, 1977).

6. Jeremy Bernstein, "Eastern Mysticism and the Alleged Parallels with Physics," *American Journal of Physics* 57, no. 8 (1989): 687.

7. Stephen W. Hawking, *A Brief History of Time* (Toronto: Bantam, 1988), p. 121.

8. Freeman Dyson, *Disturbing the Universe* (New York: Harper & Row, 1979), p. 250.

9. John D. Barrow and Frank J. Tipler, *The Anthropic Cosmological Principle* (Oxford: Clarendon, 1986).

10. E.g., John Leslie, in Clifford N. Matthews and Roy Abraham Varghese, eds., *Cosmic Beginnings and Human Ends: Where Science and Religion Meet* (Chicago and La Salle, Ill.: Open Court, 1995), pp. 337–53; Errol E. Harris, *Cosmos and Theos: Ethical and Theological Implications of the Anthropic Cosmological Principle* (London: Humanities Press, 1992), chap. 13.

11. Heinz R. Pagels, *Perfect Symmetry: The Search for the Beginning of Time* (New York: Simon and Schuster, 1985), p. 359.

12. Stephen Hawking, in Stephen Hawking and Roger Penrose, *The Nature of Space and Time* (Princeton: Princeton University Press, 1996), chap. 5.

13. Victor J. Stenger, *The Unconscious Quantum: Metaphysics in Modern Physics and Cosmology* (Amherst, N.Y.: Prometheus Books, 1995), pp. 218–23. Alan H. Guth, *The Inflationary Universe: The Quest for a New Theory of Cosmic Origins* (Reading, Mass.: Addison-Wesley, 1997).

14. Andrei Linde, Dmitri Linde, and Arthur Mezhlumian, "From the Big Bang Theory to the Theory of a Stationary Universe," *Physical Review D* 49, no. 4 (1994): 1783; Martin Rees, *Before the Beginning: Our Universe and Others* (Reading, Mass.: Addison-Wesley, 1997). For criticism, see Stephen W. Hawking and Neil Turok, "Open Inflation without False Vacua," *Physics Letters B* 425 (1998): 25.

15. Lee Smolin, "Did the Universe Evolve?" *Classical and Quantum Gravity* 9 (1992): 173; Lee Smolin, *The Life of the Cosmos* (New York: Oxford University Press, 1997).

16. Questions concerning the adequacy of current theories of inflation, whether there is a cosmological constant, and possible deviations from flatness continue to be debated. An accessible review is Ronald Ebert, "A Bang or a Whimper? Cosmology at the Beginning of a New Millennium," *Skeptic* 8, no. 3 (2000): 48. No doubt things will change further by the time this book is published.

17. Paul Davies, *The Mind of God: The Scientific Basis for a Rational World* (New York: Simon & Schuster, 1992), p. 222.

18. Hawking, *The Nature of Space and Time*, p. 90.

19. Nicholas Humphrey, *Leaps of Faith: Science, Miracles, and the Search for Supernatural Consolation* (New York: Basic, 1996), chap. 12, uses a similar argument he calls an "Argument from Unwarranted Design" against claims of psychic powers.

20. Georges Lemaître, *The Primeval Atom: An Essay on Cosmology*, trans. B. H. Korff and S. A. Korff (New York: D. Van Nostrand, 1950).

21. Robert Jastrow, *God and the Astronomers* (New York: Warner, 1980), p. 125.

22. William Lane Craig, in William Lane Craig and Quentin Smith, *Theism, Atheism, and Big Bang Cosmology* (Oxford: Clarendon, 1993), p. 4. The emphasis is mine.

23. Ibid., chap. 1. For classical Islamic versions of the argument, see Majid Fakhry, *A History of Islamic Philosophy*, 2d ed. (New York: Columbia University Press, 1983), pp. 74–77, 224.

24. Smith, *Theism, Atheism, and Big Bang Cosmology,* chap. 2. See also Michael Martin, *Atheism: A Philosophical Justification* (Philadelphia: Temple University Press, 1990), pp. 105–106.

25. This is not to say all physicists are satisfied with the big bang; see Geoffrey Burbridge, Fred Hoyle, and Jayant V. Narlikar, "A Different Approach to Cosmology," *Physics Today* 52, no. 4 (1999): 38, and Andreas Albrecht's reply following it. However, such views are very much in the minority.

26. Charles W. Misner, Kip S. Thorne, and John Archibald Wheeler, *Gravitation* (San Francisco: W. H. Freeman, 1973), p. 934.

27. Stephen W. Hawking, "Breakdown of Predictability in Gravitational Collapse," *Physical Review D* 14 (1976): 2460; Smith, *Theism, Atheism, and Big Bang Cosmology*, pp. 195f.

28. For example, to circumvent randomness, Craig relies on the classical Islamic notion that the outcome of an indeterminate situation is to be explained

as a choice made by a personal agent, in this case a personal God. William Lane Craig, *The Kalam Cosmological Argument* (New York: Barnes & Noble, 1979), pp. 149–52. This is a misunderstanding of randomness, as well as archaic psychology.

29. Lee Smolin, *Three Roads to Quantum Gravity* (New York: Basic, 2001).

30. Hawking, *A Brief History of Time*, pp. 133–41. A somewhat forced interpretation of a no-boundary universe as divine creation from nothing is possible; e.g., Chris J. Isham, in Robert J. Russell, William R. Stoeger, S.J., and George V. Coyne, S.J., eds., *Physics, Philosophy, and Theology: A Common Quest for Understanding* (Vatican: Vatican Observatory, 1988), pp. 397–402. However, this derives its substance from the classical cosmological argument, not physics.

31. James Rachels, *Created From Animals: The Moral Implications of Darwinism* (Oxford: Oxford University Press, 1991), pp. 107–10.

32. Craig, *Theism, Atheism, and Big Bang Cosmology,* p. 57.

33. Hawking, *The Nature of Space and Time*, chaps. 1, 3. Hawking also explains how black holes are also white holes, due to their thermal radiation (pp. 124–27). I would extend Hawking's argument. Black holes destroy our knowledge of both macroscopic properties and microscopic states. Thermal radiation creates new microscopic states unconnected to previous states; however, this contains practically no macroscopic information. In other words, black holes are a gravitational agent of thermodynamic irreversibility, even though the physics remains CPT-invariant.

34. Smith, *Theism, Atheism, and Big Bang Cosmology*, makes this analogy; Craig objects (pp. 143–44) by pointing out that the quantum vacuum is a dynamical state, hardly nothing. However, the very complexity of the vacuum is a result of virtual particle creation processes. Creation out of absolute nothing is a metaphysical quagmire anyway, since nothing must at least have the potentiality for becoming something (Peter A. Angeles, *The Problem of God: A Short Introduction* [Amherst, N.Y.: Prometheus Books, 1980], pp. 59–60). Since we are stuck with potentiality, it might as well be something like a quantum vacuum.

35. Victor J. Stenger, *Timeless Reality: Symmetry, Simplicity, and Multiple Universes* (Amherst, N.Y.: Prometheus Books, 2000), p. 207. Stenger also uses time-symmetry to give an attractive explanation of contextuality in quantum mechanics—the fact that the entire experiment, in future and past, must be taken into account in making predictions; Stenger, *The Unconscious Quantum*, pp. 204–206. See also Huw Price, *Time's Arrow and Archimedes' Point: New Directions for the Physics of Time* (New York: Oxford University Press, 1996), chaps. 7–9.

36. Judea Pearl, *Causality: Models, Reasoning, and Inference* (New York: Cambridge University Press, 2000).

37. See Patrick Grim, ed., *Philosophy of Science and the Occult*, 2d ed. (Albany: State University of New York Press, 1990).

38. Catherine L. Albanese, "The Magical Staff: Quantum Healing in the New Age," in *Perspectives on the New Age*, ed. James R. Lewis and J. Gordon Melton (Albany: State University of New York Press, 1992); Deepak Chopra, *Ageless Body, Timeless Mind: The Quantum Alternative to Growing Old* (New York: Harmony, 1993). For a critique, see Douglas Stalker and Clark Glymour, in Douglas Stalker and Clark Glymour, eds., *Examining Holistic Medicine* (Amherst, N.Y.: Prometheus Books, 1989), pp. 107–25.

39. Ken Wilber, in Ken Wilber, ed., *The Holographic Paradigm and Other Paradoxes: Exploring the Leading Edge of Science* (Boulder, Colo.: Shambhala, 1982), p. 159. Mystical misinterpretations of physics attract liberal Abrahamic believers as well; e.g., Richard Elliott Friedman, *The Hidden Face of God* (San Francisco: Harper, 1995), sees God as a kind of progressively revealed cosmic consciousness, though he distorts the big bang to make it fit the mysticism of the Kabbalah, rather than playing with quantum mysticism like New Agers.

40. Gerald M. Edelman, *Bright Air, Brilliant Fire: On the Matter of the Mind* (New York: Basic, 1992), pp. 212–18. Daniel C. Dennett, *Consciousness Explained* (Boston: Little, Brown, 1991), chaps. 6, 9.

41. Eugene P. Wigner, *Symmetries and Reflections: Scientific Essays of Eugene P. Wigner* (Bloomington: Indiana University Press, 1967), p. 172.

42. David Hodgson, *The Mind Matters: Consciousness and Choice in a Quantum World* (Oxford: Clarendon, 1991), pp. 383–87; Danah Zohar, with I. N. Marshall, *The Quantum Self: Human Nature and Consciousness Defined by the New Physics* (New York: William Morrow, 1990), p. 69; John Hitchcock, *The Web of the Universe: Jung, the "New Physics," and Human Spirituality* (New York: Paulist Press, 1991). See also the "panexperientialism" of David Ray Griffin, *Parapsychology, Philosophy, and Spirituality: A Postmodern Exploration* (Albany: State University of New York Press, 1997), chap. 3.

43. The quantum state $|\psi\rangle$ is not analogous to a classical single-system state; treating it as if it were a classical physical quantity is misleading; Stenger, *The Unconscious Quantum*, pp. 86–91, 165–70. $|\psi\rangle$ changes, or "collapses" after a measurement; Heisenberg suggested this blurred the Cartesian boundary between knower and known (Werner Heisenberg, *Physics and Philosophy: The Revolution in Modern Science* [New York: Harper & Brothers, 1958], pp. 78–83). This seems more profound an observation than it really is, because we tend to think of the "collapse" like a change in a classical physical entity.

44. Wojciech H. Zurek, "Decoherence and the Transition from Quantum to Classical," *Physics Today* 44, no. 10 (1991): 36. See also Gregory R. Mulhauser, "Materialism and the 'Problem' of Quantum Measurement," *Minds and Machines* 5, no. 2 (1995): 207–17.

45. For a defense of the fundamental role of decoherence, see Roland Omnés, *The Interpretation of Quantum Mechanics* (Princeton: Princeton University Press, 1994), chap. 7. A nontechnical exposition is Murray Gell-Mann, *The Quark and the Jaguar: Adventures in the Simple and Complex* (New York: Freeman & Co., 1994), chap. 11. For a critique and alternative approach, see Sheldon Goldstein, "Quantum Theory without Observers—Part One," *Physics Today* 51, no. 3 (1998): 42.

46. E.g., V. B. Braginskii and F. Y. Khalili, *Quantum Measurement* (Cambridge: Cambridge University Press, 1992).

47. David Bohm, *Wholeness and the Implicate Order* (London: Ark, 1983); Stenger, *The Unconscious Quantum*, chaps. 4, 5.

48. This is not to say a Bohmian approach is not useful on occasion. Some still think it is the best conceptual framework; e.g., Sheldon Goldstein, "Quantum Theory without Observers—Part Two," *Physics Today* 51, no. 4 (1998): 38; plus, it might be a starting point to develop theories which go beyond

conventional quantum mechanics. Hard-core realists in philosophy who demand interpretations without wavefunction collapse also are attracted to Bohm; Jeffrey Bub, *Interpreting the Quantum World* (New York: Cambridge University Press, 1997). Many physicists are skeptical of all such efforts; Christopher A. Fuchs and Asher Peres, "Quantum Theory Needs No 'Interpretation,' " *Physics Today* 53, no. 3 (2000): 70.

49. Philippe H. Eberhard and Ronald R. Ross, *Foundations of Physics Letters* 2 (1989): 127. Taking *information* as fundamental is a promising way to approach quantum mechanics; Anton Zeilinger, "A Foundational Principle for Quantum Mechanics," *Foundations of Physics* 29 (1999): 631.

50. Murray Gell-Mann and James B. Hartle, "Quantum Mechanics in the Light of Quantum Cosmology," in *Complexity, Entropy, and the Physics of Information*, ed. Wojciech H. Zurek, Santa Fe Institute Studies in the Sciences of Complexity, vol. 8 (Redwood City, Calif.: Addison-Wesley, 1990); Stenger, *The Unconscious Quantum*, chap. 5.

51. Ken Wilber, ed., *Quantum Questions: Mystical Writings of the World's Great Physicists* (Boulder, Colo.: Shambhala, 1984). Note that even here, no *direct* support by physics is claimed. Also see contributions by religious physicists in Henry Margenau and Roy Abraham Varghese, eds., *Cosmos, Bios, Theos: Scientists Reflect on Science, God, and the Origins of the Universe, Life, and* Homo Sapiens (La Salle, Ill.: Open Court, 1992).

52. E.g., Detleff Dürr, Sheldon Goldstein, and Nino Zanghi, *Journal of Statistical Physics* 67 (1992): 843. The authors argue for a Bohmian pseudo-deterministic interpretation of quantum mechanics, and support this by a God's-eye appeal to rational design.

53. Albert Einstein, *Ideas and Opinions* (New York: Crown, 1954), p. 40. On Einstein's cosmic religion and Spinozan God, see Max Jammer, *Einstein and Religion* (Princeton: Princeton University Press, 1999). Today's religious physicists continue to see God in the "rational" order of basic physics; e.g., Roland Omnès, *Quantum Philosophy: Understanding and Interpreting Contemporary Science* (Princeton: Princeton University Press, 1999), chap. 16; Mariano Artigas, *The Mind of the Universe: Understanding Science and Religion* (Philadelphia: Templeton Foundation Press, 2000).

54. Zohar, *The Quantum Self*, pp. 17–20, 219–20.

55. Edmund D. Cohen, *The Mind of the Bible-Believer* (Amherst, N.Y.: Prometheus Books, 1988), p. 104. Cohen criticizes fundamentalism from a psychoanalytic perspective; a case of the pot calling the kettle black if there ever was one.

56. Local or gauge symmetries are the most conspicuous example of building fundamental physics on symmetry; Elliot Leader and Enrico Predazzi, *An Introduction to Gauge Theories and Modern Particle Physics* (New York: Cambridge University Press, 1996).

57. Some physicists, notably Roger Penrose, think otherwise: that irreversibility is such a profound fact about our world that it must have a more "fundamental" explanation; Penrose, *The Nature of Space and Time*; Roger Penrose, *The Emperor's New Mind* (Oxford: Oxford University Press, 1989), pp. 302–47. Since the initial smoothness of our universe is so improbable, Penrose suspects a time-asymmetric law of nature is behind this. This is certainly possible; however, as I noted in

the previous chapter, statistical physicists see no need to build irreversibility into our fundamental theories. Price, *Time's Arrow and Archimedes' Point*, chap. 4.

58. David Lindley, *The End of Physics: The Myth of a Unified Theory* (New York: Basic, 1993).

59. T. D. Lee, *Symmetries, Asymmetries, and the World of Particles* (Seattle: University of Washington Press, 1988), p. 41. The importance of experiment in particular is sometimes clearer with failed physical ideas; see Allan Franklin, *The Rise and Fall of the Fifth Force: Discovery, Pursuit, and Justification in Modern Physics* (New York: American Institute of Physics, 1993).

60. Even within physics, what is "fundamental" is subject to debate; the elementary particle physicists' is not the only perspective. Silvan S. Schweber, "Physics, Community and the Crisis in Physical Theory," *Physics Today* 46, no. 11 (1993): 34; Philip W. Anderson, "More Is Different," *Science* 177 (1972): 393.

61. John Polkinghorne, *Science and Providence: God's Interaction with the World* (Boston: Shambhala, 1989), p. 34.

62. Michael J. Buckley, in Russell et al., *Physics, Philosophy, and Theology*, p. 99. Also, Michael J. Buckley, S.J., *At the Origins of Modern Atheism* (New Haven, Conn.: Yale University Press, 1987).

Chapter Four
History and Holy Writ

The merest accident of microgeography had meant that the first man to hear the voice of [the Great God] Om, and who gave Om his view of humans, was a shepherd and not a goatherd. They have quite different ways of looking at the world, and the whole of history might have been different.

For sheep are stupid and have to be driven. But goats are intelligent and need to be led.

—Terry Pratchett, *Small Gods* (1992)

Sometimes it seems religion is the urge to thump very hard on a suitable Book. Our traditions tell us a God directs our history, and even communicates with us, though often he—usually a "he"—says some very peculiar things. We then collect stories about our Gods in Holy Books, and demand everyone treat them with utter solemnity. But we can also study history with a more critical eye. When we do this, we find that religion is a very human creation, and that our history does not have an overall purpose any more than the evolution of life.

SPECIAL REVELATION

Traditionally, the wonders of the natural world were supposed to convince us there is an awesome creative force who designed it all. Somehow, though, this "general revelation" was not very satisfying—it never seemed to say much about God. If, for example, the universe required a creating cause, there was no guarantee this cause was a personal deity. And even if science or philosophy could lead us to God, this was a dry, lifeless knowledge which could not substitute for an encounter with the Word of God. God must reach out to us, otherwise we fall short. Protestants often go so far as to declare that our very intellects are darkened by sin, making remedial revelation necessary for a saving knowledge of God.[1]

Special revelation—especially scripture—is central to the Abrahamic religions; much more so than the God of the philosophers. Children grow up hearing tales from the Bible or the Quran; the stories of our religion become templates for our culture. We absorb lists of thou-shalt-nots, examples of moral rectitude, everything Deeply Significant with some entertaining sex and violence thrown in. We learn that "God has power over all things" not from metaphysicians, but from the Quran's story about a man whom God let die, then brought back to life a hundred years later. We learn that the dead can rise again from God's demonstration to Abraham, instructing him to tame four birds and then kill them and scatter the pieces on separate hills; they then came flying back when Abraham called (2 Al-Baqarah 259, 260).

To religious conservatives, such wonders are supposed to have taken place in real history, not once-upon-a-time. They are not just strange happenings out of the blue but divine messages, part of a religious community's relationship with God. Life is not one damn thing after another; events fit into a story—a divine plan. God might be a cosmic power, but with divine acts in history, cosmic purposes come down to earth, touching ordinary lives. In the words of a popular Old Testament textbook,

> biblical faith, to the bewilderment of many philosophers, is fundamentally historical in character. It is concerned with events and historical relationships, not abstract values and ideas existing in a timeless realm. The God of Israel is known in history—a particular history—through his relations with Abraham, Isaac, and Jacob.[2]

Prophets and patriarchs speak for God, calling nations to repentance. In the history of Yahweh's chosen people, Jews and Christians see God acting in the world. For Muslims, Muhammad was the last prophet, bringing the final and perfect revelation. Human history is a religious drama, beginning with Creation and ending in Judgment.

Christians in particular used to believe in an elaborate salvation history. From the time of the church fathers to the Renaissance, historians regularly began their accounts with Creation, followed the biblical story up to Jesus, brought the reader to the present, and even told of the Apocalypse and Judgment which would bring the story to its climax. The whole universe was a stage for the drama of human salvation. Philosophers such as Augustine wrote that history would go through six ages corresponding to the six days of creation, leading into a seventh age in which the faithful rest eternally. Such explicit religious patterns remained part of traditional historical writing for many centuries.[3]

Another way of seeing God's plan in history was observing how prophets foretold the future. Christians read the Jewish scriptures as a prophecy of Jesus,[4] though somehow God never allowed the Jews themselves to see this. To this day, conservative Christians believe in biblical prophecy, even that we may be living in the End Times. Harold Camping, who rashly predicted the world would end in 1994, nevertheless gives voice to a common expectation:

> The return of Israel to its land, the potential for massive worldwide destruction by nuclear war, and the rapid increase in communication technology (permitting the Gospel to penetrate everywhere in the world), are some of the phenomena that cause serious students of the Bible to wonder if the end of time is at hand.[5]

Though such beliefs remain popular, they have disappeared from educated circles. The European Enlightenment changed how we think of history, as well as the universe. Historical writing became secularized; history itself began to be seen not as a cosmic drama, but a narrative of human events; "the scope of histories became ever more limited as the focus of attention gradually shifted from the supernatural to the natural, from the universe to nation, principality, city."[6] Historians became increasingly critical, evidence-oriented, and naturalistic. And as with the new natural sciences, critical history became a major headache for religion. Instead of divinely inspired scriptures, we were left with stories which inspire religious communities. Religion itself became an object of secular historical study, no longer a window into the supernatural purposes behind human events.

Today, conservatives still steadfastly defend their scriptures, saying their religion is supported by facts like "the unity of the Bible, its fulfilled prophecies, its remarkable accuracy as determined by archaeology and historical investigation."[7] But the claims which Abrahamic religions have made about history are as dubious today as creationism in biology. What was once obvious has now become ridiculous. So liberal religious people have learned to read stories of prophecy or Noah's Ark as morality tales, not historical truths.

However, like creationism, we still have to take traditional historical claims seriously. For most believers, the idea of God is inseparable from the particular special revelation they put their trust in. Beliefs like prophecy fulfillment survive for good reason; they have always served to confirm that there is a divine plan of salvation, that scripture is true, even that there is a God at all.[8] A robust sense of revelation requires a religious pattern in history; a bunch of moral fables is not enough. If we found that something like Augustine's six ages were true, we might see that a supernatural power influences the course of events. If students of prophecy such as Camping were able to tell us what lies ahead, we could construct a religious theory with a predictive power secular historians would envy.

Augustine was wrong, and we can no longer believe that the Bible or the Quran is holy writ straight from God. But it still might be possible to salvage a sense of purpose in history from old-time religion. The Bible might not be free of error, but it may still testify reliably to a supernatural pattern in ancient Israel's history. Prophets such as Muhammad may bear witness to the divine intruding in ordinary time. For if there is a God offering salvation to sinners, we should expect *some* signs of special revelation in our religious history. The Abrahamic God whom so many have so fervently believed in must speak to us—religion needs a God who reaches out to its creatures, not a creator permanently aloof from our predicaments. There is no point in looking behind the vast silences of an indifferent universe just to find an equally silent God.

YAHWEH'S PROMISE

Yahweh, the story goes, starts out by creating the world and the first human pair. Humans, however, turn out to be a disappointment; they are a rebellious bunch and continually do evil. Frustrated, Yahweh destroys his creation with a great flood, saving only a chosen few. Unfortunately, humans multiply and begin acting like pests again. So Yahweh decides to cultivate a special nation from the seed of Abraham. He gathers the Jews out of Egypt, and leads them to a promised land which becomes theirs after some ethnic cleansing. He also promises that if the Jews try and become more godly, obey his laws, be nice to their neighbors, and slaughter whom he dislikes, they will become a privileged people. Yahweh, the Lord of All Creation, will be present in Jerusalem, and the temple-state will prosper. In time, the Jews will lead the other nations into righteousness. But meanwhile, being the chosen people is also a heavy burden; the Jews will transgress against Yahweh and be severely punished for doing so.

At face value, this is an overly ambitious national epic. But in various mutated forms, Yahweh's promise ended up capturing the imagination of billions. Part of its appeal is that it is supposed to be real, historical—not a tale of a timeless mythic realm. With revelation in history, divinity is manifested in change as well as a sanctified social order.

So the orthodox often claim the Hebrews developed a unique historical consciousness. Biblical narratives are supposed to be infused with a linear sense of history, a concreteness lacking in the cyclic myths of their pagan neighbors. Dying and rising agricultural deities were not for the Jews; time marched on driven by a divine purpose from Creation to the Day of the Lord. As a result, theologians such as Paul Merkley argue, "in the Jewish tradition standards of faithfulness in preserving and perpetuating received accounts of past events were incomparably higher than ever became the case among the Hellenes, and were of longer standing by many centuries."[9] From trustworthy sources, we learn of the prophets, and that the fate of the Jews depended on how faithful they were to Yahweh. The biblical writers were set apart by their awareness of the transcendent forces behind history. Even their dogged monotheism was unique compared to pagan beliefs. Left to their own devices, humans set up idols of their own contrivance, worshiping social or natural forces. The Jewish sense of a peerless God transcending all creation can only arise from revelation.

From a secular, critical point of view, however, the Jewish epic is spectacularly indifferent to history. As Thomas L. Thompson puts it,

> Here and there, the Bible uses data gleaned from ancient texts or records. It often refers to great figures and events of the past . . . at least as they are known to popular tradition. But it cites such "historical facts" only when they may serve as grist for one of its various literary mills. The Bible knows nothing or nearly nothing of the great, transforming events of Palestine's history. Of historical causes, it knows only one: Palestine's ancient deity Yahweh. It knows nearly nothing of the great droughts that changed the course of Palestine's world for centuries, and it is equally ignorant of the region's great historical battles at Megiddo, Kadesh and Lachish. The Bible tells us nothing directly of four hundred years of Egyptian presence. Nor can it [teach] us anything about the wasteful competition for the Jezreel in the early Iron Age, or about the forced sedenterization of nomads along Palestine's southern flank.[10]

The Bible is a religious work, centered on themes like the errors of old Israel and the making of the new, which will follow God's will rather than that of humans. Both the old Israel and the new are largely mythic, not part of history. The Bible's interest is to demand unconditional loyalty to Yahweh, not to recount the past.

Still, using the bits of history in the Bible, and independent textual and archaeological evidence, we can outline what happened. In the beginning, Yahweh was a local weather god. Exactly how he became the chief god in Jerusalem and Samaria is unclear. But even in the Bible, we can see a Yahweh very similar to his neighboring gods. He spoke through thunder, threw around lightning like arrows, fought great monsters and cut them up, directed battles as the Lord of Armies, received sacrifices. He had his holy mountains, his throne, cultic objects, a chosen people.[11] We also find hints of Yahweh as a son of the Most High god, as a national deity in charge of Israel. Expressing nationalist conceit, he was the most powerful among the sons of El Elyon. And then we find Yahweh, "a great king above all gods" (Ps. 95:3), in the position of supreme god himself, a creator on his throne, surrounded by a heavenly court of lesser gods.

Traces of Iron Age religion appear only as fossil fragments in the Bible, where Yahweh, though ancient in name, has become a more universal, transcendent God—as much a Platonic philosophical abstraction as a distinct local deity. This development was brought on by the empires, particularly Assyria, which displaced Near Eastern peoples and promoted a concept of divinity reflecting the centralization of political power. Local gods were consolidated, located in hierarchies, and became part of a more universal religious perception.

This is not to say that local elites lost interest in their national religions; if anything, their unstable circumstances provided greater motivation to impose meaning on their world through theological reflection. Yahwists also found plenty of traditions, including different versions of the same stories, to assemble into a literary epic, and plenty of ancient wisdom about the will of their gods to construct tales about how God directed their lives.

For example, ancient Near Eastern religions explained national calamities by divine anger. A Moabite inscription from about 830 B.C.E reveals that their national fortunes fluctuated according to the whims of their chief god. Chemosh—"the abomination of Moab" to Yahweh's partisans (1 Kings 11:7)—is angered with Moab, and uses the Israelites as an instrument to punish Moab. Mesha, "son of Chemosh, king of Moab," regains the favor of Chemosh and strikes back:

> Omri was king of Israel, and he oppressed Moab for a long time, for Chemosh was angry with his land. And his son succeeded him and he too said: "I will oppress Moab." In my days Chemosh spoke thus, and I looked down on him and his [Omri's] house.

Like Yahweh of old, Chemosh instructs his human instruments in battle and proper extermination, so Mesha says:

> And Chemosh said to me, "Go, take Nebo against Israel!" And I went by night and fought against them from dawn until midday. And I took it and slew them all: seven thousand warriors and old men—together with women and old women and maidens—for I had consecrated it for Ashtar-chemosh. And I took from there the ram of Yahweh and dragged it before Chemosh. And the king of Israel had built Jahaz and remained in it while fighting against me. But Chemosh drove him out before me, when I took from Moab two hundred men, all its brave men.[12]

The mythic history running from Deuteronomy to 2 Kings uses this theme of disloyalty and divine anger. Written in or after the exile of the elites of Jerusalem to Babylon, its story looks forward toward the time when the Jews will be driven out of their land. The Jews are not properly attentive to Yahweh, so Yahweh hands them over to oppressors for a while, rescues them, and then the cycle repeats itself. Yet the Deuteronomist uses the disloyalty and anger theme in a new way. He appears to represent the views of a faction which, after the exile, reformed the religion to preserve it in the absence of royalty—as the rabbis would later adapt Judaism to life without a Temple. A historical relationship with Yahweh would substitute for the ancient cult mediating the people's relationship with a timeless divine realm. This view is pressed in a theological fiction, written in light of the displacement caused by regional empires, preserved by a religious establishment who liked its theme. They also kept other versions of the same story, such as Chronicles, which revises Kings to fawn over the figures of David and Hezekiah.

The result is no doubt a compelling myth, especially when thought to be real. Yahweh's promise came to be reflected in a pattern believers expected to see in Jewish history. Later, Christians adapted this pattern to themselves, usually by making their own nation the focus of divine attention. Even today, many Americans imagine themselves part of a chosen nation, rising or falling by the way they carry out biblical mandates.[13]

Of course, not everyone submitted to this vision. While the religious establishment preferred one god, one people, one temple, the people themselves often had other ideas. The Bible continually complains about Jews who follow gods lacking the official seal of approval, so these idols must have been popular. Even the belief that Yahweh was a second God, the great angel of Israel, mightiest of the sons of the Most High, held on for long; the early rabbis still had to denounce heretics believing in "two powers in heaven."[14] Established Jewish religion distinguished itself by its condemnation of freelance contact with the supernatural—magic. Folk beliefs, as always, remained different; Jews also acquired a reputation for producing powerful magicians.

Nevertheless, the Yahwists succeeded. More and more, their epic came to be etched in stone, so that when Jewish communities reimagined their myths, they would start with themes from the official texts. After the Babylonian exile, not only was Yahweh brought more firmly into history, but apocalyptic visions of the end of history flourished as never before. The Book of Daniel, for example, advertises itself as the prophecies of a Jewish hero during the exile, laying out the future of the chosen people. From the exile to the reign of Antiochus IV over the Jews some centuries later, Daniel seems to have had a clear vision except for some minor errors. During that time, nationalist Jews erupted in the eventually successful Maccabean rebellion. Daniel is aware of Antiochus's policies, and explains foreign domination over the Jews as punishment for Israel's sins against Yahweh. However, with the predictions of Antiochus's death and beyond, its narrative becomes increasingly bizarre.

Daniel was written under a false name, during the rebellion against Antiochus.[15] With the outcome uncertain, the writer encouraged his side in a prophetic manner which stretched out into apocalyptic fantasy. Burton L. Mack, a historian of early Christianity, points out that apocalyptic scenarios emerge from religious communities in trouble. "It is a people's investment in their traditions, institutions, culture, and ideals that determines what is 'good.' Only when the good is in danger of being overwhelmed by external circumstances does an apocalyptic rationale begin to make sense."[16] Because history *does not* follow the expected pattern, the apocalyptic imagination vindicates the goodness of God by setting history aright at a cataclysmic end. So the writer of Daniel produced a piece of wartime propaganda, disguised as ancient prophecy. Like other apocalyptic writers, he borrowed the authority of a revered figure of legend. Daniel expresses an inspiring vision of history under God's control; it is also a pious fraud.

We tend to think history is more straightforward than physics—a matter of telling a story. But critical history does not come any more naturally to us than science. We are mythmakers, not historians. Competing nations in today's Near East still tell their histories to bolster nationalist and religious ideologies. In the United States, schools teach a national epic of unflagging progress, with a few misunderstandings like slavery along the way.[17] And when we want to tell the story more accurately, we end up in the quagmire of having to avoid upsetting the self-images of every tribe.

For centuries, Yahweh's promise has been a myth central to our cultures, so we have wanted to believe that the ancient Jews were reliable historians. But, as historian Robin Lane Fox observes, "The people who have been described as 'obsessed with history' had not a single historian among them with a critical idea of evidence."[18] Stories of God's will and theologies of covenants and retributions are continually rewrought myths of religious communities, telling us much about the communities but little about history.

It is because the Bible is story and sermon that it remains so vital to communities. Myths invite retelling, reimagining. Liberal theologians, finding that the Bible was a theological enterprise from the beginning, can produce layers upon layers of new interpretations. Realizing that grand patterns in history are mythical can liberate the religious imagination, inviting theologians to contemplate the deep hopes and fears expressed even in dark apocalyptic fantasies. And despite their claim to be faithful to the original revelation, conservatives also interpret their myths creatively. Evangelical Christians, for example, make excuses for the fact that we are still here despite Daniel's promise that the world would end seventy sevens of years after exilic times. Since the last block of seven occurs after an "anointed leader" disappears and vanishes (Dan. 9:26), and this, with some stretching, falls in the ballpark of Jesus' crucifixion, many believe that the prophetic timetable is interrupted to postpone the last seven years indefinitely.[19] Such flexibility is the rule; believers in prophecy have seen the figure of Antichrist as the emperor Nero revived from death, a Jewish false messiah, a Muslim Turk, the pope, Martin Luther, Napoleon III, and more[20]—all supported by impressive exegesis and detailed calculations.

Where, then, does all this leave revelation? The liberal response to history seems like a retreat before modern knowledge, but there is more to it. Part of being religious is engaging a myth, working within a tradition to create new meanings for it. So perhaps, by a kind of progressive evolution, our religious beliefs are continually refined toward higher truths. We are not merely custodians of a past revelation which can only degenerate—communities of faith produce revelation as well as preserve it.

This is an attractive idea. However, the conservatives also have a point. Inspirational literature is not enough; religions must make contact with a reality which is more than the human imagination. And the present scholarly history of religion does not challenge naturalism. Even a secular person who identifies with the tradition rooted in a religion, or who just enjoys myths as literary creations, can find it worthwhile to continue rewriting the myths for our times. Claims of grand patterns of reward and retribution and prophecies foretelling the future seem crudely fundamentalist today, but if they were true, they would point to something beyond nature. With a liberal, scholarly approach, religious traditions still provide comfortable metaphors to express feelings of transcendence or cosmic meaning, but the course of events no longer reveals a God. Religious experiences within history, seen in the context of tradition, may, of course, still tell us of God. Prophetic messages or the founding events of great religions may still demonstrate a radical novelty intruding on the natural flow of history. Nevertheless, these are much weaker claims than the traditional robust sense of purpose in history. As with natural science, God starts to vanish from the public stage, retreating to the individual experiences of religious people.

THE MESSENGER OF GOD

Inhabitants of what used to be Christendom are still trying to come to terms with the fact that history does not follow the biblical epic. Islam proclaims a different story, with different difficulties. Muslim history unfolds between Creation and Judgment as well, but what happens in the meantime need not follow a grandiose pattern of salvation history. Occasionally, God sends messengers to the different nations, punishes a disobedient tribe, or performs miracles through a prophet. But such stories in the Quran are told as "a collection of interesting anecdotes about persons who had lived at some period in the past—a collection not in any way chronologically ordered."[21] For example, 34 Saba 12–14 speaks of jinns, lesser supernatural beings of Arab tradition, enslaved to build things for Solomon:

> We (subjugated) the wind to Solomon. Its morning journey took one month, and the evening's one month. We made a spring of molten brass to flow for him; and many jinns labored for him by the will of his Lord. Anyone of them who turned from Our command was made to taste the torment of blazing fire. They made for him whatever he wished, synagogues and statues, dishes as large as water-troughs, and cauldrons firmly fixed (on ovens; and We said): "O House of David, act, and give thanks." But few among My creatures are thankful. When We ordained (Solomon's) death, none but the weevil, that was eating away his staff (on which he rested), pointed out to them that he was dead. When he fell down (dead) the jinns realized that if they had full knowledge of the Unknown they would never have suffered demeaning labor.

The Quran does not place such stories in a historical narrative; they illustrate the ways in which God acts, but they are not events in a great historical drama. Even the exact point of the stories is often unclear. Their original audience no doubt knew tales about Solomon, and could fill in the background which is lost to us; this particular story may derive from legends about Solomon building the Temple. Many such stories appear in the Quran, and sometimes make sense only when compared to Jewish and Christian sources from outside the Bible.[22]

Since Muslims construct their myths around the Quran, they have a different vision of history than Christians and Jews. In Islam, history testifies to God through the revelations sent to the nations. Muhammad, the Messenger of God, the final prophet, speaks the words of God, and from then on everything is radically different. There was an Age of Ignorance before the Prophet, which was set aside, if not erased, by the coming of Islam. Even non-Arab Muslim peoples tend to see their history before Islam as a time of darkness, as if history only began for them once they received Muhammad's message. The sacred history of a chosen people dominates Judaism, Jesus stands at the center of the Christian imagination, but Islam is focused on a divine message—the Quran itself. All the Abrahamic religions tend to take their scriptures to be the inerrant word of God, but for Muslims, "The Qur'an is unique. It embodies the word of God—unchanged, unabridged and uncompromised. It does not contain any element that is a product of a human mind."[23]

Among Jews and Christians, few aside from scholars have come to accept that their holy writings are very human documents. For Muslims, the difficulty is even greater: admitting a flawed Quran is like a Jew disavowing the idea of a chosen people, or a Christian saying Jesus was not divine. Nevertheless, the Quran was not dictated by any God. The text is a badly edited mess, full of repetition, interpolation, and arbitrary arrangement of material. It gives contradictory rulings, forcing orthodox interpreters to claim that later revelations abrogate earlier verses. And it is too obviously set in a world corresponding to the beliefs of seventh-century Arabs. God creates the universe in a span of days, with seven layers of skies anchored on the earth (41 Ha Mim As-Sajdah 9–12). God fashions Adam from clay, breathes spirit in him, and commands the angels to adore him (38 Sad 71–72). The world of the Quran is familiar from ancient Near Eastern beliefs, and from the Bible, except that the Quran lacks a coherent narrative. This is not a minor matter—God condescending to use the language of seventh-century Arabs to reach them with a moral message. The Quran depends on that ancient context to present itself as revelation, to establish the authority of God.

If the Quran is a rather implausible message for a God to dictate, how do we account for the origin of Islam? There is, of course, the traditional story. In the Age of Ignorance, the Arabs did many awful things; chiefly, they worshiped idols alongside the One True God. Muhammad was a merchant from Mecca, a town organized around a pagan shrine originally built by Abraham. At age forty, he was called to prophethood and began to recite revelations. He invited his fellow Arabs to join the pure religion of Abraham he was proclaiming, but the Meccans were not convinced. They persecuted him when he insisted. So the Messenger of God packed up with his followers and migrated to Medina, a Jewish and pagan oasis. Soon, after some ethnic cleansing, he became the master of Medina. He raided Meccan caravans, fought a few battles with mixed success, finally took control of Mecca itself. He successfully united the nomadic Arab tribes, and was getting ready to invade other lands, when he suddenly dropped dead. The revelations he uttered until his death were carefully preserved in memory, and later assembled in a holy book.

In outline, the traditional story makes sense. Unfortunately, Muslim traditions about the Messenger are far from perfect. Muslim interest in the historical Muhammad began late, and it was motivated by more than curiosity: the Prophet's life could answer questions which arose in the Islamic empire, questions the Quran was silent on. Often a story was fashioned not out of independent evidence, but from interpretations of the Quran and a sense of how things *must* be to fit theological and political views. Many of the traditions orthodox Muslims rely on are sheer inventions, few are trustworthy at face value. Even the chronology of Muhammad's life is murky; for example, Muhammad's prophetic career probably began long before the ideal age of forty, but the story was told according to Near Eastern expectations for an ideal prophet's biography. Historian F. E. Peters expresses the difficulty Muslim sources present:

> At every turn, then, the historian of Muhammad and early Islam appears betrayed by the sheer unreliability of the sources. One confronts a community whose interest in preserving revelation was deep and careful, but who came to history, even to the history of the recipient of that revelation, too long after the memory of the events had faded to dim recollections over many generations, had been embroidered rather than remembered, and was invoked only for what is for the historian the unholy purpose of polemic.[24]

The unreliability of our sources casts doubt on the traditional history, so much that some scholars have come up with very radical accounts of Islamic origins. Patricia Crone and Michael Cook, for example, depart from the few, but intriguing, non-Muslim sources we have, and reconstruct a version of early Islam according to which the traditional story is mostly a later fabrication.[25] Muhammad was an Arab merchant who proclaimed the religion of Abraham, but the immediate focus of his message was that as children of Abraham through Ishmael, Arabs also had a claim on the Promised Land and Jewish messianic hopes. The first thrust of the Arab tribes was into Palestine—allied with Jews. Islam was fashioned out of that initial ideology, political developments including a break with the Jews, the influence of Samaritan beliefs, and the need for a myth of common identity for the peoples thrown together by the first wave of Arab conquests. The traditional biography of Muhammad was shaped by such concerns, along with the legend of Moses which was a model for a prophetic life.

Revisionists like Crone and Cook have not convinced most historians of Islam; most believe the traditional framework of events is still serviceable. But the fact that such radical scenarios are serious possibilities highlights how much of the story of early Islam was shaped by mythmaking. As with the Bible, we have a mythic salvation history—a story fashioned in sectarian debate, not a record of events. Even if we assume a historical core was correct, Muslims fleshed out the traditional framework with detailed stories about Muhammad, his sinless nature, his revelations, and his miracles. They invented these myths for good reasons: to convince their audience that something supernatural happened, that *their* Prophet was as good as and even better than those of the Jews and Christians. But today, all we have is an obscure messenger of God and a very dubious message.

Religious liberals, of course, can still find revelation in Islam. Muhammad may well have revealed some truths about God, had an experience of the divine, and spiritually transformed his community. We perhaps do not have the uncorrupted word of God in the Quran more than any other scripture, but the hand of God can still be discerned in the origins of Islam. In Muhammad's mystical experience, moral reforms, and inspiring effect on a worldly and corrupt community, we see a radical change toward a God-centered way of perceiving the world.

We do not know which verse of the Quran came first; Muslim tradition preserves different views. But tradition agrees that Muhammad first had a visionary experience during an obscure seclusion ritual, probably while asleep.[26] The Quran refers to his vision of God, "He had surely seen Him on the clear horizon" (81 At-Takvir 23), though later doctrine would identify this presence as the angel Gabriel. Karen Armstrong elaborates, imagining how revelation "invaded his being with fearful force, doing violence to his natural self which was not built for such a divine impact. What all these prophets had experienced was transcendence, a reality that lay beyond concepts and which the monotheistic faiths call 'God'."[27] Muhammad quite possibly had a visionary experience induced by ritual procedures, though the traditional account which impresses Armstrong owes much to stereotyped stories about how a prophet ought to receive his call, validating his message by casting him in the role of ideal prophet. In any case, Muhammad did not attract followers by pontificating about Transcendence or the Ineffable Other. The Messenger brought *words* from a personal God.

Muhammad produced revelations throughout his prophethood, often in response to particular circumstances. Disbelievers seem to have questioned this gradual process, and the Quran explains that "it was sent thus that We may keep your heart resolute" (25 Al-Furqan 32). A later passage declares that the Quran had indeed been revealed in its entirety during a holy night (97 Al-Qadr), thus conforming to the Jewish belief that the Torah had been revealed on a single occasion.[28] The early revelations were similar to the speech of poets and seers, though the Quran assures listeners that it is inspired by God. Like Christian miracles, which were distinguished from magic by the operation of the Holy Spirit rather than demons, the Quran's words were special because they were brought down by the "trusted spirit" of God (26 Ash Shuara 193). The

Quran also declares itself impossible to imitate, though poets composing in a Quranic style have generally been deterred by worldly punishments and not lightning bolts from above.

Some traditions describe Muhammad as speaking new verses after something vaguely similar to an epileptic fit.[29] However, it is hard to say how much these stories owe to prophet-stereotypes. More interesting is how Muhammad's circumstances shaped the Quran, though again, it is hard to tell if "occasion of revelation" traditions are historical or inventions to explain obscure passages. The most notorious example is the Satanic Verses. According to a Muslim tradition, Muhammad once announced some verses praising three goddesses worshiped in Mecca. Muhammad was a messenger of the Most High God who Meccans already acknowledged, but the goddesses were intermediary beings who could intercede with the High God. This gave Muhammad some breathing room in Mecca. However, he soon came to reject the idea of lesser gods. So he withdrew the Satanic Verses and replaced them by others denouncing the goddesses. Tradition blames Satan for the earlier version; since the story became embarrassing to later orthodoxy, modern Muslims vehemently deny the whole incident ever happened.

The Satanic Verses are just one case where, even if we accept the traditional story, we catch a glimpse of Muhammad's evolving convictions. Orthodox Islam did not come fresh out of the box; we can see signs of how the Quran's theology developed.[30] Late in Muhammad's traditionally Meccan period, he came to deny all lesser gods; subsequent Medinan revelations stress the uniqueness of God. Even beings like jinns totally disappear.

Philosophers admire this pared-down, uncompromising monotheism; ironically, this late period is also the time when the Quran endorses Muhammad's violent political career. When Islam was a small movement in a hostile environment, the Quran pleaded for tolerance between faiths, and disavowed compulsion in matters of conscience. Verses from Medina, when Muhammad was a stronger leader with military ambitions, demand that non-Muslims submit and pay a protection-tax (9 At-Taubah 5, 29). Muhammad's personal circumstances would also shape his revelations. For example, Muslim tradition tells of Muhammad's adopted son divorcing his wife so that Muhammad could marry her. Adopted sons being considered equivalent to sons in every respect, this was bound to raise eyebrows. Immediately, God sent a verse (33 Al-Ahzab 37) to make it known that Muslims may marry the daughters of their adopted sons. Muslims regarded their prophet as a moral model, so his personal life being an occasion for revelations was only to be expected. The adopted son incident, however, looked uncomfortably self-serving even to the faithful.[31]

Though there is little that is certain about early Islam, we can tell its revelations emerged through a creative process; they were not zapped down from heaven. In the Quran we see evolving theologies, influence of Jewish ideas, even changes in response to criticism. The Quran is like a work of literary art, written in interaction with a human environment, including messy politics and personal life.

In recent times, we have had similar examples of religions started by a charismatic leader. Mormon founder Joseph Smith, for example, received new scriptures endorsing nineteenth-century American superstitions, was involved in politics, and tended to receive revelations solving immediate problems with his developing church and his personal life.[32] The United States has had its share of prophets: Joseph Smith, Mary Baker Eddy who started Christian Science, Ellen White of Seventh-day Adventism, and many more. Even now, New Age channelers speak new revelations from supernatural beings, maybe even new scriptures.[33] History shows us that some people have visionary experiences, and that many more are inclined to see this as a sign of holiness. Never, however, do our prophets reveal credible divine messages. We invariably find false and superstitious claims, signs of creativity and adaptation to external circumstances, often even pious fraud. This is not to say the words of our prophets are worthless. Particularly for members of disadvantaged groups, like women, claiming a supernatural voice is sometimes the only way to speak up and be taken seriously.[34] But prophets are no evidence for any supernatural reality. The voices that inspire us are only human.

GOD'S EMPIRES

Whatever the origins of Abrahamic religions, they have clearly been very successful. And if divine providence shapes history, the faithful have reasoned, it should ensure that the True Faith flourishes. Jews imagined that Yahweh had once granted them a Davidic mini-empire. When their sect was promoted to become the Roman state religion, Christian writers such as Eusebius saw this as confirmation that Jesus was the Messiah. After all, God was in control of history. And when their turn came, Muslims took the phenomenal expansion of the early Islamic empire to be a sign of Islam's truth. With imperial success, the One God became our cultural background, an actor in all the stories dear to us.

Since God is now tied up with so much that is important, we tend to think that a prophet establishing monotheism ushers in progress, or at least responds to the deep spiritual needs of his community. For many of the devout, squashing idolatry is progress enough. Liberals are more tolerant, so they like to think prophets proclaim a more humane moral order. Prophetic success, however, owes as much to power and political violence as to an inspiring moral vision.

Western scholars often portray Muhammad as a moral reformer. According to this story, Muhammad's hometown, Mecca, was a city based on trade. Its residents, however, were recently desert nomads, and their inherited social order no longer fit their newfound capitalist prosperity. There was a spiritual malaise in Mecca, indeed, throughout Arab lands. Commercial success led to moral degradation and a cynical disregard for the less fortunate. Muhammad restrained the worldly attitudes of the settled Arabs, and so effectively overcame the malaise that the Arabs were able to unite and achieve their magnificent conquests.[35]

This story is largely a modern myth. To begin with, Mecca was not a prosperous trading town. It was a desert settlement with no resources, in the middle of nowhere, at a time when the spice and incense trade had collapsed even if it were implausibly to go through Mecca. What Mecca had was a prominent pagan shrine. F. E. Peters observes that "Mecca may have indeed begun to construct for itself, out of the matrix of its shrine influence and revenues, and in the face of the rapidly receding economic and political influence of both the Sasanians and the Byzantines, a modest trading zone along the now unpatrolled frontiers of Syria and Iraq." This trade, however, was "on a far more modest level than is generally thought."[36]

Mecca surely had its share of social tensions, even without large-scale trade. Still, there is no solid evidence for Arab malaise. Patricia Crone points out that Muslim sources portray disbelieving Meccans as depraved precisely because Islam was unattractive to them. If Islam had a specifically ideological advantage, it was its program of forming an Arab nation, enabling the desert nomads to unite in their plunder of neighboring lands. God "told the Arabs that they had a right to despoil others of their women, children, and land, or indeed that they had a duty to do so: holy war consisted in obeying. Muhammad's God thus elevated tribal militance and rapaciousness into supreme religious virtues."[37]

Crone sounds a jarring note in an environment where scholars of Islam are concerned about Western culture once again demonizing Islam to fashion a substitute for the Red Menace. But whether we are as blunt as Crone or not, there is no escaping the fact that early Islam is full of violence.

Muhammad appears to have been invited to Medina as a holy man who could act as a judge and bring peace to the fractious oasis. He was as good as his word, and eventually achieved peace by getting rid of the local Jews and melding everyone else into one Muslim community. This was a difficult process. Muhammad seems to have been especially concerned about poets who, in an age without television, could be influential in shaping public opinion. Fortunately, he was usually able to have them assassinated. Orthodox tradition describes how Muhammad dealt with Kab ibn al-Ashraf, Jewish leader and irritating satirical poet:

> The Prophet said, "Who is ready to kill Ka'b bin Al-Ashraf who has really hurt God and his Apostle?" Muhammad bin Maslama said "O God's Apostle! Do you like me to kill him?" He replied in the affirmative. So, Muhammad bin Maslama went to him (Ka'b) and said, "This person (the Prophet) has put us to task and asked us for charity." Ka'b replied "By God, you will get tired of him." Muhammad said to him, "We have followed him, so we dislike to leave him till we see the end of his affair." Muhammad bin Maslama went on talking to him in this way till he got the chance to kill him.[38]

Orthodox Muslims think Kab deserved his fate. He had plotted against the Islamic State and the Prophet, and had insulted the honor of Muslim women with immoral songs.

Like any competent politician, Muhammad was willing to use greater displays of force when necessary. For example, according to Muslim tradition, Muhammad dealt with the last remaining Jewish tribe in Medina by a method straight out of the Bible. The Jews, it seems, had assisted Muhammad's enemies, though their side of the story has not been preserved. So the Muslims, led by Muhammad, killed all the men, divided the property among themselves, and enslaved the women and children.[39]

After Muhammad's death, according to the traditional story, the Arabs burst out of the desert, and bestowed salvation on their neighbors—though to an untrained eye, this looked a lot like slaughter and slavery. They conquered Coptic Egypt, Berber North Africa, Spain, Persia, Turkish Central Asia. God had a shiny new empire. And this empire meant that Islam was to take shape as an imperial ideology, enjoining obedience to the rulers, sanctifying the continuing wars of conquest, and prescribing just what to do with the subject peoples. This imperial ideology was not, however, a corruption of an originally pure revelation. The jurists and theologians of the early empire were elaborating on a theme sounded by the Quran itself.

Political violence is not unique to Islam. Judaism used to be a temple-state ideology which was as bloodthirsty as anything in Muslim doctrine when it came to smiting idolaters. For centuries the universal Church was too small to persecute anyone but dissident Christians, but after the emperor Constantine converted, the new imperial religion could deal with infidels in typical monotheistic fashion. Christian empires destroyed natives of the American continents and assimilated their cultures into the imperial ideology just as surely as Islam swallowed Egypt, Iran, or Central Asia. It is curious that a benevolent God would choose such violent institutions to spread revelation. Idolatry must be a terrible offense indeed, if God's empires are its punishment.

Monotheism, it seems, serves very nicely as an imperial ideology. And if a faith can help organize an effective use of violence, it gives its adherents a competitive advantage against their rivals. Even so, there is more to monotheism than authoritarian ideology; stopping here would be as much a corruption of history as reverent epics about the Triumph of God's Truth. While Islam would not have succeeded without political violence, the Islamic world was not carved out just by the sword. Such an empire would have been as ephemeral as Genghis Khan's. The conquered peoples resisted, but eventually genuinely converted, made Islam their own, even formed the next generation of God's soldiers.

More importantly, the fortunes of empire changed, and Islam along with them. Orthodox Sunni Islam, which claims the allegiance of the vast majority of Muslims today, crystallized after the first five centuries of Islam, driven by—interestingly enough—developments in Iran. Sunni Islam, with its popular consensus on doctrine and social forms, and under the leadership of a rabbinical class of *ulema*, has found remarkably uniform expression throughout the Muslim world. But this did not come about through governmental imposition, nor as a direct reflection of early Islam. It developed in response to hard times and political instability—urban overpopulation and then the Mongol invasions. The result is an ideology that famously claims authority over *all* aspects of life, making rulings on bathroom habits as well as the characteristics of an acceptable ruler. And Sunni Islam was still suitable as an instrument of state control, in places like the later Osmanlı empire. Yet it has a peculiar political impotence about it at the same time. The Islam of the ulema took shape in the absence of a stable state; it is more concerned with fencing off a private realm from arbitrary exercises of power. With God taking over all law and social control, the legitimacy of state power remains ever in doubt, useful only so far as it promotes Islam.[40]

And of course, the success of Islam at the state and popular levels meant that oppositional beliefs came to be expressed within Islam, if only as heterodoxies. Muhammad's cousin and fourth successor, Ali, was murdered—a common fate of caliphs—during the civil warfare of the early empire; Shii Muslims opposed Sunni rule by creating myths around Ali and his martyrdom. Turks migrating to Anatolia created a mystical, heterodox Islam out of their pre-Islamic beliefs and the folk religions of their new land, the conquering message they were exposed to, and messianic hopes for heaven on earth when all oppression passed away. Turkish popular Islam centered not on the Quran, but saint cults, magical beliefs which monotheistic purists held in suspicion, and the poetic wisdom sayings of mystics. Stories about a merciful God became the voice of the downtrodden, as well as the dogma of empire. While the Islam of empire was bent on smiting infidels, while the ulema drew up rules for godly practice in everything, heterodox wise men declared that it did not really matter whether one was a heretic or unbeliever, and that God had a place in Hell for the hard-hearted persecutor.[41]

So Islam, like any world religion, soon came to include a whole family of constantly mutating and mutually contradictory myths. Many Christians have quoted the Bible to support slavery, but African Americans have found stories of release from bondage in the same pages. Similarly, Islam supported violent interests, but also expressed hopes for a world without violence. Modern secularists in the Muslim world sometimes portray the heterodox saints and poets of popular Islam as representatives of a humanist morality beyond theism, but this is historically false. The mystic poet Yunus Emre said, "we love the Creation on account of the Creator"—God is still at the center of heterodox beliefs, even though this God has a different character than the imperial God.[42] Our religions encompass everything human, both our humane and our murderous sides. If we create God in our own image, it is no surprise that God should turn out to have so many different faces.

The history of religions makes God implausible, but not just because God's empires are unlikely instruments for a loving God. It is because religions seem very human, not the product of gods and demons. We have hopes and hates, interests and fears. We tell stories. We seek supernatural sanction for our ideals, hoping this will make them something more than *just* our ideals. All this does not, by itself, mean the claim of One God is false. But religion seems so much a human creation that we do not need revelation to explain it. The One God succeeded and flourished for reasons which had nothing to do with revealed truth. Monotheism was violent enough to ensure growth, and flexible enough to contain both desire for power and hope for freedom.

THE MEANINGS OF HISTORY

When all else fails, try philosophy.

Our religions appear to be human creations. But reaching this conclusion depends on treating history like a science—examining claims about the origin of Islam, asking whether events show a pattern explained by a contract between Jews and a God. We start with theories, throw evidence at them, make adjustments; take what seem to be historical facts, throw theories at them; stir well, keep going around, and hope to converge on something sensible. Since human societies are very complex, history is a messy business; nevertheless, it still seems to be a kind of science of figuring out the human past.

But not all historians believe their work is continuous with natural science. With the European Enlightenment, science and the humanities both gained independence from theology, but only to go their separate ways. Natural science appeared impersonal, concerned with reproducible events understood as specific instances of general laws. But if nature was the realm of necessity, history was that of freedom. Historians dealt with the personal, the unique and unpredictable. They wondered about the *meaning* of our past, not just what once happened. They produced narratives, not equations; interpretations, not experiments. Hence historian Jeffrey Burton Russell thinks his discipline is different: "its goal is always the understanding of the individual. In this way history is transcendent, almost religious."[43]

Revelation is supposed to be an intrusion of divinity into the ordinary world. And if history is concerned with individual, meaningful occasions, it can present and validate revelation while natural science cannot. Religions begin to look like they erupt from singular events—Muhammad's call to prophethood, Jesus' Resurrection, Moses' visit to Mount Sinai—which confound the natural order. We recognize such moments of radical novelty as we attune our interpretive skills to the meanings of history. Theologian Langdon Gilkey says, "Spirit transcends history by surveying its past and in that light envisioning its possible future . . . eternity, transcending time but a transcendence united with time, invade[s] personal and communal life in historical understanding and in political action alike." This presumably means that we can be aware of the past and act accordingly. But there is more, since "Some deep assumed principle of interpretation is always at work whenever we think of temporal passage or seek to act within it."[44] After all, philosopher Eric Voegelin tells us:

> Philosophy is the event of being on the occasion of which the logos of realization becomes transparent. Philosophy, therefore, is that phenomenon in the field of history, with the help of which not only are other phenomena recognized as phenomena of realization but also is the field of history recognized as the field of tension between the phenomena.[45]

Among all these oracular pronouncements about spirit and realization, there is the inconvenient fact that the ordinary work of historians tends to deflate claims of revelation. But this is no great challenge. If history is about meaning and interpretation, theologians can now charge that critical history has assumed a naturalistic principle. It is no wonder we find that religions are only human, if after the Enlightenment, we rigged the process to exclude a transcendent viewpoint.

This conception of history, centered on individuality and interpretation, has a rather Protestant character. In the history of religions, it serves to portray Christianity as unique, incomparable to pagan beliefs.[46] Besides specific apologetic uses, however, it is a response to a broader concern. We no longer live in tight-knit communities with a single tradition demanding our allegiance. So theologians have to deal with modern notions of historical evidence, and confront a plurality of religious options, while preserving the integrity of their tradition.

Followers of the Abrahamic religions did not always have this difficulty, because they did *not* read their scriptures as accounts of the past. Instead, they found stories which imposed meaning on the chaos of present experience. Modern theologians, even if they read the myth of Israel as a description of actual events, have to worry about how to find an "essential message" about unconditional faith, and make the story speak to today, to a different culture. Premodern Christians, not to mention Muslims or Jews, did not have to bridge this gap between past and present; the meanings of their world were already organized according to these stories. Readers of the words of the sages of blessed memory, or the traditions of the Prophet and his Companions, did not find a linear narrative of history, but encountered the timeless structures of the universe and of society. Interpreting their tradition, rabbis and ulema could rule on everyday matters. Their scriptures always spoke to them in the "eternal present."[47]

This way of thinking is quite natural; it certainly imposes structure on social life. However, it operates in a closed system. Using timeless scriptures as templates to organize experience produces a static world of meaning. Rabbis and ulema will disagree with their colleagues on details, but their overall framework cannot be criticized, only disrupted. And since the system is not subject to nonsocial reality checks, interpretations vary—just not in the same social unit. We end up with a landscape dotted with villages of Druze, Monophysites, Nestorians, Sunni, and Shii. As long as the villages or neighborhoods are socially reasonably isolated, conflicts between the different timeless truths do not surface too often.

Claims to historical truth surface when this order is disrupted—when communities seek converts, when they have active messianic hopes, or just when they find themselves in an empire with a mix of people and many revelations as live options. History then serves to show that one tradition is right and its rivals wrong. Christianity claimed a historical messiah; by claiming to be the One True Faith, it became the One Compulsory Myth for an empire. Rabbinic Jews retreated further into their eternal present, keeping faith in the face of overwhelming worldly failure.

The Protestant view of history came about through another such disruption. Against Catholic tradition, Protestants brought forth Christianity's submerged millennialism, and the idea of purifying truth by going back to the time of revelation. Thus they invoked history; to reform the present, they reopened the gap between present and past.

This might have been manageable—the social order could have convulsed and settled down on a different eternal present. Unfortunately, a cognitive revolution took place. Along with a new, potent science, critical history became available, and got out of control. Protestants separated myth from history, bringing their religion into the modern world. If history reproduced the myth, this trick would have worked. But conservatives insisted that the Bible is history, and got stuck in a sterile and patently false literalism. Liberals embraced the myth and the meaning, but lost any reason to call their tradition truth, with a God revealed in an objectively real history.

If theologians have to worry about bringing myth and history back together, however, the Enlightenment also has unfinished business. We still have a quasi-religious dualism if we separate history from nature. This keeps us in Protestant territory, arguing about history and the individual, and lets some philosophers claim narrative is an "irreducible form of understanding."[48] Plus, we always confront competing interpretations of history, and events for which we seek meaning. Consider the Holocaust. An Orthodox Jew might say this was Yahweh's latest punishment of a straying Jewish nation. Sherwin Wine, a humanistic rabbi, disagrees:

> The Holocaust is dramatic testimony to the sublime indifference of the universe to undeserved human suffering—and to undeserved human pleasure, for that matter. In the twentieth century, the Jews learned that the reward for faith and humility is despair and humiliation. If they did not resist reality, they were confronted by an ironic truth. The Jewish people, whose official establishment proclaimed for over two thousand years that Jewish history is testimony to the presence of God is, indeed, the strongest testimony to the absence of God.[49]

Historians alone cannot tell us how to respond to the Holocaust, or what its moral significance is, though those questions may be just what motivates our interest in history. We certainly bring our hopes and allegiances, and perhaps our myths, to the task of understanding an event. But we also want to *explain* our history, to address questions of fact. Having separated myth and history, some of us still try to see where the evidence might take us, and to set our understanding in the context of our knowledge about the world—our science.

The most obvious way to connect science and history would be to find some general laws. Carl Hempel once argued that history could capture the unique individual quality of an event, but only because it described an event in terms of general concepts.[50] So history was no different than the sciences which presented their phenomena in terms of general laws.

Unfortunately, we know no general laws of history. When we want to know about the origins of Islam, we try and reconstruct the social environment of seventh-century Arabia, and then tell a story about Muhammad in that setting. This is not an arbitrary narrative; we are guided by the evidence we have. Our picture of Muhammad is not totally unique; we understand him better by comparing him to other examples of charismatic prophets in precarious political circumstances. We do not, however, explain Muslim history by calling on a Fifth Law of Prophets, describing how a prophet will solve domestic strife between his wives by announcing an appropriate revelation.

History seems stubbornly contingent: the course of events could have been otherwise, and the only way we can see why our history took place is to recount the individual events. We can make some general statements; for example, we can say a missionary faith has a competitive advantage. But which specific faith succeeds is largely a matter of accidents. Muhammad's authority in Medina was enhanced when the Muslims won, against the odds, the battle of Badr against Meccan forces. Later, the Muslims were defeated at Uhud, and were lucky the Meccans did not press the advantage and destroy them. The Quran presents Badr as a sign of divine favor—God sent down three thousand angels to fight alongside the Muslims. Uhud, of course, was a test of faith.[51] Events could have taken a different turn. A parallel universe where Muslims were soundly defeated, and consequently Islam was only a religious movement which flared up briefly and disappeared, would not violate any law of history. We explain the initial success of Islam by some advantageous features of the religion, Muhammad's political talents, and lots of historical accidents. Judaism as we know it would not exist today if not for the Babylonian exile, and then political ruin in Roman hands. Christianity might not have succeeded, or at least would have evolved very differently, if Constantine had not converted.[52] Many religious movements do not get the lucky breaks and are barely remembered in history.

This contingency, however, does not set history apart from nature. Complex physical systems exhibit chaotic dynamics or an element of randomness, forcing us to list what actually took place when describing their histories in detail. Many different futures are compatible with the information we possess, and events in the past could have been otherwise if not for some accidents that we can do no more than describe. Modern physics lays the groundwork for an accidental, historically contingent world. The supposed contrast between science with its laws, and history with its unique individual events, is based on a very outdated picture of science. Historical sciences such as evolutionary biology are all about teasing out individual histories from the evidence we have. In a complex system like a human society, unpredictable behavior which forces us to describe individual events is exactly what we expect.[53]

Of course, there is a long way from the physics of accidents to a historical narrative, with its reliance on commonsense causality and on folk psychology. We do not just string facts together but assemble a story. However, the reason we use narrative is our brains. Consider a geographer and her maps. A map is like a spatial narrative, displaying information about individual features of an area. Like history, geography is not completely haphazard; there are definite patterns to things like mountain formations. But as a rule, a geographer has to individually depict features of the terrain we are interested in, like a mountain pass, or the location of a city we want to visit. Maps depict complex systems we cannot capture with a few elegant equations. Moreover, a map, like a historical narrative, is not just a dumping ground of information. We are visual animals who can take in colorful two-dimensional pictures at a glance. We do not read the raw data off a satellite, even when this expresses the same information. On a map, we recognize patterns; we can see the shortest route to our destination, or, if we know enough geology, make a good guess at what processes formed the mountains we see. A map, in other words, is a good way of presenting information to our kind of brain. Similarly, we best assimilate historical information as narratives: stories with a theme we can remember, and see patterns of personal and social relationships within. Telling a story about Muhammad helps us make sense of the Prophet as a person in his social circumstances. We do narrative history because of the way our brains have evolved, not because human events transcend physics, and certainly not because a religious transcendence is built into the fabric of history.

The practice of critical history is also continuous with that of science. As in science, the results of historical inquiry are not predetermined by principles of interpretation. Historians could have ended up supporting the Bible; we could have discovered that Jewish history, for example, follows a pattern of disloyalty and retribution. We could even have found something about a transcendent force behind events, and changed how we do history accordingly. We did not. In the modern world, we had to risk putting sacred history to the test; unless revelation is an objective historical fact, as conservative theologians fear, God itself increasingly becomes a mere psychological metaphor.[54] And if critical historians had confirmed the claims of our religions, theologians would surely be celebrating. Instead, we have found that our history fits the naturalistic world of science. The only purposes shaping the course of events appear to be our own. If now theologians cry foul, this shows the depth of their loyalty to their myth, but no more. We have learned something about how to do history, as well as natural science; theological spin-doctoring under the name of "interpretation" is not part of either.

THE END OF REVELATION

There used to be no accidents. History flowed toward salvation; everything had its place in the divine purpose. We were all actors in a drama beginning at Creation and ending with Judgment. Every decision we made took meaning from its place in the story, and how it affected our eternal fate. Even the punishments of Hell were a way of highlighting the significance of our acts; we would choose our spiritual condition in our finite earthly life, and God would honor this choice throughout eternity.[55]

Many of us still think this way. Even the most conservative Abrahamic religions do not lack for followers, perhaps precisely because they heroically embrace the old stories. Nevertheless, times have changed. Islam is no longer a world unto itself, and both modernizers and revivalists now feel confined by the orthodox religious consensus.[56] The Western world is no longer Christendom; countries once dominated by their churches are now populated by individualist religious consumers.[57] We have become relativists about religion—though we usually affirm the virtue of faith, it does not seem to matter what particular Eternal Truth we trust in. Even Protestants no longer believe in the Bible, but believe in believing in the Bible. Muslims may yet succeed in walling themselves off from modernity, but conservative Christians have to hope they will succeed in the religious marketplace as they once prevailed in the religious chaos of the Roman world.[58] Maybe another Constantine will come looking for an imperial ideology. Meanwhile, our One True Faiths are not as confident as they once were.

Of course, a changed religious climate does not mean we no longer have saving truths; perhaps many of us have just grown deaf to God's words. But which words are we supposed to listen for? Religious feeling alone does not take us far; even in pagan times, people had profound personal experiences with their local gods. And today, our religious marketplace is overrun by promises of salvation. There seem to be too many revelations, eloquently testifying to human religiosity but not to the truth of any single tradition. We once may have hoped God's gift of reason would help illuminate the true path to salvation, but nowadays this looks like a hollow promise. A critical approach taught us that our holy epics are more works of fiction than records of events. Reason could have vindicated our gods and demons; instead, it demoted them to myths.

Since old-time religion is out of touch with our new smorgasbord of faiths, and intellectually disreputable to boot, philosophers and theologians need a new kind of God. Maybe the half-zillion revelations swirling around us are not all wrong, but all correct. God reaches out to everyone; we receive the outpouring of divine love through the religious experiences we channel through diverse traditions. There is no *one* sacred history, no single chosen people. The core of religion is encountering fundamental reality as something deeply personal, which is totally beyond, yet within, us. Some of us attend Mass to express our dependence on this mystic God, some of us pray toward Mecca; but these are the disciplines of spirituality, not the substance.

Such a liberal religion is perfect for our time. It is wonderfully tolerant and undogmatic, so full of sweetness and light as to kill any self-respecting cynic of nausea. It allows scholars to rip apart traditional doctrines during the week and join their congregations on the Sabbath. The old verities get reinterpreted; apocalyptic prophecies become psychological metaphors of our internal struggle with evil, messengers of God become vaguely left-wing moral reformers. But liberal religion protects itself from historical criticism so well that history can no longer reveal God. If our religious traditions express revelation, we have to take their specific claims seriously—we cannot dissolve them in an amorphous spiritual soup. All too often, religious liberals bury the fire-breathing God under inoffensive platitudes. Mohammed Arkoun says, "Revelation is the accession to the interior space of a human being—to the heart, the *qalb*, says the Qur'an—of some novel meaning that opens up unlimited opportunities or backcurrents of meaning for human existence." His examples include "the teachings of Buddha, Confucius, African elders, and all the great voices that recapitulate the collective experience of a group in order to project it toward new horizons and enrich the human experience of the divine."[59] This is a vapid broadmindedness, ignoring the content of new ideas. Early Buddhist writings, for example, were shaped by criticism of Hindu theism; it does these teachings no honor to redefine them as divine revelations. Teachers make real claims, sometimes they even say other teachers have been dead wrong. And if their ideas are something new under the sun, so they are. In an accidental world, people will have new ideas, even inspiring ideas. Declaring these to be supernaturally granted vistas of deep meaning is sheer superstition.

There is a peculiar irony in defining revelation as radical novelty. The most radical idea of modern times is the skepticism which comes with science. The followers of a jealous God taught us to take truth seriously, to reach beyond timeless myths enshrined in custom and community. Some of us were bound to go so far as to reject the authority of revelation as well. And as Jacques Monod suggests, this is perhaps too radical for our taste:

> Modern societies accepted the treasures and the power that science laid in their laps. But they have not accepted—they have scarcely even heard—its profounder message: the defining of a new and unique

> source of truth, and the demand for a thorough revision of ethical premises, for a total break with the animist tradition, the definitive abandonment of the "old covenant," the necessity of forging a new one. Armed with all the powers, enjoying all the riches they owe to science, our societies are still trying to live by and to teach systems of values already blasted to the root by science itself.[60]

Monod is no prophet, he heralds the end of revelation. And if Monod sounds naively evangelistic these days, it is because we are too busy complaining that science gives us only truth when what we want is myth.

So the quest for God in history continues. After all, here in the ruins of Christendom, we still tell stories of Jesus. The Jews only prepare the way, and Muhammad we ignore. Maybe, if we place Jesus at the center of history, we will then see the purposes of God. Theologians such as Wolfhart Pannenberg say that history is revelation, that Jesus Christ makes everything come together, and that this revelation is open to everyone who honestly examines history.[61] This Jesus cannot be the God-man predicted by the Hebrew prophets, who descends from above, gets born of a virgin, saves us from Satan with his sacrifice, gets resurrected, and flies up to heaven to wait for his return in glory. This story is too fantastic, even nonsensical, to salvage. We are too aware now of how religious communities spin myths. Still, liberal Christians claim that even after critical history does its worst, we are left with a new and improved Jesus who answers our mature religious needs. Let us seek, and see who we shall find.

NOTES

1. Calvinists in particular. Bernard Ramm, *Special Revelation and the Word of God* (Grand Rapids, Mich.: William B. Eerdmans, 1961), p. 19. However, even such theologies do not deny that general revelation is available.

2. Bernhard W. Anderson, *Understanding the Old Testament*, 3d ed. (Englewood Cliffs, N.J.: Prentice-Hall, 1975), p. 12.

3. C. A. Patrides, *The Grand Design of God: The Literary Form of the Christian View of History* (London: Routledge and Kegan Paul, 1972), chaps. 2, 3.

4. Burton L. Mack, *Who Wrote the New Testament? The Making of the Christian Myth* (San Francisco: Harper, 1995), chap. 10.

5. Harold Camping, *1994?* (New York: Vantage, 1992), p. 29. For criticism of prophecy beliefs, see Tim Callahan, *Bible Prophecy: Failure or Fulfillment?* (Altadena, Calif.: Millennium, 1997).

6. Patrides, *The Grand Design of God*, p. 58.

7. Jack W. Cotrell, in Charles R. Blaisdell, ed., *Conservative, Moderate, Liberal: The Biblical Authority Debate* (St. Louis, Mo.: CPB Press, 1990), p. 33. For a philo-

sophical defense of biblical inerrancy, see Norman L. Geisler, ed., *Biblical Errancy: An Analysis of Its Philosophical Roots* (Grand Rapids, Mich.: Zondervan, 1981).

8. Paul Boyer, *When Time Shall Be No More: Prophecy Belief in Modern American Culture* (Cambridge: Belknap Harvard, 1992), pp. 293–96.

9. Paul Merkley, *The Greek and Hebrew Origins of Our Idea of History*, vol. 32, Toronto Studies in Theology (Lewiston, N.Y.: Edwin Mellen, 1987), p. 76. Merkley extends his confidence in Israelite historical competence to the compilers of the Christian canon, naturally.

10. Thomas L. Thompson, *The Mythic Past: Biblical Archaeology and the Myth of Israel* (New York: Basic, 1999), p. 99.

11. Diana Vikander Edelman, ed., *The Triumph of Elohim: From Yahwisms to Judaisms* (Kampen, The Netherlands: Kok Pharos, 1995); Robin Lane Fox, *The Unauthorized Version: Truth and Fiction in the Bible* (New York: Alfred E. Knopf, 1991), pp. 54–63; Gerd Theissen, *Biblical Faith: An Evolutionary Approach* (Philadelphia: Fortress, 1985), pp. 56–57.

12. From "The Mesha Inscription," in Walter Beyerlin, ed., *Near Eastern Religious Texts Relating to the Old Testament*, trans. John Bowden (Philadelphia: Westminster, 1975), pp. 238–39.

13. Mack, *Who Wrote the New Testament?* pp. 303–306. This is a historically common belief; Patrides, *The Grand Design of God*.

14. Margaret Barker, *The Great Angel: A Study of Israel's Second God* (Louisville, Ky.: Westminster/John Knox, 1992); Alan F. Segal, *Two Powers in Heaven: Early Rabbinic Reports About Christianity and Gnosticism* (Leiden, The Netherlands: Brill, 1977).

15. André LaCocque, *Daniel in His Time* (Columbia: University of South Carolina Press, 1988); G. A. Wells, *Who Was Jesus?* (La Salle, Ill.: Open Court, 1989), pp. 160–68.

16. Mack, *Who Wrote the New Testament?* p. 109. See also Stephen O'Leary, *Arguing the Apocalypse: A Theory of Millennial Rhetoric* (New York: Oxford University Press, 1994).

17. James W. Loewen, *Lies My Teacher Told Me: Everything Your American History Textbook Got Wrong* (New York: New Press, 1995).

18. Lane Fox, *The Unauthorized Version*, p. 116.

19. R. Ludwigson, *A Survey of Bible Prophecy* (Grand Rapids, Mich.: Zondervan, 1951), p. 46. The interrupted prophecy belief among fundamentalists derives largely from the nineteenth-century dispensationalism of John Nelson Darby; Ernest R. Sandeen, *The Roots of Fundamentalism: British and American Millenarianism, 1800–1930* (Grand Rapids, Mich.: Baker, 1978), pp. 62–64.

20. Bernard McGinn, *Antichrist: Two Thousand Years of the Human Fascination with Evil* (San Francisco: Harper, 1994); Tom McIver, *The End of the World: An Annotated Bibliography* (Jefferson, N.C.: McFarland, 1999).

21. W. Montgomery Watt, *What Is Islam?* (New York: Frederick A. Praeger, 1968), p. 79.

22. Heribert Busse, *Islam, Judaism, and Christianity: Theological and Historical Affiliations,* trans. Allison Brown (Princeton: Markus Wiener, 1998). Critics can often embarrass orthodox Muslims by doing little but print stories from the Quran, note that they look like fairy tales, and compare them to Jewish

antecedents; e.g., İlhan Arsel, *Şeriat'tan Kıssalar* (İstanbul: Kaynak, 1996). The closest English example I know of is Ibn Warraq, *Why I Am Not a Muslim* (Amherst, N.Y.: Prometheus Books, 1995), pp. 49–65.

23. Thomas Ballantine Irving, Khurshid Ahmad, and Muhammad Manazir Ahsan, *The Qur'an: Basic Teachings* (London: Islamic Foundation, 1979), p. 29.

24. F. E. Peters, *Muhammad and the Origins of Islam* (Albany: State University of New York Press, 1994), p. 265.

25. Patricia Crone and Michael Cook, *Hagarism: The Making of the Islamic World* (Cambridge: Cambridge University Press, 1977); Ibn Warraq, ed. and trans., *The Quest for the Historical Muhammad* (Amherst, N.Y.: Prometheus Books, 2000). Cook further develops his argument about the unreliability of Muslim sources in *Early Muslim Dogma: A Source-Critical Study* (Cambridge: Cambridge University Press, 1981). For an argument against such comprehensive fabrication, see Clinton Bennett, *In Search of Muhammad* (London: Cassell, 1998), pp. 54–65.

26. Peters, *Muhammad and the Origins of Islam*, pp. 141–52. Mostafa Vaziri, in *The Emergence of Islam: Prophecy, Imamate, and Messianism in Perspective* (New York: Paragon, 1992), points out another possibility: that Muhammad's vision was triggered by his sighting Halley's comet (pp. 10–12).

27. Karen Armstrong, *Muhammad: A Biography of the Prophet* (San Francisco: Harper, 1992), p. 84.

28. Peters, *Muhammad and the Origins of Islam*, pp. 150, 205–206.

29. Western critics of Muhammad have often seized on epilepsy as an explanation of his revelations; e.g., Paul Kurtz, *The Transcendental Temptation: A Critique of Religion and the Paranormal* (Amherst, N.Y.: Prometheus Books, 1991), pp. 211–15. This accusation goes back to the Byzantines; Bennett, *In Search of Muhammad*, p. 80.

30. Peters, *Muhammad and the Origins of Islam*, chap. 6.

31. Frederick Mathewson Denny, *An Introduction to Islam* (New York: MacMillan, 1985), pp. 87–88.

32. Ernest H. Taves, *Trouble Enough: Joseph Smith and the Book of Mormon* (Amherst, N.Y.: Prometheus Books, 1984), pp. 64–65.

33. Suzanne Riordan, "Channeling: A New Revelation?" in *Perspectives on the New Age*, ed. in James R. Lewis and J. Gordon Melton (Albany: State University of New York Press, 1992). Western religion is increasingly becoming more cultlike and individualistic (Steve Bruce, *Religion in the Modern World: From Cathedrals to Cults* [Oxford: Oxford University Press, 1996]), so modern prophets are increasingly likely to bring new revelations rather than affirm the doctrines of an organized religion.

34. Ann Braude, *Radical Spirits: Spiritualism and Women's Rights in Nineteenth-Century America* (Boston: Beacon, 1989); Catherine Wessinger, ed., *Women's Leadership in Marginal Religions: Explorations Outside the Mainstream* (Urbana and Chicago: University of Illinois Press, 1993); Keith Thomas, *Religion and the Decline of Magic: Studies in Popular Beliefs in Sixteenth and Seventeenth Century England* (New York: Oxford University Press, 1971), pp. 138–40.

35. This seems to be the standard story from a liberal religious perspective, e.g., Armstrong, *Muhammad*; William E. Phipps, *Muhammad and Jesus: A*

Comparison of the Prophets and Their Teachings (New York: Continuum, 1996). It derives mainly from the work of W. Montgomery Watt.

36. Peters, *Muhammad and the Origins of Islam,* p. 75.

37. Patricia Crone, *Meccan Trade and the Rise of Islam* (Princeton, N.J.: Princeton University Press, 1987), pp. 234–45. The aggressive element in classical Islam is indubitable, though modernist apologists often argue that early Muslims fought only in self-defense. Rudolph Peters, *Jihad in Classical and Modern Islam* (Princeton, N.J.: Markus Wiener, 1996); Reuven Firestone, *Jihad: The Origin of Holy War in Islam* (New York: Oxford University Press, 1999).

38. From the tradition collection of Bukhari, Jihad 270. A more detailed account is in Sahih Muslim, Kitab-al-Jihad chap. 744, 4436. The earliest source for the story is Ibn Ishaq, *The Life of Muhammad*, trans. A. Guillaume (Oxford: Oxford University Press, 1955), pp. 364–69. For a modern justification of the assassination as a legitimate covert operation, see Bennett, *In Search of Muhammad*, pp. 117–18.

39. Peters, *Muhammad and the Origins of Islam*, pp. 222–24.

40. Richard W. Bulliet, *Islam: The View from the Edge* (New York: Columbia University Press, 1994); Olivier Roy, *The Failure of Political Islam* (Cambridge: Harvard University Press, 1994); Ahmet Yaşar Ocak, *Türkler, Türkiye ve İslâm: Yaklaşım, Yöntem ve Yorum Denemeleri* (İstanbul: İletişim, 1999); L. Carl Brown, *Religion and State: The Muslim Approach to Politics* (New York: Columbia University Press, 2000).

41. Ahmet Yaşar Ocak, *Türk Sufîlîğine Bakışlar* (İstanbul: İletişim, 1996), p. 46. There is controversy over how much of Turkish heterodox Islam derives from pre-Islamic Turkish traditions or from Anatolian Kurdish beliefs; e.g., Cemşid Bender, *Kürt Tarihi ve Uygarlığı* (İstanbul: Kaynak, 1995) defends the Kurdish view.

42. Ocak, *Türk Sufîliğine Bakışlar,* pp. 126–36. The translation from Yunus Emre is mine.

43. Jeffrey Burton Russell, *The Devil: Perceptions of Evil from Antiquity to Primitive Christianity* (Ithaca, N.Y.: Cornell University Press, 1977), p. 38. See also R. G. Collingwood, *The Idea of History* (London: Oxford University Press, 1956); criticized by G. A. Wells, *Belief and Make-Believe: Critical Reflections on the Sources of Credulity* (La Salle, Ill.: Open Court, 1991), pp. 25–34.

44. Langdon Gilkey, in C. T. McIntire and Ronald A. Wells, ed., *History and Historical Understanding* (Grand Rapids, Mich.: Wm. B. Eerdmans, 1984), pp. 5, 11.

45. Eric Voegelin, *Anamnesis*, ed. and trans. Gerhart Niemeyer (Notre Dame: University of Notre Dame Press, 1978), p. 136.

46. Jonathan Z. Smith, *Drudgery Divine: On the Comparison of Early Christianities and the Religions of Late Antiquity* (London: University of London, School of Oriental and African Studies, 1990), chap. 2.

47. Jacob Neusner, *Rabbinic Judaism: Structure and System* (Minneapolis: Fortress, 1995).

48. Louis O. Mink, "Narrative Form as a Cognitive Instrument," in Louis O. Mink, *Historical Understanding*, ed. Brian Fay, Eugene O. Golob, and Richard T. Vann (Ithaca, N.Y.: Cornell University Press, 1987).

49. Sherwin T. Wine, *Judaism Beyond God* (Hoboken, N.J.: KTAV Pub.

House; Milan, Kan.: Milan Press, 1995), p. 98. See also Yehuda Bauer, "The Holocaust, Religion and Jewish History," reprinted in *Judaism in a Secular Age: An Anthology of Secular Humanistic Jewish Thought*, ed. Renee Kogel and Zev Katz (Hoboken, N.J.: KTAV Pub. House; Milan, Kan.: Milan Press, 1995).

50. Carl G. Hempel, "The Function of General Laws in History," in *Theories of History: Readings from Classical and Contemporary Sources*, ed. Patrick L. Gardiner (Glencoe, Ill.: Free Press, 1959).

51. Peters, *Muhammad and the Origins of Islam,* chap. 9.

52. Robin Lane Fox, *Pagans and Christians* (New York: Alfred E. Knopf, 1986).

53. It might then be possible to use ideas from physics, like chaotic dynamics, to understand history better. See Michael Shermer, "The Chaos of History: On a Chaotic Model that Represents the Role of Contingency and Necessity in Historical Sequences," *Nonlinear Science Today* 2, no. 4 (1993): 1; "Exorcising Laplace's Demon: Chaos and Antichaos, History and Metahistory," *History and Theory* 34, no. 1 (1995): 59. However, that history is based in physics does *not* mean that physical techniques should be immediately transportable to history; I remain somewhat skeptical.

54. David F. Wells, *No Place for Truth, or, Whatever Happened to Evangelical Theology?* (Grand Rapids, Mich.: Wm. B. Eerdmans, 1993), p. 270.

55. Merkley, *The Greek and Hebrew Origins of Our Idea of History*, pp. 4–5; Gary R. Habermas and J. P. Moreland, *Immortality: The Other Side of Death* (Nashville, Tenn.: Thomas Nelson, 1992), pp. 162–64.

56. Don Peretz, "Islamic Revival or Reaffirmation," in Don Peretz, Richard U. Moench, and Safia K. Mohsen, *Islam: Legacy of the Past, Challenge of the Future* (New York: New Horizon/North River, 1984); Bassam Tibi, *The Crisis of Modern Islam: A Preindustrial Culture in the Scientific-Technological Age* (Salt Lake City: University of Utah Press, 1988).

57. Bruce, *Religion in the Modern World*; Wade Clark Roof, *Spiritual Marketplace: Baby Boomers and the Remaking of American Religion* (Princeton, N.J.: Princeton University Press, 1999).

58. Wells, *No Place for Truth*, pp. 259–64.

59. Mohammed Arkoun, *Rethinking Islam: Common Questions, Uncommon Answers*, ed. and trans. Robert D. Lee (Boulder, Colo.: Westview, 1994), p. 34.

60. Jacques Monod, *Chance and Necessity* (New York: Alfred E. Knopf, 1971), p. 171.

61. Wolfhart Pannenberg, "Revelation as History," in *God, History, and Historians*, ed. C. T. McIntire (New York: Oxford University Press, 1977).

Chapter Five
God Incarnate

Christianity is the bastard progeny of Judaism. It is the basest of all national religions.

—Celsus, *True Discourse* (c. 178)

Jesus saves. At least that is what evangelical Christians say. Apparently he saves them from evil spirits, from the terrible sins that taint their souls, even from death. He also manages to help out more liberal Christians. He saves them from despair, the sense that nothing really matters. Jesus leads to God, our loving Father, though whether he is *the* way to God does not seem so clear anymore. Not bad, for a dead Jewish preacher. Unfortunately, all we really know about Jesus are uncertain bits and pieces we dig up from under Christian myths. Scholars assemble these to draw many different portraits of Jesus; in all of them, he appears too human to be the Son of God.

IN SEARCH OF JESUS

According to traditional Christianity, Jesus is our Redeemer, the Perfect Man, the Sacrifice for our sins, our Judge when we come to account for our deeds. In fact, Jesus *is* God, fully human and completely divine at once. This is like saying something is both fully a fruitcake and a complete fruitbat; nonsense to the infidel, a profound and fruitful doctrine to theologians. To further sow confusion, orthodox Christians also insist that God, Jesus, and a Holy Spirit are three, but yet one.

Whether this scheme makes sense or not, Western culture has certainly treated Jesus as God. He appears as the leading figure in our myths, even as we keep rewriting them to fit our time and place.[1] And for all the variations, we have usually believed that a historical core is the gospel truth; that Jesus, as declared in the Apostle's Creed, was "born of the Virgin Mary, suffered under Pontius Pilate, was crucified, dead, and buried; . . . the third day he rose again from the dead." Jesus went around Palestine preaching and performing miracles, and died a horrible death in order to save from Satan all who believe in him. It seems Jesus was at the center of an awesome outpouring of supernatural power; furthermore, his life confirmed and completed the revelations given to the Jews. Jesus was at least a great prophet, perhaps the focal point of God's plan for creation, maybe even God Incarnate.

Orthodox Christians also tell us why we should trust their story. After Jesus rose from the dead and ascended to heaven, his apostles started proclaiming the good news, faithfully preserving the memory of Jesus' words and deeds. This first generation passed on the true understanding of Jesus as both God and man, which was expressed in the early creeds of the holy Catholic Church. The gospels were based on eyewitness testimony, written by disciples or their close companions.

Like so much in old-time religion, the Christian story began to fall apart a few centuries ago. Through the eyes of the Church, the canonical gospels were a harmonious whole, complementing one another with their varying points of view. As historical documents, however, scholars noticed they told significantly different stories. In the gospel we call Mark, for example, Jesus is a wandering miracle worker who talks of the coming Kingdom of God. Curiously, he also wants to keep his message and his identity a secret. Even his disciples are a bumbling lot, unable to grasp what is going on. In the gospel of John, Jesus is a different character—an unearthly redeemer come down from heaven. This Jesus does not keep quiet about his divinity; though the Jews seem quite incapable of understanding him. Mark tells us the end of the world is near, John stays away from an immediate apocalypse. Even some common events in their stories occur differently. Both say Jesus caused an incident in the Jerusalem Temple by overturning the tables of the moneychangers. In Mark, this happens at the end of Jesus' public career, shortly before his crucifixion. In John, it comes at the very beginning.

Conservative Christians still try to harmonize their gospels, especially where important miracles like Jesus' resurrection are at stake.[2] Mainstream scholars have long given up on this; the gospels look different because they *are* different stories. Our gospels are heavily edited collections of traditions; they were origin myths for different Christian communities, not historical biographies. They are not carefully recorded memories of disciples but anonymous works dating from 70–120 C.E., long after any crucifixion in the early 30s. Mark is the earliest gospel, loosely stringing together stories about Jesus to form a narrative. Matthew and Luke use Mark as a source, along with a sayings collection known as "Q," and also throw in other traditions and their creative imaginations. John spins a different tale, put together mostly from independent traditions.[3] There were many early Christianities, all producing myths to justify their existence. Our gospels survived because they fit the emerging orthodox ideology; others were suppressed or forgotten. Like Islam, Christianity has murky origins; sorting out actual history from political and theological fictions is very difficult.

To recover hints of history, scholars have to perform a kind of textual archaeology, uncovering different layers of composition, trying to find material expressing earlier beliefs rather than the needs of the author's community. The results are often quite curious. For example, among the earliest statements in the New Testament about the nature of Jesus, some say Jesus was adopted as the Son of God at his Resurrection. In Rom. 1:3–4 the apostle Paul says, "on the human level he was a descendant of David, but on the level of the spirit—the Holy Spirit—he was proclaimed Son of God by an act of power that raised him from the dead." Paul is here probably quoting an already existing creed, which suggests that the Resurrection story began as a tale about how God vindicated a martyr by raising him from the dead.[4]

Critical history, in other words, has made the orthodox story unbelievable. But Jesus is still central to how we understand God in our culture. So conservative Christians continue to proclaim the old stories, decrying the philosophical prejudices they detect behind Enlightenment science and history. More liberal theologians no longer accept the claim of God Incarnate, but they still want to capture the significance of Jesus expressed in such a myth.[5] They may not agree that a corpse got up and walked, but they still say a mysterious event beyond the ordinary must have propelled the radical vision of Christian faith into the world. Even infidels hostile to Christianity feel compelled to ask who Jesus really was, if only to find out where Western culture went wrong.

Now we run into a problem. Our search for Jesus produces few definite results. As John Dominic Crossan observes in his own biography of Jesus, "Historical Jesus research is becoming something of a bad joke" because of "the number of competent and even eminent scholars producing pictures of Jesus at wide variance with one another."[6] Crossan's Jesus is a peasant sage; we also find apocalyptic prophets, faith healers, magicians, Jewish nationalists, bandits, rabbis, and more. Scholars have a disturbing tendency to come up with portraits of Jesus which reflect their own circumstances. John C. Mellon, for example, argues that Mark was produced by a community of alcoholics, and describes Jesus as a humbled person on the road to recovery. Mellon is himself a recovering alcoholic, and sees signs of the spirituality of modern twelve-step recovery movements in Mark.[7] It seems we have too many ideas about Jesus but too little evidence to converge on a single picture.

So we will not find a clear-cut Jesus who satisfies all our historical curiosity. But we can at least look at some of his more plausible portraits, and try to understand how the traditional story emerged. There is much we do not know about Jesus, but we know enough to see he was quite human—not someone who manifested God on earth.

AN APOCALYPTIC PROPHET

Some gospel stories are probably historical. Mark begins with John the Baptizer dunking Jesus in the Jordan River. It would be peculiar for the eternal Son of God to start out with a baptism for the forgiveness of sins, subordinating himself to John. The later gospels alter the story to emphasize Jesus' superiority; Matt. 3:14 makes John say, "It is I who need to be baptized by you." When we find potentially embarrassing material in our gospels, there is a good chance it was not invented to glorify Jesus. The story of Jesus' baptism probably has historical roots, or is at least an early tradition too well known to be censored by the gospel writers.

Some gospel stories are myths. Matthew and Luke start with two different fairy tales about Jesus' birth of a virgin and his precocious infancy. We know these tales are fiction because they are fantastic to begin with, they contradict one another, and they get their history wrong when they make claims we can check independently. When Muslims invented stories about how Muhammad's birth and childhood foreshadowed his prophethood, they were following a long tradition of imagining their leader's life as that of an ideal prophet. Christians did the same.[8]

We want to strip away the layers of Christian myth to get at the historical Jesus. Unfortunately, there is so much myth that finding the actual Jesus is an infuriatingly difficult task. We are not completely in the dark; it is fairly obvious that stories like Jesus' baptism are much more credible than the Virgin Birth. But we cannot just start digging in gospel texts armed with some rules about what is more credible. For example, if we collect embarrassing stories which go against the sensibilities of early Judaism or the emerging church ideology, we will end up with some episodes which have a reasonable claim to be historical. But if we stop there, we will be left with a biased picture of Jesus, making him look unique in his time and place. Surely Jesus said many things which were perfectly in line with the Judaism of his time, and early church ideology may well have been based in part on the memory of Jesus.[9]

To hold our picture of Jesus together, we need to see Jesus as a man of his times, inhabiting a social and cultural environment. Of course, we can fit Jesus into many social roles: prophet, teacher, revolutionary, healer, and so on. So we have to look at the alternatives, and see if any one best makes sense of the historical evidence and the fact that early Christians told so many fantastic tales about Jesus.

The most straightforward starting point is to think of Jesus as a Jew. The New Testament was written in Greek, when Christianity was starting to become a non-Jewish religion. Myths like the virgin birth of a savior-god have a Greek flavor. But the first Christians were Jews, and the gospel stories take place in a Jewish environment. The stories are steeped in Jewish beliefs about God, angels and devils, prophecies, and apocalyptic upheavals. So it is reasonable to suppose that Jesus was a wandering Jewish preacher and healer who was put to death by the Romans occupying Palestine, quite possibly with the collusion of Jewish priestly elites. The gospel stories would then be correct in outline, though much embellished by a mythic imagination and the apologetic concerns of later Christian communities.

This bare outline does not tell us much of Jesus, what he taught, or why he was crucified. But upon looking closer at the New Testament stories, many scholars have encountered an apocalyptic prophet—not the ancient equivalent of a wild-eyed man bearing a "Repent, The End Is Near" sign; rather, a prophet of Jewish restoration. Restoration theology proclaimed the imminent redemption of Israel and a Gentile world which would recognize the sovereignty of Yahweh. The Bible's prophecies for Israel would be fulfilled. A messiah would come—perhaps to restore the Davidic monarchy. The Temple in Jerusalem would be destroyed and reerected by a miracle of God, the dead would be resurrected, and much more. Jesus believed God's Kingdom on earth was near, and that he was charged with announcing its coming.

E. P. Sanders describes Roman Palestine in Jesus' time as a buffer zone for an empire; not heavily occupied, not close to revolt. Jews were not on the edge of their seats with messianic hopes, though many did expect some sort of divine intervention. In this environment, Jesus was "an individual who was convinced that he knew the will of God."[10] Sanders's Jesus starts out being baptized by John, and like John, proclaims a message squarely in the tradition of Jewish restoration theology:

> Jesus saw himself as God's last messenger before the establishment of the kingdom. He looked for a new order, created by a mighty act of God. In the new order the twelve tribes would be reassembled, there would be a new temple, force of arms would not be needed, divorce would be neither necessary nor permitted, outcasts—even the wicked—would have a place, and Jesus and his disciples—the poor, meek and lowly—would have the leading role.[11]

Signs of restoration theology are all over the New Testament. For example, take the story about Jesus overturning the tables of the money changers and pigeon sellers in the Jerusalem Temple. Judaism, according to Christian tradition, was trapped in ritual rigidities, and even Temple worship was corrupted by crass commercial activity. But this does not make sense. Pilgrims from all over needed to purchase unblemished animals for their ritual sacrifices. Money changers and pigeon sellers were performing a vital service, not corrupting the Temple. And Jesus' disciples continued to worship in the Temple after his death—not exactly an indication that Jesus condemned normal Temple observance. Jesus thought the Temple's days were numbered, not that it was corrupt.[12] In Mark 14:58, Jesus says, "I will pull down this temple, made with human hands, and in three days I will build another, not made with hands." Overturning tables was an act symbolizing imminent destruction, exactly what Jesus threatened would happen to the Temple. This destruction was a prelude to a miraculous rebuilding at the coming of God's Kingdom.

Other actions of Jesus also make sense when we see them as part of his hope for Israel's restoration. Paula Fredriksen lists:

> his calling twelve disciples to represent all the tribes of eschatological Israel; his intensification of the ethical norms embodied in Torah; his journey to Jerusalem to greet the coming Kingdom; his prophetic gesture at the Temple, in anticipation of a Temple not made by hands; his prophecy of the imminent fulfillment of God's promises to Israel revealed in Torah.[13]

Such a restoration obviously requires divine intervention. So Jesus and his followers expected to be delivered by a miracle, not by military force. Jesus may have entered Jerusalem preaching apocalypse and restoration, which could well mean that Jewish elites saw him as yet another troublemaker stirring up people and inviting a violent Roman response. He would have been tried and executed in an offhand manner—like John the Baptizer—but his disciples would be left alone, as they were harmless without a leader.

What about Jesus' teachings, which Christians have thought to be a radical novelty, a revelation of how God wants us to live with one another? Sanders argues that Jesus thought that "His special mission was to promise inclusion in the coming kingdom to the outsiders, the wicked, if they heeded the call."[14] But though Jesus was distinctive in the way he emphasized the wicked and downplayed *national* repentance, this did not place him in opposition to Judaism. The legalistic, externalistic Judaism rejected by Jesus in favor of grace and inner experience is a Christian myth, not history; the notion that Jesus was killed by the Jews for preaching God's grace is sheer nonsense. Christians have traditionally thought their God Incarnate must stand out by his uniqueness, but Sanders observes that "we cannot say that a single one of the things known about Jesus is unique: neither his miracles, non-violence, eschatological hope or promise to the outcasts."[15]

As an apocalyptic prophet, Jesus is believable; a man of his time and place. With such a Jesus, we can be conservative about the New Testament. Though garbled and filtered through theological perspectives Jesus would not have shared, most of the Christian story is based on actual history. But now the problem is explaining Christianity. How does a Jewish apocalyptic prophet who was dead wrong about the coming Kingdom and who was crucified as a pest end up starting a Greek religion?

Since we are being conservative, we can continue with the gospel stories. These indicate that shortly after Jesus' crucifixion, his disciples had visions of him. These accounts are confused, and a belief like the Resurrection would acquire a very heavy overlay of myth, no matter what the initial experiences were. Still, we might be able to make some psychological sense of the story. Gerd Lüdemann points out that the disciple Peter's Easter experience, including seeing a ghost, is plausible as part of a mourning process. Conversion visions, often associated with guilt, are well-known psychological phenomena.[16] Visions alone would not generate a movement; the disciples had a strong investment in the coming Kingdom, and Jesus' unexpected execution would seem to make the prospects for this rather dim. But interestingly, committed believers often respond to failed prophecy by reaffirming their beliefs with an increased supernatural emphasis, sometimes even becoming zealous missionaries.[17] The disciples began to proclaim that their apocalyptic hopes would be realized when Jesus came again, which was sure to take place very soon. If Jesus had expected the Son of Man announced by the book of Daniel to arrive soon, the early Christians would begin to think Jesus was actually referring to himself.

There is still a long way to go between revived hopes of Jewish restoration and the increasingly Greek Christianity of the New Testament. Apocalyptic beliefs are conservative: they promise supernatural vindication for social values which are under attack. Why would Gentiles be attracted to ideals of Jewish restoration?

This is a difficult question, and some resort to extreme measures to answer it. Hyam Maccoby argues that Gentiles were in fact not attracted to the historical Jesus. Jesus and his disciples were observant, conservative Jews; Jesus claimed to be a messiah who would restore the Davidic monarchy. The followers of Jesus made up the early Jerusalem church, which the New Testament describes as strictly obedient to Jewish religious laws. The apostle Paul, whom Maccoby thinks was only superficially a Jew, invented Christianity. He founded a libertine Hellenistic religion combining features of a mystery cult and gnostic belief in saving knowledge brought by a heavenly redeemer, and made it more attractive and permanent by claiming that the Jewish religious epic was really all about Christ. The diverse theologies we see in the New Testament were generated by the rivalry of Pauline and Jewish Christianity.[18]

Paul was certainly very influential in shaping Christianity and in bringing the new religion to the non-Jewish world. But there is little reason to depart from a conventional picture of Paul,[19] and Paul's creativity alone does not explain how hopes for Jewish restoration started a Greek religion. Restoration theology included an expectation that at the End Times, the Gentiles would also come to know God and be redeemed. A missionary group proclaiming restoration may have tried to bring Gentiles into the Kingdom as well. Furthermore, Greek and Jewish cultures were not radically separate; there were plenty of Hellenized Jews and "God-fearers" among Gentiles who were interested in Jewish religion. So an apocalyptic message may well have been a starting point for a religious movement bringing together Jews and Gentiles at the margins of society.

Jesus' message had failed with his execution, but his disciples had resurrected their hopes and sought new converts. However, in a few decades the small but expanding Christian communities faced failure once again, when Jesus neglected to return. Christians made up marginal social groups expecting an imminent miraculous change of the world, and they were beginning to get into serious conflicts with mainstream Jews. The gospel stories began to take shape in such an environment. Different Christian communities responded to apocalyptic failure in diverse ways, making myths which departed far from restoration theology. To Paul, the second coming was imminent; then Christians began to wait for the first generation who had witnessed Jesus to die off. Writing well into the second century, the author of 2 Peter assured his readers that prophecy had not failed, and told them that "In the Lord's sight one day is like a thousand years and a thousand years like one day." Christian mythology began to reflect the existence of an organized church, and started falling into an orthodox mold. Jesus started out as a failed prophet, and ended up as God Incarnate who founded the universal Church.

This is a compelling account of Christian origins. An apocalyptic prophet fits the evidence, and allows us to treat the gospels as largely historical. Of course, there are many points at which this theory is strained. While discussing the diversity of early Christian beliefs, proponents of an apocalyptic prophet often resort to speculation, or call upon Christian experiences of the Risen Lord as a catch-all explanation solving every difficulty. Still, Jesus may well have preached that restoration was at hand—history is full of accidents, and when our evidence is poor in quality sometimes all we can do is speculate about the accidents which could have led from Jesus to Christianity. But these difficulties do mean we should look at other portraits of Jesus, and see if they do a better job explaining Christian mythmaking.

MYTHMAKING

Early Christians did not invent haphazardly; the life of Jesus had to reflect his many mythic roles. He was, for example, the Messiah. So Luke and Matthew list different genealogies to show that Jesus was descended from King David—even though they also say he had no human father. In other words, Christians made up historical fictions. They were not out to deceive people, but they followed a mythic sense of how history ought to have been. As Randel Helms remarks, they knew what sort of story to produce:

> When the author of Mark set about writing his Gospel, circa 70 A.D., he did not have to work in an intellectual or literary vacuum. The concept of mythical biography was basic to the thought-processes of his world, both Jewish and Graeco-Roman, with an outline and a vocabulary already universally accepted: a heavenly figure becomes incarnate as a man and the son of a deity, enters the world to perform saving acts, and then returns to heaven.[20]

New Jewish myths had to start from the Jewish religious epic. So Christians composed their tales with an eye on scripture, convinced that Jesus fulfilled Old Testament promises. Many Jews outside of Palestine used the Septuagint, a Greek version of the Old Testament. As a result, our gospel stories often derive from the Septuagint, sometimes showing close verbal parallels, even following mistakes in translation. For example, Matt. 1:23 quotes Isaiah out of context to prophesy the Virgin Birth, saying, "a virgin will conceive and bear a son." The Septuagint passage speaks of a virgin, but the Hebrew says only that a young woman is with child.

The gospels tell miracle stories about Jesus raising dead people, or feeding multitudes with a few loaves and fishes. These parallel Old Testament miracles; the gospels even lift phrases straight out of the Septuagint. Mark contains two sets of five miracle stories, each beginning with a sea crossing, continuing with exorcisms and healings, and ending with feeding four or five thousand people. Miraculous sea crossings and feeding of multitudes allude to the Israelites' exodus from Egypt. The healings, including Jesus raising Jairus's daughter, recall the prophets Elijah and Elisha. It seems that before Mark was put together, some followers of Jesus began to imagine themselves as a new Israel. So they made myths which cast Jesus as a figure similar to Moses and Elijah, except that his miracles were even more powerful.[21]

Our examples can be multiplied. Christians created stories based on the Jewish epic, and began to fashion a more Hellenistic myth as Christianity became more Gentile. However, this now brings up an interesting question. Christianity seems to have become a dumping ground for so many myths that it becomes very difficult to say much about the historical Jesus. Even the evidence that Jesus was an apocalyptic prophet is not rock-solid. A savior figure, for Jewish Christians, could well include stereotyped expectations about a prophet of restoration. Plus, early Christianities were precarious social experiments which would respond to hard times by calling on the Jewish apocalyptic tradition. So perhaps they cast Jesus as an apocalyptic prophet only to fill yet another mythic role. What, then, can we say about the historical Jesus at all?

Many rationalist critics of Christianity have suspected that the Jesus stories are entirely mythical—that we cannot even say that Jesus was a preacher in first-century Palestine who was executed by the Romans. Today, almost all scholars of early Christianity think this is preposterous. But G. A. Wells, though he has come to accept that the New Testament contains some memories of Jesus, still argues it is mostly myth.[22]

Wells points out that non-Christian evidence concerning Jesus is virtually nonexistent. Josephus, the Jewish historian, mentions him twice, but his text has obviously been altered by Christians, probably because the original was not complimentary to Jesus. Wells, however, argues that all this material is interpolated, and that in any case, Josephus was writing at a late date and could have gotten his information from Christian sources.[23] A few Roman historians mention Jesus, but their information may also derive from Christians. For the supernatural Savior of the Christian story, this lack of reliable independent documentation is curious.

This would not mean much if garbled historical memory was still the best explanation for the bulk of the gospel stories. But Wells also argues that we see signs of an originally unhistorical myth being elaborated over time, even acquiring a historical setting for apologetic purposes. We suspect Muslims fabricated many traditions, giving them a concrete historical setting, though they addressed later controversies in the Islamic Empire. Something similar may be true for the New Testament.

Aside from traditions incorporated into the gospels, Paul's letters are our earliest Christian documents. And these letters are remarkably silent about the historical Jesus. Paul loves supernatural wonders as a sign of divine power, but he shows no knowledge of Jesus' miracles. He was struggling to assert his authority against the church in Jerusalem, led by Peter, but he does not bring up the story of Peter denying Christ. He has plenty of advice for his readers, but he almost never mentions that the earthly Jesus did or taught the same thing. Paul's Jesus is not a preacher and miracle worker recently crucified in Jerusalem by the Romans, but a supernatural being who existed before the world was made. He had descended into a world dominated by hostile supernatural powers, lived in humble obscurity, redeemed us through his death, and would come again very soon to usher in the apocalypse. In other words, he was a figure of myth similar to Wisdom in Judaism. In the ancient anthropomorphic tradition, Wisdom was a name of Yahweh as the great angel, the first creation of God, who participated in making the world, and who came down to earth to seek a place among humans. In 1 Cor. 24, Paul calls Jesus "the power of God and the wisdom of God," in other words, a manifestation of the Lord.[24]

Other early letters besides those of Paul are also silent regarding the historical Jesus. Only after the gospels are composed do letter writers begin to sprinkle biographical references in their exhortations. Since Christian letter writers had no objection to detailed historical claims, their early silence suggests that they did not know much about the historical Jesus.

Wells argues that Christians started giving their founder figure a concrete history, incorporating memories of a failed Galilean preacher, when their cosmic Savior failed to return as promised. They began to emphasize how the first visit by Jesus was effective for salvation. His sacrifice atoned for our sins, but Christians also began to think he fulfilled scripture in a more comprehensive way, gave them ethical instructions, and instituted church practices. The Christ cult had accumulated teachings formulated by their leaders and by prophets who spoke when possessed by Jesus' spirit. Attributing these to Jesus gave them a more definite authority. The emerging Church could also use a more concrete Jesus, since the forerunners of orthodoxy were constantly battling heretics. There were Jewish Christians who treated Jesus as no more than a great prophet; and gnostics who denied Jesus was human and interpreted their myth in individualist, magical terms. The orthodox preferred a man who walked the Palestinian soil, yet who was obviously much more than a man.

Wells is correct to point out the massive mythmaking in early Christianity. However, it still is difficult to account for the gospel stories by wholesale invention. Why *our* stories? Why speak of Jesus of Nazareth and cook up far-fetched tales about how he must have been born in Bethlehem—would it not have been easier to avoid the Nazareth business in the first place? The Jesus of the gospels may well be largely a myth, but we need a better approach to the mythmaking.

Paul's Christians had no interest in history, only a supernatural savior; they practiced cultic rituals like baptism and the Eucharist; they engaged in faith healing and became possessed by "the Spirit." This Christ cult, however, was not the only early Christian group. In fact, its focus on the story of Christ's death and Resurrection did not take over Christianity until much later. Jonathan Z. Smith observes that from our nonliterary evidence, early Christians appear as

> a heterogeneous collection of relatively small groups, marked off from their neighbors by a rite of initiation (chiefly, adult baptism), with their most conspicuous cultic act a common meal, and a variety of other activities that would lead a scholar to classify these groups as being highly focused on a cult of the dead.[25]

They seem to have cultivated the memory of their dead and believed in a continuing relationship; Jesus' atoning death does not appear in the picture. We find signs of Christianities which were not interested in Jesus' death in our texts as well. For example, many scholars believe Matthew and Luke incorporate an older text called "Q," which consists mainly of Jesus' sayings. Q's Jesus is not a messiah, a supernatural being, or a resurrected savior; he is a teacher who inspired a novel social formation. Burton L. Mack points out that such "Jesus movements" did not reflect on Jesus' death the way the Christ cult did. Q only places Jesus' death in the tradition about Israel killing its prophets, reassuring the people of Q that they were correct in following Jesus; nowhere does Jesus' death have any significance for salvation.[26]

In fact, Mack finds a bewildering variety of early Christianities, all which imagined Jesus differently. Paul tells us of the Jerusalem Church, made up of Orthodox Jews. Mark's miracle stories with Old Testament themes might come from a community whose members thought of themselves as a new Israel. Mark also includes stories which picture Jesus as a founder of a philosophical school, perhaps produced by a movement which found its identity in debate with Orthodox Jews. In the Gospel of Thomas, which might incorporate early elements, we encounter a Jesus who was a revealer of cosmic wisdom. Though Mack perhaps too easily assumes that each textual tradition belongs to a distinct community, these traditions have little in common beyond Jesus as a founder figure.

With a prophet of restoration, diversity was a result of apocalyptic failure. Christianity survived Jesus because of an obscure event which underlies our Resurrection stories. Mack, however, points out that the variety of early Christianities strains such conservative theories. To understand Christian origins, we must explain the mythmaking. And this takes place together with social experimentation: people tell new stories to sanctify a new social ideal. The Jesus movements remembered Jesus as a teacher who called for a Kingdom of God in which people from all ethnic and social backgrounds would come together. He did not proclaim a detailed ideology; it fell to the movements themselves to imagine what the Kingdom meant and to flesh it out in their lives. So we got diverging social groups which made different Jesus myths.

Some scholars compare Jesus to the Cynics, itinerant Hellenistic philosophers who made a career out of poking fun at social conventions. They behaved so as to upset the usual order; and they relied on witty retorts and aphorisms, not detailed ethical principles and teachings. John Dominic Crossan's Jesus is "a peasant Jewish Cynic" who opposed the rigidly stratified order of an agrarian society under imperial control; criticizing relationships of patronage and rank, even the patriarchal family.[27] Being from Galilee, a province with a mix of peoples, Jesus also encouraged experiments bringing together Jews and Greeks. Into the Greek Cynic example, he infused Jewish concerns about an ideal community. As Mack puts it,

> two themes mark the genius of the [Jesus] movement. One is a playful, edgy challenge to take up a countercultural lifestyle. . . . The closest analogy for this kind of invitation to live against the stream is found in Cynic discourse of the time. It does appear that Jesus was attracted to this popular ethical philosophy as a way for individuals to keep their integrity in the midst of a compromising world. The other theme is an interest in a social concept called the "kingdom of God." This concept was not worked out with any clarity, but the ways it was used show us that something of a social vision appeared in the teachings of Jesus. The kingdom of God referred to an ideal society imagined as an alternative to the way in which the world was working under the Romans. But it also referred to an alternative way of life that anyone could take at any time.[28]

No wonder, then, that different communities, all of which engaged in mythmaking, burst on the scene. Christianity emerged out of this process, not memories of Jesus' life, not the circumstances of his death. The Romans would have killed Jesus offhandedly, without a trial, as a potential source of trouble—the brutal execution of a peasant nonentity would not attract much attention outside the circle of his followers. Crossan argues that Jesus' body was lost, cast into a pit to be eaten by dogs.[29] Jesus did not return in a resurrected body; what happened was that Jesus' followers remained committed to the ideal of a Kingdom where no one would be oppressed. The movements Jesus inspired soon attracted followers from the literate "retainer class," who would flesh out their Jesus stories by reflecting on Jewish scripture.

The myths about Jesus quickly mutated. Mack argues that we can see stages in the composition of Q, corresponding to the changing circumstances of the Q people. Their experiment brought them into conflict with traditional society, so they began to demand loyalty to the movement, setting themselves apart from unbelievers. Eventually, their conflict with mainstream Jewish identity led them to take up apocalyptic rhetoric, and to make an inept attempt to appropriate the Jewish epic in order to show that the conventional Jews were wrong. On top of the aphorisms and wisdom sayings by which they had remembered Jesus, they added a layer of apocalyptic threats.[30]

The Christ cult evolved differently. They proclaimed Jesus was a martyr vindicated by being raised from the dead to God's side; and they imagined a supernatural Kingdom of God, present and valid no matter what the circumstances of this world. Their social ideal was still Jews and Gentiles coming together in fellowship. So they avoided concrete claims about Jesus' execution and Resurrection, as this would needlessly implicate priestly Judaism or the Romans as oppressors. The myth of the Risen Lord worked precisely *because* it ignored history.

Mack believes that our narrative gospels, particularly the story of Jesus' passion and execution, came about because the author of Mark wanted to join the myths of the Jesus movements to the Resurrection story of the Christ cult. The Jesus movements had started out with an open invitation to everyone, including sinners and the wicked. Things did not go as they had hoped; unable to reform conventional society, they ended up as marginal groups which had to exist independently. Mark gave a rationale for this independent existence in a new origin myth which used ideas from the Christ cult. The Christ myth supplied martyr themes and a link to Wisdom traditions. Mark toned down the cultic aspects of the Christ myth by historicizing it, portraying Jesus as more than a teacher, more than an apocalyptic prophet, but as one whose life and death had cosmic implications.[31]

Jesus may well have been a teacher after the Cynic mold. If so, his vague ideas about how people should live caught fire, then fizzled out; only to survive in the radically different form which is the orthodox Christian myth. Of course, no theory of Jesus we have is so strong as to be *the* answer. A modern cynic will be suspicious of this utopian-socialist Jesus; he seems too conveniently relevant to today's socially engaged liberal churches. Christianity was made by communities at odds with the normal social order, and scholars defending the Cynic hypothesis rightly focus our attention on social history and mythmaking. Yet a teacher calling for social experimentation seems unlikely to impress people so much that they make bizarrely inflated myths about the Son of God. We need to add something more to the mix.

A MIRACLE WORKER

As an apocalyptic prophet or a Cynic teacher, Jesus is a man with a message. He speaks words which inspire his followers to make a religion in his name. But this is quite an effect for words alone. Though Muhammad impressed the faithful with his words, this was because he convinced them these words had a supernatural source. He had esoteric experiences which put him in touch with God. Jesus, the gospels say, knew the will of God and taught it. But he also demonstrated that his message was certified by supernatural forces: he performed miracles. Whether he was announcing the end of the world or a Kingdom where all could live as equals, he impressed people first of all with his supernatural connections.

Conservative Christians quite reasonably believe that Jesus' miracles are the best sign he was sent by God. They even say Jesus demonstrated the kind of control over nature only a god could pull off. He stilled storms, walked on water, created food. When he died, there was an earthquake and many of the dead were resurrected; or, if we trust Luke rather than Matthew, a darkness fell over the land at midday. But these are obviously mythical. No evidence exists for them outside of Christian propaganda; not even all early Christian movements testify to them. And there are plenty of reasons to invent miracle stories. Miracles work wonders for origin myths and stories enhancing Jesus' stature. Old Testament-like miracles were supposed to resume in Messianic times, and a myth paralleling the Jewish epic would need them in any case. Plus, miracle stories were also useful for legitimating the authority of the emerging Church.

Miracle stories about religious leaders are not unusual. According to Philostratus, a religious wonder worker called Apollonius who lived in the first century C.E. performed miracles, raised the dead, and exorcized demons. He even rose from his own death, appeared to his disciples, and ascended to heaven.[32] Muslims remembered Muhammad as a channel for God's words rather than a miracle worker, and they rarely thought of him as a supernatural being. But even he started appearing in miracle stories. His most striking nature miracle was temporarily splitting the moon to demonstrate God's power to infidels.[33]

Jesus' healing miracles, however, are different than this sort of mythic invention. Jesus often deals with just the sort of psychological ailment a faith healer and exorcist can cure. Modern psychiatry recognizes many such problems, including "conversion disorders" which lead to real loss of physical functioning. Ancient Galileans would not see these as any different from more straightforward organic disorders.[34] Jesus lived in a time when everyone thought magic worked, and that the world was infested with spirits and demons. The Mediterranean world was familiar with people who practiced magic, did exorcisms and healings, and trafficked in esoteric experiences. In Mark, Jesus appears as just such a faith healer. In his hometown, where people knew him too well, he did not find enough faith to work with (Mark 6:4–6). Later he healed a blind man by spitting on his eyes, which was not fully effective, and so he had another go (Mark 8:22–26).

So Jesus practiced at least some small-time healing magic and exorcism. This would not be unusual for a teacher or prophet. Muslim tradition tells of Muhammad healing people by using the magical power inherent in the words of the Quran. Indeed, as a prophet, Muhammad would have found it hard to avoid being a healer on occasion. Magical healing was probably much more important to Jesus' career. John Dominic Crossan argues that Jesus was a faith healer and folk magician, and that this was part of his stand against established authority. After all, magic is "subversive, unofficial, unapproved, and lower-class religion."[35] Someone who had a holy spirit, exorcized demons, and healed through magic would have been remembered as a prophet as well. Both Hellenistic holy men and Near Eastern prophets had ecstatic experiences and emanated healing power. Old Testament prophets were filled with the spirit of God.

Magic and ecstatic experience can help flesh out our picture of Jesus. But we should also explore whether we should think of Jesus primarily as a magician or healer rather than a man with a message. After all, esoteric spirituality usually promotes an individual approach to the supernatural realm, not a social ideology. If people believe a man has supernatural power, they will be impressed regardless of whether he taught much of anything. And if Jesus practiced magic, we can more easily understand the distinctive supernatural beliefs and cultic practices of the Christ cult.

According to Morton Smith, Jesus was a magician. He believed he had acquired a supernatural spirit which gave him powers of healing and exorcizing demons. He also practiced techniques of ascending into heaven, and taught these to his disciples. In effect, he started a mystery cult with baptism as a rite of initiation. Smith argues that the Kingdom of God was a supernatural realm accessed through magic: "Initiates were given what they thought was an experience of entering the heavens and they were thus trained to have such visions as those reworked in the transfiguration and resurrection stories." Such journeys into the heavenly kingdom are known from magical texts, even the Dead Sea Scrolls.[36] In Mark 4:11, Jesus says, "To you the secret of the kingdom of God has been given; but to those who are outside, everything comes by parables." If Smith is correct, passages like these refer to an esoteric mystery granted only to an inner circle, perhaps "a magical rite, by which initiates were made to believe that they had entered the kingdom and so escaped from the realm of Mosaic Law."[37] The Christ cult and early gnostic groups continued with personality-altering esoteric practices and the theme of being released from the Law.

Ancient magicians performed rites to acquire a spirit and thus become a god. Magical papyri describe how the heavens open up, and the spirit comes down in the form of a bird. In Mark 1:10–11, Jesus becomes the Son of God after his baptism, when "he saw the heavens break open and the Spirit descend on him, like a dove." Smith uses the papyri to illuminate other obscure Christian passages and rituals. For example, the Eucharist, where the disciples eat Jesus' body and drink his blood in the form of bread and wine, is very similar to rites magicians used to bind their disciples to themselves.

Just a set of parallels between magic and Christian scriptures would be suggestive, but not much more. Smith argues that other evidence also leads to a magician. If early Christians presented Jesus as a miracle worker, outsiders would accuse Jesus of being a magician. Magic was a private relationship with supernatural beings, with antisocial implications. A big issue would be how to separate a positive "divine man" from a shady magician. Thus much of the apologetic material in the gospels is there to respond to charges that Jesus used magic. Unbelievers must have accused Jesus of having a demon—being a magician—so the gospels present Jesus as producing wonders through his own power. In Mark 3:22–30, we find that "the scribes" accused Jesus of being possessed by the prince of demons; Mark then explains why this was not so. Writings by rabbis and the pagan philosopher Celsus indicate that non-Christians thought Jesus must be a magician. Early Christians were sporadically persecuted because they were thought to belong to a magical association founded by Jesus. And we know of other cases where a magician's followers responded to accusations of magic by portraying their master as a holy man—like Apollonius, that other first-century man who rose from the dead and ascended to heaven. Smith says such items of evidence "fit together and give a coherent picture of a magician's life and his work."[38]

Jesus the magician has not convinced scholars. Smith's magical papyri do not reflect the world of folk magic, his parallels between them and the gospels are somewhat selective, and there are equally impressive alternative explanations for rites like the Eucharist.[39] However, Smith forces us to take the esoteric experiences of early Christians more seriously. Whether conjured up by ritual magic or not, Jesus' "spirit" probably had to do with strange psychological experiences and altered personalities. It is plausible that such experiences—Jesus' spirit—remained available to the disciples after Jesus' death, which led to a belief in his spiritual resurrection.[40] The spirit-filled prophets of the Christ cult certainly did not pick up their habit from out of the blue. We need, if not ritual magic, another way of understanding such experiences; and we need, if not a magician, another intelligible social type for Jesus explaining how he used such experiences in his healing.

Recently, Stevan L. Davies has argued that Jesus was a spirit-possessed healer. Smith's Jesus is a magician who thought he possessed control *of* a spirit, while Davies's Jesus believed he was possessed *by* a spirit. Spirit-possessed healers are common in the religions and magical practices of traditional peoples. In modern Western societies, we have Pentecostals who continue the Christian tradition of spirit possession and faith healing, and New Age trance channelers who speak in the voice of a being of light from a distant galaxy.[41] It is very natural to interpret such experiences in terms of folk theories which affirm that an outside spirit has taken possession of the host body.

Many psychologists describe spirit possession as an altered state of consciousness; a state where part of the mind takes over the whole stream of consciousness and adopts characteristics of a different personality. Others emphasize how the possessed person acts out a social role appropriate to her ritual circumstances.[42] In any case, nothing from outside descends on the brain. A woman may use her role as someone possessed by a powerful spirit to accuse her husband of adultery, but she does not turn into a research chemist. Beings of light from far galaxies never give us new astrophysical knowledge, only metaphysical verbiage. Even so, spirit possession is a powerful, impressive experience.

Davies suggests that Jesus believed himself to be possessed not by any old homeless ghost, but the spirit of the Jewish God. As described in Mark, his first possession took place after his ritual baptism by John. Being possessed by the spirit of God is quite compatible with his acting as a prophet—he must somehow have got the idea that he spoke God's will. This made him very effective in healing psychological disorders, since Jesus "could present himself as the manifestation of God on earth (through spirit possession), announce forgiveness and set in motion a comprehensible set of psychological factors that would lead to an elimination of the presenting symptoms."[43] Ancient Galileans would explain all this supernaturally. Today we can compare Jesus' practice with what we know of anomalistic psychology. Some modern therapeutic techniques induce a susceptible state in patients, and then adjust the way they see themselves within a shared mythic context. Jesus may have done the same. He induced spirit possession in others, and forgave their "sin" in the context of Jewish mythology. Naturally, he ministered to the sinners, the wicked, and the mentally ill.

In Davies's theory, as in Morton Smith's, the Kingdom of God was originally an ecstatic state induced by Jesus in his disciples. This state, and Jesus' spirit, was available after Jesus' death. Such a theory makes us see the gospels differently. Jesus' paradoxical, aphoristic speech might not be Cynic wordplay but part of the possession process. The weird, ethereal theology in the gospel of John may come from trying to make sense of possession experience. So Davies argues that the earliest Jesus movements were driven not by a social ideal of equality but by spirit possession.[44]

Christianity may well have started out as a cult based on magical or possession experiences. It would have attracted marginal, lower-class people; it would have been a chaotic cult, not organized around any real teaching or social ideal. Soon, more literate people would join. Some of these would try and fit the loose, chaotic movement into the Jewish religious epic. As distinct communities formed, ideology and mythmaking would follow. Others, like the early gnostics, would intellectualize possession in more individualist terms, tying in esoteric experiences to revealer myths.

The history of early Christianity is in fact that of a cult moving toward middle-class respectability. As Christianity became an organized religion, spirit possession and baptismal mystery themes were toned down. Even the idea of Jesus' Resurrection became controversial because of its cultic associations, until orthodox doctrine stabilized on a bodily resurrected Jesus who instituted Church authority.[45] Meanwhile, ecstatic prophecy kept breaking out in heretical movements.[46] We expect a religion starting with ecstatic experiences among marginal groups to grow this way. It becomes institutionalized, more mainstream, attracts educated and higher-class adherents. Eventually, the incoherent pronouncements of spirit-filled prophets become embarrassing—the religion settles down on older revelations sanctified by official doctrine. Nonsense becomes respectable theology.

Once more, we have a compelling account of Christian origins, which should at least be part of the puzzle. Of course, we do not have the kind of evidence necessary to be certain about Smith's or Davies's scenarios. Scholars will continue debating whether we should see Jesus primarily as a prophet or a teacher or something else, and how much weight we should give to esoteric experiences. We have a lot more to learn about possession-like experiences; the modern therapies to which Davies compares Jesus' behavior derive from trial and error, with little solid theoretical understanding behind them as yet. Nevertheless, early Christians cultivated experiences they thought brought them in contact with a supernatural realm. This was an important part of how they understood God, and how they came to believe that Jesus manifested God. And from a modern perspective, we have to conclude that they were mistaken—perhaps understandably mistaken, even inspired to lofty ideals by their mistake, but still mistaken.

THE RISEN LORD

Jesus was human. He was not a god, he was not the center of a great supernatural display of power. In the Christendom of old, this was a strange, even dangerous opinion; for today's serious historical scholarship, it is commonplace. Our evidence is sparse, so we can fit it with many different portraits of Jesus. But in every case, we can make sense of Jesus as a man of his time and place. We do not need gods and demons to explain the man or the myth.

This puts theologically minded historians in a tight spot. Christians need a God revealed in history, so they would like to take historical scholarship seriously. As Paula Fredriksen says,

> If history, for the church, is important, then undistorted history is very important. Only by meeting this obligation with intellectual integrity can the church, with integrity, continue to witness to that message proclaimed by the first apostles, expounded by Paul, and reflected in the gospels: that the horizontal plane of the human and the vertical plane of the divine met at the cross of Jesus of Nazareth.[47]

So liberal Christian scholars have to dismantle the Christian myth, and at the end of their efforts, turn back and reassure believers that Jesus is still where we meet God. This is a hard task, when Jesus appears all too human. Many historians reconcile their work with their religion by emphasizing the dogma that Jesus was completely human as well as fully God. He must have been a man acting in mundane history, not a timeless celestial being; to say otherwise is nothing but an old heresy.[48]

But if we are to find God through Jesus, never mind God Incarnate, there must have been *something* special about him. What is left, once we get rid of the Virgin Birth, the Resurrection, all the supernatural signs and wonders? If Jesus was a spirit-possessed healer, we have no reason to believe he was actually possessed by God. If he was a man with a message, he was not too different from prophets like Muhammad. All that we learn about Jesus reinforces the way we understand religion as a natural human phenomenon. If Jesus was a mere man, we have no reason to think he must have had some connection with a God. A charismatic figure who inadvertently started a religious movement which happened to succeed is not enough.

Conservative Christians are correct to insist that where the human meets the divine, supernatural sparks must fly. They want to reaffirm the old doctrines, particularly Jesus' bodily resurrection. Perhaps modern scholars are overly impressed with how science seems to have exorcized the supernatural from our world. Perhaps if we just let historical testimony speak to us directly, we will find that in Jesus' life, miracles happened. Christian scriptures record eyewitness testimony about the Risen Lord—resurrection appearances, an empty tomb, disciples energized with missionary fervor after his seemingly shameful death. We should take these at face value, not discard them because of philosophical prejudices against miracles.[49]

Modern historians are not fond of divine intervention. It is plausible that some resurrection experiences underlie the story of the Risen Lord, and that these experiences were important for the rise of Christianity. But in that case, what exactly happened is lost to us. Something like the psychology of spirit possession might help us fill some of the blanks, but bridging the gap by divine intervention is too great a leap—it is like deciding Santa Claus exists because we cannot figure out who bought one of the presents under our Christmas tree. In natural science we find many gaps, but we also have excellent reasons to believe these are gaps in our knowledge, not discontinuities in an otherwise natural world. The same is true with history. Historical inquiry does not exist in a limbo independent of our other sciences; nonmiraculous explanations which comfortably fit our overall body of knowledge are much more credible than stories of divine intervention.

So we must have extraordinarily solid evidence to accept a claim like Jesus' Resurrection. This is appropriate even from a Christian point of view. Jesus' miracles must be truly extraordinary if they are to stand apart from the half-zillion other religious miracle claims. If the gospels claim Jesus appeared to his disciples after his death, well, Philostratus tells a similar tale about Apollonius. Mormon prophet Joseph Smith found three eyewitnesses to the golden plates he supposedly translated, and later added eight more.[50] The Roman historian Tacitus thought the healing miracles of Sarapis, a recently invented god, were based on reliable eyewitness testimony.[51] Neither the veracity nor the significance of miracle tales are all that clear, even if we take them at face value.

As it stands, the Resurrection is a fantastic claim, supported by conflicting and inadequate testimony, buried under obvious myths. The traditional story is a mess. Plus, we can make sense of Jesus without miracles. If Jesus was a failed apocalyptic prophet, a teacher, or a faith healer, it is hard to see why a God would bother resurrecting him anyway. All this is more than enough to dismiss the Resurrection. But we can do even better. Our evidence gives us incomplete but intriguing hints of how the Resurrection story began.

We are used to thinking Christianity is all about proclaiming the Risen Lord—as Paul says in 1 Cor. 15:14, "if Christ was not raised, then our gospel is null and void, and so too is your faith." Not all Jesus movements thought so. The communities who produced Q and the Gospel of Thomas did not make much of a Resurrection; they quite possibly did not even know the story.

Still, we have evidence that belief in a Risen Lord existed very early in the Christ cult. In 1 Cor. 3–7, Paul says:

> First and foremost, I handed on to you the tradition I had received: that Christ died for our sins, in accordance with the scriptures; that he was buried; that he was raised to life on the third day, in accordance with the scriptures; and that he appeared to Cephas, and afterwards to the Twelve. Then he appeared to over five hundred of our brothers at once, most of whom are still alive, though some have died. Then he appeared to James, and afterwards to all the apostles.

Here, Paul is quoting an earlier, carefully composed creed. Now, Paul wrote only a couple of decades after Jesus' death, in which time the Resurrection belief arose and became the polished creed Paul quotes. Christian apologist Gary Habermas argues that here we have eyewitness testimony to Jesus' appearances, which brings us, "for all practical purposes, back to the original events." What better testimony could a historian ask for?[52]

Paul, however, does not preserve a pristine testimony. The list of appearances, which has an obvious apologetic function, is most likely a later elaboration, possibly based on spirit-possession experiences like those described in the Pentecost story in Acts 2:1–13.[53] And the gospels do not seem to know much about these specific appearances. The creed before the appearance list is more interesting, since it tells us what sort of belief the very early Resurrection was.

Burton L. Mack argues that the creed combines a martyr myth with a Wisdom tale. Jesus dies for our sins—in other words, the Christ cult believed Jesus was a martyr, like in legends of Jewish martyrs whose deaths washed away the sins of their community. "Sin," of course, was a deviation from the will of God, or conventional Jewish Law. The raising of Jesus, on the other hand, recalls the Wisdom tale of a sage who was persecuted by a foreign power, but who was vindicated after his ordeal. Jews of Jesus' time had begun to imagine that God's justice would be realized in an afterlife, since it was obviously not happening on earth. So the Wisdom tale also changed to let the vindication scene take place after death. In the Christ-cult version, God endorses Jesus—who is a Wisdom figure in Paul's letters—by raising him to a position of honor in heaven.[54]

So a movement seeking significance in Jesus' death had plenty of mythic material to draw on. Those involved in the movement were concerned about sin, perhaps because they interpreted Jesus' teaching to bring Jews and Gentiles—sinners—together in one community, or because they found, as in the Pentecost story, that people from all nations could be possessed by the Holy Spirit. A myth like Paul's creed, where Jesus died for their sins and was vindicated by being raised up, would nicely justify the practices of the Christ cult. And if they practiced spirit possession, their self-conception was probably already permeated by dying-and-resurrection themes, since dying in the spirit and being born again would be a very natural way to describe some of their experiences.

It may seem incredible that a group could be so immersed in their mythic imagination as to believe their fictions. But the Jesus whom Paul found in the Christ cult was only vaguely historical; he proclaimed a resurrection, but not a detailed story. This fit the myths they already thought described the world. Believing in a peculiar but deeply meaningful creed is not unusual for a fringe religious group. Later Christians would be convinced the Jewish scriptures *had to* be an oracle of Jesus' life, and wrote his story accordingly. Members of the early Christ cult were convinced Jesus *had to* be vindicated by their God, and converted the disaster of his death into a reason for missionary fervor.

It does not take much to trigger fantastic stories. Consider the UFO myth.[55] In the 1940s and 1950s, the idea of flying saucers piloted by aliens burst on the scene. Some began to interpret strange lights in the sky as an alien presence, fitting this ambiguous information into a story with themes from science fiction and cold war paranoia. Soon, "contactees" appeared, who personally communicated with the space brothers visiting us. The mythic theme of superhuman saviors from heaven acquired a space-age veneer. Today, UFO beliefs center on alien abductions and government cover-ups. Strange experiences like sleep paralysis, timeless religious hopes, sexual frustration, and paranoia about the government now drive the myth. Believers invent their stories in collaboration with hypnotherapists, not prophets; they get their story templates from the media, not scripture; but we still have a classic case of mythmaking here. Perhaps lights in the sky were the "spaceship appearances" for flying saucer cults, but these do not go very far in explaining the UFO myth; they certainly do not show there were actual spaceships in our skies.

Jesus' appearances, then, are not the key to the Resurrection story, though it is quite plausible that the disciples saw a ghost. Particularly if they were cultivating spirit possession and visions, they probably did have the sort of experience which fed into the Resurrection myth. Indeed, we have some indication that the Christ cult encouraged visionary encounters with Jesus, as well as possession by his spirit. After his list of appearances in 1 Corinthians, Paul adds, "Last of all, he appeared to me too." Resurrection appearances were not limited to the time between Jesus' death and ascension, and a latecomer like Paul could also see Jesus raised in a "spiritual body." We do not know what exactly happened; it is hard to say what is behind Paul's words aside from themes like a vindicated Savior, a sign of a coming resurrection of the dead, and Paul's claim of an authority to preach.

The Resurrection story developed in different ways. The orthodox, respectable party insisted on a more concrete, bodily resurrection. Some Gnostics, adopting a more individualist, esoteric spirituality, spoke of a spiritual resurrection and tied it to their myth of Jesus as a cosmic Revealer. As Elaine Pagels points out, "Some say that the person who experiences the resurrection does not meet Jesus raised physically back to life; rather, he encounters Christ on a spiritual level. This may occur in dreams, in ecstatic-trance, in visions, or in moments of spiritual illumination."[56] The forerunners of orthodoxy were not comfortable with the chaos of continual individual revelation. No respectable religion can live with spirit-filled prophets continually speaking with supernatural authority and contradicting one another. In the process of moving from cult to church, orthodoxy needed a myth to legitimate the Church as an institution faithful to Jesus. Revelation had to be disciplined; its essentials had to be confined to the time of the Lord himself. Authority would derive from the few apostles Jesus personally appeared to, not the spirit-filled prophets. A bodily resurrection would help: Jesus would rise by another impressive miracle, establish the Church by delegating authority to his disciples, and then ascend and get out of the way. Christians facing martyrdom could follow the example of a Lord who physically suffered and rose bodily.[57]

This may look like the early Church consciously set out to create a self-serving myth. Not at all; the actual process was probably rather Darwinian. Our UFO myth is spread by the mass media, so the stories of UFO-believers quickly fall into a stereotyped outline. Without the media, Christian stories could not spread very fast, but this also meant that a stereotyped story could not suppress the variant myths of separate Christian groups. So very early on, there was a diversity of Christian stories. Communities which promoted a stable, less individualistic practice had a better chance of passing on their myths in the long run. And when a universal Church authority began to crystallize, it became another selective force by promoting those stories congenial to itself and suppressing Gnostic and Jewish-Christian stories.

The gospels are part of this process. Mark ends with Jesus' empty tomb, and women leaving afraid to tell anyone—it does not even include resurrection appearances. Paul did not know of the empty tomb story; it may have been invented by the author of Mark. After all, then and now, saint's tombs are important shrines in the Near East; Jesus' tomb would have been remembered if known. As myth, however, the empty tomb and the lack of appearances tones down the cultic aspects of the Resurrection. While the Christ cult celebrated the presence of Jesus' spirit, the empty tomb highlights Jesus' absence until his imminent second coming. Only the women come up to the tomb, and are the first witnesses to the

Risen Lord in later gospels. Jesus' disciples fall away as his crucifixion approaches—finally only the lowliest are truest to Jesus.[58]

Matthew, Luke, and John add appearances, though none of their stories agree with each other. Interestingly, their Risen Lord is still not quite a material body; some of the appearance stories are designed to support a bodily resurrection, while in others Jesus is insubstantial, hard to recognize, as if in a vision. In any case, he authorizes the disciples and zooms back up to heaven. The gospels following Mark also elaborate the story to provide more solid proof of the Resurrection. Matthew makes sure the tomb is guarded and sealed. The noncanonical Gospel of Peter adds seven seals on the tomb, soldiers keeping watch two by two, a magnificent angelic descent to open the tomb, and many witnesses to the whole event.

The Resurrection started as a myth of martyrdom and vindication, supported by visions and spirit possession. By the time the universal Church was in place, it had become an elaborate tale of a bodily resurrection and appearances legitimating apostolic authority. At every step along the way, the Resurrection was a myth. Elvis did not fake his death because of the burdens of stardom; he does not appear once in a while to his faithful fans. Hitler was not spirited away by Nazi allies from within the Hollow Earth; he does not wait plotting his revenge. Jesus did not rise from the dead; he will not come again.

THE BAD NEWS

As critical history took hold, scholars—most of them devoutly Christian—set out to examine the good news about Jesus. For centuries now, they have been encountering myths. It would have been nice if God really took on flesh, but for Christians who take history seriously, God Incarnate must become a poetic way of claiming reality is human-centered at its deepest level. The Resurrection used to be a miracle, even evidence for eternal life; now it is just a symbol of hope. Yet theologians cannot, of course, banish Christ to the never-never land of an ancient religious imagination. Liberal Christians still confess that Jesus is Lord, though they are no longer quite sure what they mean by that.

One way to discover God anew is to look for some divine mystery, some radical novelty, in the ruins of traditional Christianity. Bishop John Shelby Spong, a popular spokesman for liberal Christianity, denies Jesus was bodily resurrected. He manages to unsay what he said, however, by claiming Easter was still a mystic event where God touched our world:

> If . . . Easter and resurrection are aspects of a human experience, timeless but always subjective, breaking through our barriers now and again

> in mind-altering, consciousness-raising revelations, then we can use the words of our forebears in faith to journey toward the experience in which their lives were changed. . . . our ultimate goal is not objectivity, certainty, or rational truth. It is rather life, wholeness, heightened consciousness, and an expanded sense of transcendence.[59]

Modern, open-minded Christians like Spong too often reduce their religion to psychological hand waving and existentialist exhortation. Perhaps this has its place; all of us could do with a bit of inspiration. But it is still strange to watch Christianity dissolve into a vapid verbiage and contempt for truth which would be more at home in a Californian psychotherapy cult.

Alternatively, we might ignore history. Paul did not care about history, he just charged ahead proclaiming the good news. So Christians may say the Christ of faith is more important than the Jesus of history.[60] In the Christian message, however mythical, we can still hear Jesus offering us God's mercy. Even better, the mythic nature of Christianity might be a positive. After all, theologians such as Reinhold Niebuhr assure us that myths point to realities mere scientific rationality is unable to grasp, helping us recognize the divinity and unity in all of our experience. In the end, it seems, reality "can be revealed and expressed only in mythical terms."[61]

No doubt myths say much about our hopes and fears. And since myths are invented and worked over by communities, they can tell us a lot about social reality. Somebody interested in the anxieties of Americans today can learn much from our UFO myths. In that sense, myths are indeed not just false stories; they are useful devices for negotiating our social world. But once we step outside of social reality, we find an accidental world. Within this impersonal world we maintain a bubble of reality infused with personality and purpose, overflowing with human meanings. That bubble of personal reality is our creation, not of some sort of God. *We* keep our social world alive, not emanations from a mythic deep reality.

This is bad news. We do not live by truth alone; we want to feel at home in our world. Our religions and traditional philosophies claim they can bring truth, happiness, and morality together in harmony. If we could discover a meaningful personal order to the universe, we could get in tune with that order and everything would fall into place. We could find true happiness beyond the pursuit of superficial pleasures. We could pattern our social lives after the harmonies of the universe and attain justice and virtue. So, to achieve a sane society, we follow Jesus, God Incarnate and Perfect Man. Or, in today's New Age, we hope to find happiness by recognizing our divine nature. So we insist that the true physics presents us with a world seething with the sublime energies of consciousness; not impersonal matter in mindless motion.

For the philosophers of old, myth became metaphysics. Our Rational Souls could come to know God and achieve immortality by contemplating metaphysical truths and purging themselves of material entanglements. God was so far above mere human thought that all myths and metaphysical systems only imperfectly reached out toward God. Nevertheless, harmony was there for us to seek.

But in an accidental world, there is no more harmony. Jesus is not the Way, the Truth, and the Life; indeed, there is no way, truth need not deliver happiness, and life is disappointingly finite. For many of us, this is unbearable news. Somehow, there must be a way to bring truth back in harmony with our hopes. It is hard to do this if we take our sciences seriously, but what if we go back to the basics? Maybe miracles do take place—not in a special time set aside for revelation, not in any privileged religious tradition, but everywhere, all the time. Maybe we do have souls, maybe our minds can magically control matter. To find God, we have to begin with the miracles of saints, the visions of mystics, the paranormal powers of psychics. The early Christians with their miracles and magic were right: there really is a supernatural realm.

Some of us, on the other hand, are convinced that we inhabit a godless world never blessed by the Perfect Man. We think we have to muddle through without supernatural guidance, trying to make something of our lives. This is not necessarily a liberating belief; after all, it means there are no saving truths. Truth itself becomes tarnished, dragged down to earth from the heights of metaphysical purity. Of course, an accidental world is an interesting place, so truth is worth pursuing if we have a sense of wonder. But truth is also often ugly, something to shake our fists at—and change, if we are able. So be it. We can look up at the stars not to praise God, but with fascination, even some pride in what we have learned about them. It is not salvation, but it is no small accomplishment either.

NOTES

1. Jaroslav Pelikan, *Jesus Through the Centuries: His Place in the History of Culture* (New Haven, Conn. and London: Yale University Press, 1985).

2. E.g., John Wenham, *Easter Enigma: Are the Resurrection Accounts in Conflict?* 2d ed. (Grand Rapids, Mich.: Baker, 1992).

3. This is common textbook knowledge; e.g., Howard Clark Kee, Franklin W. Young, and Karlfried Froehlich, *Understanding the New Testament*, 2d ed. (Englewood Cliffs, N.J.: Prentice-Hall, 1965). Some recent proposals are quite interesting; e.g., Alan Watson, *Jesus and the Jews: The Pharisaic Tradition in John* (Athens: University of Georgia Press, 1995), identifies a Jewish, anti-Christian source in John.

4. Bert D. Ehrman, *The Orthodox Corruption of Scripture: The Effect of*

Early Christological Controversies on the Text of the New Testament (New York: Oxford University Press, 1993), p. 48; Burton L. Mack, *Who Wrote the New Testament? The Making of the Christian Myth* (San Francisco: Harper, 1995), pp. 79–87.

5. John Hick, ed., *The Myth of God Incarnate* (Philadelphia: Westminster, 1977).

6. John Dominic Crossan, *The Historical Jesus: The Life of a Mediterranean Jewish Peasant* (San Francisco: Harper, 1991), p. xxvii.

7. John C. Mellon, *Mark as Recovery Story: Alcoholism and the Rhetoric of Gospel Mystery* (Urbana: University of Illinois Press, 1995).

8. Christians told the story of Jesus' infancy to parallel Old Testament heroes like Moses; John Dominic Crossan, *Jesus: A Revolutionary Biography* (San Francisco: Harper, 1994), pp. 10–15. The apocryphal Infancy Gospels of Thomas and James are even more fantastic; see Robert J. Miller, ed., *The Complete Gospels* (San Francisco: Harper, 1994). On stories about Muhammad, see F. E. Peters, *Muhammad and the Origins of Islam* (Albany: State University of New York Press, 1994), chap. 4.

9. John Meier, *A Marginal Jew: Rethinking the Historical Jesus* (New York: Doubleday, 1991), pp. 168–84; E. P. Sanders, *Jesus and Judaism* (Philadelphia: Fortress, 1985), pp. 16–17.

10. E. P. Sanders, *The Historical Figure of Jesus* (London: Penguin, 1993), pp. 28–32, 42.

11. Sanders, *Jesus and Judaism*, p. 319.

12. Ibid., pp. 73–76; Paula Fredriksen, *From Jesus to Christ: The Origins of the New Testament Images of Jesus* (New Haven, Conn.: Yale University Press, 1988), pp. 111–14.

13. Fredriksen, *From Jesus to Christ*, p. 125.

14. Sanders, *Jesus and Judaism*, p. 227.

15. Ibid., p. 319. However, Christian historians keep slipping into portraying Jesus as unique within Judaism. E.g., James H. Charlesworth, *Jesus within Judaism: New Light from Exciting Archaeological Discoveries* (New York: Doubleday, 1988), while warning against the tendency to portray an ossified Judaism (pp. 45–46), later adopts a variant of the traditional argument (pp. 73–74).

16. Gerd Lüdemann, *What Really Happened to Jesus: A Historical Approach to the Resurrection*, in collaboration with Alf Özen, trans. John Bowden (Louisville: Westminster John Knox, 1995), pp. 93–94; Michael Goulder, "The Explanatory Power of Conversion-visions," in *Jesus' Resurrection: Fact or Figment: A Debate Between William Lane Craig and Gerd Lüdemann*, ed. Paul Copan and Ronald K. Tacelli (Downers Grove, Ill.: InterVarsity, 2000).

17. Leon Festinger, Henry W. Reicken, and Stanley Schachter, *When Prophecy Fails: A Social and Psychological Study of a Modern Group That Predicted the Destruction of the World* (New York: Harper & Row, 1964). It has since become clearer that only few among group members respond so aggressively. However, in various forms, compensating for worldly failure by making grander supernatural promises is common in religions. William Sims Bainbridge, *The Sociology of Religious Movements* (New York: Routledge, 1997), pp. 135–38, 191.

18. Hyam Maccoby, *The Mythmaker: Paul and the Invention of Christianity* (New York: Harper & Row, 1986).

19. Jerome Murphy-O'Connor, *Paul: A Critical Life* (New York: Oxford University Press, 1996); Ben Witherington III, *The Paul Quest: The Renewed Search for the Jew of Tarsus* (Downers Grove, Ill.: InterVarsity, 1998).

20. Randel Helms, *Gospel Fictions* (Amherst, N.Y.: Prometheus Books, 1988), p. 24.

21. Ibid., chap. 4; Burton L. Mack, *A Myth of Innocence: Mark and Christian Origins* (Philadelphia: Fortress, 1988), chap. 8; G. A. Wells, *The Historical Evidence for Jesus* (Amherst, N.Y.: Prometheus Books, 1988), pp. 206–10.

22. My summary is based on Wells, *The Historical Evidence for Jesus*, and G. A. Wells, *The Jesus Myth* (Chicago: Open Court, 1999). See also Robert M. Price, *Deconstructing Jesus* (Amherst, N.Y.: Prometheus Books, 2000). A historical survey of the myth theory is Gordon Stein, ed., *An Anthology of Atheism and Rationalism* (Amherst, N.Y.: Prometheus Books, 1980), pp. 178–84.

23. Scholars generally think Josephus contains a Christian alteration of a less complimentary original. E.g., Charlesworth, *Jesus Within Judaism*, pp. 91–95; and Wells, *The Jesus Myth*, pp. 200–21 in reply.

24. An alternative to Wisdom is that some Christians thought of Jesus as an angel—"son of God" was a title of angelic mediators between humans and God, and angel worship is a concern in Pauline material. See Alan F. Segal, "The Risen Christ and the Angelic Mediator Figures in the Light of Qumran," in *Jesus and the Dead Sea Scrolls*, ed. James H. Charlesworth (New York: Doubleday, 1992). The comparison to Metatron, the highest angel, is particularly interesting when read together with Margaret Barker, *The Great Angel: A Study of Israel's Second God* (Louisville, Ky.: Westminster/John Knox, 1992), which brings together the angel and Wisdom points of view, tying early Christianity to the undercurrent of the second God in unofficial Judaism.

25. Jonathan Z. Smith, *Drudgery Divine: On the Comparison of Early Christianities and the Religions of Late Antiquity* (London: University of London, School of Oriental and African Studies, 1990), pp. 129–30.

26. Mack, *A Myth of Innocence*, p. 303; Burton L. Mack, *The Lost Gospel: The Book of Q & Christian Origins* (San Francisco: Harper, 1993).

27. Crossan, *The Historical Jesus*, pp. 299–302. Richard A. Horsley and Neil Asher Silberman, *The Message and the Kingdom: How Jesus and Paul Ignited a Revolution and Transformed the Ancient World* (New York: Grosset/Putnam, 1997), also portray Jesus as a radical teacher, but they see the Cynic analogy as overly individualist. However, their Jesus looks like a modern left-wing hero standing against Jewish and Roman elites caricatured as totally exploitative and commercial-minded.

28. Mack, *Who Wrote the New Testament?* p. 40; *The Lost Gospel*, p. 120. Against the Cynic view, see Hans Dieter Betz, "Jesus and the Cynics: Survey and Analysis of a Hypothesis," *Journal of Religion* 74, no. 4 (1994): 453; and Birger A. Pearson, *The Emergence of the Christian Religion: Essays on Early Christianity* (Harrisburg, Pa.: Trinity, 1997), chap. 2. Donald Harman Akenson, *Saint Saul: A Skeleton Key to the Historical Jesus* (New York: Oxford University Press, 2000), also includes a useful critique of this approach.

29. John Dominic Crossan, *Who Killed Jesus? Exposing the Roots of Anti-Semitism in the Gospel Story of the Death of Jesus* (San Francisco: Harper, 1995),

chap. 6. Crossan's arguments are partly based on his dubious claim that an early layer of the Gospel of Peter contains the earliest form of the Passion story; Crossan, *The Historical Jesus*, pp. 385–87; Crossan, *Who Killed Jesus?* pp. 223–27.

30. Mack, *The Lost Gospel*, pp. 140–47. However, finding compositional layers in a text which is a reconstruction to begin with is risky business; it is still possible that Q remembered Jesus as an apocalyptic prophet from the beginning. See Roland A. Piper, ed., *The Gospel Behind the Gospels: Current Studies on Q* (Leiden, The Netherlands: Brill, 1995).

31. Mack, *A Myth of Innocence*, pp. 111, 319–24, 354.

32. Philostratus, *The Life of Apollonius of Tyana, the Epistles of Apollonius, and the Treatise of Eusebius* (Cambridge: Harvard University Press, 1969).

33. Clinton Bennett, *In Search of Muhammad* (London: Cassell, 1998), pp. 45–54.

34. Stevan L. Davies, *Jesus the Healer: Possession, Trance, and the Origins of Christianity* (New York: Continuum, 1995), chap. 5; Morton Smith, *Jesus the Magician* (New York: Barnes & Noble, 1978), pp. 8–10. On the psychological basis for faith healing, see Leonard Zusne and Warren H. Jones, *Anomalistic Psychology: A Study of Magical Thinking*, 2d ed. (Hillsdale, N.J.: Lawrence Erlbaum, 1989), pp. 42–53.

35. Crossan, *The Historical Jesus*, pp. 305, 336–37.

36. Morton Smith, "Two Ascended to Heaven—Jesus and the Author of 4Q491," in Charlesworth, *Jesus and the Dead Sea Scrolls*.

37. Smith, *Jesus the Magician*, pp. 134–35. If Smith is correct, perhaps Gnostic Christianity preserved rites like baptismal initiations accompanied by visions; see Helmut Koester and Elaine Pagels, introduction to "The Dialogue of The Savior," in *The Nag Hammadi Library*, rev. ed., ed. James M. Robinson (New York: Harper & Row, 1988), p. 244. Also interesting is the presence of a "double baptism" rite among Gnostics; Giovanni Filoramo, *A History of Gnosticism*, trans. Anthony Alcock (Oxford: Basil Blackwell, 1990), pp. 179–80.

38. Smith, *Jesus the Magician*, p. 152.

39. E.g., as based on the Passover meal. Watson, *Jesus and the Jews*, pp. 71–74.

40. Morton Smith, *The Secret Gospel: The Discovery and Interpretation of the Secret Gospel According to Mark* (New York: Harper & Row, 1973), p. 117.

41. Alice B. Child and Irvin L. Child, *Religion and Magic in the Life of Traditional Peoples* (Englewood Cliffs, N.J.: Prentice Hall, 1993), pp. 90–94; Clarke Garrett, *Spirit Possession and Popular Religion: From the Camisards to the Shakers* (Baltimore: Johns Hopkins University Press, 1987); Michael F. Brown, *The Channeling Zone: American Spirituality in an Anxious Age* (Cambridge: Harvard University Press, 1997).

42. Psychologists are divided on how to interpret altered states. Hypnosis has been best studied under laboratory conditions, and the debate over altered state and cognitive-behavioral perspectives is long-running. See Steven Jay Lynn and Judith W. Rhue, eds., *Theories of Hypnosis: Current Models and Perspectives* (New York: Guilford, 1991); Nicholas P. Spanos, *Multiple Identities And False Memories: A Sociocognitive Perspective* (Washington, D.C.: American Psychological Association, 1996).

43. Davies, *Jesus the Healer*, p. 75.

44. Ibid., chaps. 9–11. The synoptic Gospels are usually thought more reliable, but see Robin Lane Fox, *The Unauthorized Version: Truth and Fiction in the Bible* (New York: Alfred E. Knopf, 1991), pp. 205–209. Lane Fox believes John is based on a primary source.

45. Smith, *Drudgery Divine*, pp. 110–12.

46. Especially the Montanists of the second century; Robert M. Grant, *Gods and the One God* (Philadelphia: Westminster, 1986), pp. 145–47.

47. Fredriksen, *From Jesus to Christ*, p. 215.

48. Ibid.; James H. Charlesworth, "The Dead Sea Scrolls and the Historical Jesus," in Charlesworth, *Jesus and the Dead Sea Scrolls*, pp. 5, 40.

49. C. Stephen Evans, *The Historical Christ and the Jesus of Faith: The Incarnational Narrative as History* (New York: Oxford University Press, 1996). Raymond Martin, *The Elusive Messiah: A Philosophical Overview of the Quest for the Historical Jesus* (Boulder, Colo.: Westview, 2000) critiques both Evans and historians who restrict themselves to naturalistic accounts. However, a naturalistic approach is no more problematic than physicists assuming matter is particulate when modeling new phenomena. Successful overall theories should be abandoned only for extraordinarily compelling reasons.

50. Ernest H. Taves, *Trouble Enough: Joseph Smith and the Book of Mormon* (Amherst, N.Y.: Prometheus Books, 1984), pp. 46–48.

51. Grant, *Gods and the One God*, p. 35.

52. Gary Habermas and Antony Flew, *Did Jesus Rise from the Dead? The Resurrection Debate*, ed. Terry Miethe (San Francisco: Harper & Row, 1987), p. 23. Against Habermas's arguments, see Michael Martin, *The Case Against Christianity* (Philadelphia: Temple University Press, 1991), pp. 87–96.

53. Mack, *A Myth of Innocence*, pp. 103–104. Gerd Lüdemann, *What Really Happened to Jesus*, pp. 95–101, identifies the appearance to over five hundred with Pentecost. Robert M. Price, "Apocryphal Apparitions: 1 Corinthians 15:3–11 as a Post-Pauline Interpolation," *Journal of Higher Criticism* 2, no. 2 (1995): 69, argues that the whole creed is late.

54. Mack, *Who Wrote the New Testament?* pp. 79–87.

55. Curtis Peebles, *Watch the Skies! A Chronicle of the Flying Saucer Myth* (Washington, D.C.: Smithsonian Institution Press, 1994); James R. Lewis, ed., *The Gods Have Landed: New Religions from Other Worlds* (Albany: State University of New York Press, 1995); Benson Saler, Charles A. Ziegler, and Charles B. Moore, *UFO Crash at Roswell: The Genesis of a Modern Myth* (Washington, D.C.: Smithsonian Institution Press, 1997).

56. Elaine Pagels, *The Gnostic Gospels* (New York: Random House, 1979), p. 5. Similar ideas were pervasive within early Christianity, including in Paul's writings. Gregory J. Riley, *Resurrection Reconsidered: Thomas and John in Controversy* (Minneapolis: Fortress, 1995), also argues that spiritual resurrection beliefs preceded bodily resurrection beliefs.

57. Pagels, *The Gnostic Gospels*, p. 75; Crossan, *Who Killed Jesus?* p. 203.

58. Mack, *A Myth of Innocence*, pp. 111–13, 308–309; Lüdemann, *What Really Happened to Jesus*, pp. 19–24; Jeffery Jay Lowder, "Historical Evidence and the Empty Tomb Story: A Reply to William Lane Craig," *Journal of Higher Criticism* (forthcoming).

59. John Shelby Spong, *Resurrection: Myth or Reality?* (San Francisco: Harper, 1994), p. 100.

60. Howard Clark Kee, *Jesus in History: An Approach to the Study of the Gospels*, 2d ed. (New York: Harcourt Brace Jovanovich, 1977), pp. 125–31. An example of a historian taking a leap of faith is Meier, *A Marginal Jew*, pp. 197–98. Against such views, see Gregory W. Dawes, *The Historical Jesus Question: The Challenge of History to Religious Authority* (Louisville, Ky.: Westminster John Knox, 2001), and Evans, *The Historical Christ and the Jesus of Faith.*

61. Reinhold Niebuhr, in *The Nature of Religious Experience: Essays in Honor of Clyde Macintosh* (New York: Harper & Brothers, 1935), p. 135.

Chapter Six
Signs and Wonders

Everyone knows that dragons don't exist. But while this simplistic formulation may satisfy the layman, it does not suffice for the scientific mind. The School of Higher Neantical Nillity is in fact wholly unconcerned with what does exist. Indeed, the banality of existence has been so amply demonstrated, there is no need for us to discuss it any further here. The brilliant Cerebron, attacking the problem analytically, discovered three distinct kinds of dragon: the mythical, the chimerical, and the purely hypothetical. They were all, one might say, nonexistent, but each nonexisted in an entirely different way.

—Stanislaw Lem, *The Cyberiad* (1967)

Miracles are the tabloid underside of religion. Tales of levitating saints and weeping statues belong with poltergeists, spirit-summoning mediums, and psychics predicting California will slide into the sea. Still, psychical researchers hope to find real signs of the supernatural hidden among all the weirdness. We might have strange powers, we might survive death—beneath the surface, this could be a spiritual world with a God hidden in the works. But all we find behind miracle stories are the many ways we make mistakes.

WONDROUS PHENOMENA

These are troubled times for great intellectual systems built around God. But popular religion seems healthy enough. While liberal theologians set aside fantastic stories and still try to hear the voice of God, popular religions promise direct contact with supernatural powers. If the God of scholars and philosophers vanished overnight, the God of signs and wonders would remain as strong as ever. Take, for example, sociologist Mark A. Shibley's description of an evangelical Christian:

> Jack Hope is a Bible believer whose faith is displayed in the gifts of the spirit, and his highest calling on this earth is the salvation of wayward souls. He is a charismatic fundamentalist, and his missionary zeal gives character and direction to his daily life. Hope is a self-taught preacher and believes that seminaries—"cemeteries," he jokingly calls them—are spiritually dead places. His faith is nurtured not in theological reflection but through prayer to a living God who is miraculous and visible in the world and through constant reference to the Bible, the inerrant word of God.[1]

Such a faith is steeped in experiences of divine power which go beyond dry creeds. A Protestant like Jack Hope, of course, would not be impressed with weeping statues of the Virgin Mary, and he might think New Agers meddle with demonic forces. But though religions squabble over details, their Gods become apparent in supernatural experiences. In the Bible, revelation comes in dreams, visions, theophanies; through divining stones or visiting angels.[2] Miracles herald and *validate* divine messages.

There is, of course, the danger that wonders can become mere spectacle, getting in the way of a deeper spiritual knowledge. Nevertheless, our ideas of holiness are inseparable from magic. Consider a story told about a Muslim mystic, Abu Said, who started his saintly career with rigorous self-denial. Once he had broken the bonds of the self, however, he adopted a luxurious lifestyle. An ascetic who was not amused by this challenged Abu Said to a forty-day fast.

> While the ascetic, in accordance with the practice of those who keep a fast of forty days, was eating a certain amount of food, the Shaykh Abu Sa'id ate nothing; and though he never once broke his fast, every morning he was stronger and fatter and his complexion grew more and more ruddy. All the time, by his orders and under his eyes, his dervishes feasted luxuriously and indulged in spiritual concerts, and he himself danced with them. His state was not changed for the worse in any respect. The ascetic, on the other hand, was daily becoming feebler and thinner and paler, and the sight of the delicious viands which were served to the Sufis in his presence worked more and more upon him. At length he grew so weak that he could scarcely rise to perform the obligatory prayers. He repented of his presumption and confessed his ignorance.[3]

Such a story has many lessons for the devout, one which is that outward observance, even in the form of self-denial, is not the same as spirituality. The miracle also makes an essential point. God is not a philosophical abstraction but a power unlimited by material constraints. Saints partake of this power. An enlightened soul achieves what is impossible for a mere material being.

So holy figures always display supernatural powers. Christians also go without food or water; they levitate, are seen in two places at once, become invisible, control storms and floods, perform divine healings, gaze into the future. Hindu and Buddhist saints have similar bags of tricks. Indeed, in all cultures, wondrous events are part of the socially agreed-upon picture of reality. People perform feats of mind-over-matter, encounter apparitions, and get possessed by spirits. Our supernatural beliefs are based on universal human experiences.[4] And belief in supernatural powers is not only alive and well, it is growing in many places. Theologian Harvey Cox, observing the spread of Pentecostal Christianity and African spirit-possession cults, argues that a "primal spirituality" is sweeping across the world.[5]

There are good religious reasons to seek miracles. Ordinarily we inhabit a natural world: a dusty place where we live our everyday lives, kick up some dust, and return to dust when we die. But wondrous phenomena keep breaking in, showing us that *spirit* animates matter. Our minds reach beyond dust and death. The world depends on a deeper personal reality, and ultimately a great supernatural spirit we can but barely imagine. However, for all its commonsense appeal[6] and popular vitality, the supernatural is now an intellectual embarrassment. Ever since the European Enlightenment, the world seems to have lost its enchantment. Respectable religious experience has become personal, not a publicly verifiable event.

Science did much to bring magic into disrepute. Supernatural wonders started to look out of place in a world operating under strict impersonal laws. Reports of miracles had to be mistaken, or else they were due to natural phenomena which would be eventually understood as science progressed. Even politics turned against miracles. In Europe, Protestant elites tried to delegitimate both the "enthusiasm" of popular religious movements and the "superstition" of priestly Catholicism. And of course, technology gave us powers which were obviously real. Flying to the moon or picking up a phone to talk to someone half a planet away is amazing; and though popular culture sometimes treats technology as a secular miracle, everyone knows machines have nothing to do with spirits. We began to trust in medical science, visiting the faith healer only when all else failed. Among intellectuals, magical beliefs became relics of a premodern age.

Folk religion, of course, continued to be thoroughly supernatural, as with the charismatic movements that kept flaring up within Christianity. And for people dissatisfied with traditional religion but still seeking after spirit, individualist religious alternatives flourished; these were also sustained by immediate supernatural experience. In the nineteenth century, Spiritualism revolved around wondrous phenomena, as does the New Age today. Curiously, the adherents of these new religions believed they were at the forefront of the scientific spirit of the times, uniting science and religion in a blend authority-based religions could never achieve.

In fact, miracles are particularly useful for religion in a modern intellectual climate. Science is supposed to distrust received authority, deciding matters by putting them to empirical tests. Seekers after psychic phenomena hope to find exactly this sort of demonstrable evidence of a spiritual reality. This would be a much more secure basis for religious hopes like immortality than blind faith. Also, magic is common ground in a world where religious cultures are no longer isolated from one another. Everyone claims wonders, and instead of fighting over what miracle validates which creed, we can declare that everyone has a glimpse of a spiritual realm—everyone but the skeptics stuck in an outdated materialist orthodoxy.

So in the late nineteenth century, investigators concerned about encroaching materialism came together in societies for psychical research, hoping to establish new, scientifically respectable foundations for religion.[7] In the paranormal phenomena which fell through the cracks of established science, there could be signs of a spiritual reality. Mind was not reducible to matter; we had strange occult powers; our personality was immortal spirit, not a process snuffed out at death. Psychical research, known today as parapsychology, took its place at the fringes of science—between mainstream research and occult craziness.

To seek wondrous phenomena which might lead to God, then, we must examine psychical research. The debate on miracles can no longer center on questions like whether Jesus or Catholic saints performed wonders long ago. We should ask if psychics and gurus can work miracles as well, and see if the experiments of parapsychologists unveil a spiritual reality. The culture of mainstream science, as well as its theories, stands against the supernatural. But parapsychology also demands our attention. If we can recover the spirit in a world which seems all too material, perhaps we can eventually come to God as well.

A SPIRITUAL SCIENCE

Speaking of wonders opens a famous philosophical can of worms. Some, for example, suspect that since miracles would violate the lawful regularity of nature, we might have to choose between science and the miraculous. But many of the important philosophical debates on miracles began before the scientific revolution really gathered steam, and were driven by theology.[8] The Catholic church supported its claims to authority by pointing to the continuing miracles associated with the

Roman church and its saints. Protestants sought to cast doubt on miracles outside of scripture, but also asked whether God would operate in the Catholic manner in the first place. Their debate left us with questions philosophers still struggle with today. Which is more perfect: a God who sets the world up and leaves it alone, or a God who keeps tinkering with creation? If God has been revealed in history, are further wonders superfluous? Should God intervene to prevent evil?

These difficulties, though, arise with a generic metaphysical God. It is hard to say what the bare God of the philosophers should have to do with signs and wonders, so a straightforward connection between that grand abstraction and popular beliefs is bound to be questionable. But actual religions present a more concrete picture of the supernatural. Their God is the zenith of a top-down, hierarchical reality. In other words, the Catholics were on the right track. *If* the church could mediate between the mundane and supernatural worlds, *if* it produced a better class of miracle than its rivals, and *if* Catholic doctrine provided the best explanations of the supernatural, they would have a good case. Though wonder-working saints and Marian apparitions need not be the primary evidence for Catholicism, these can be important "motives of credibility."[9]

Establishing this credibility depends more on convincing evidence for miracles than ways around philosophical difficulties. For this, testimony preserved in religious tradition is not good enough—it is too easy to suspect mythmaking. We need a more scientific approach. Of course, science has tended to give us an uncomfortably material reality. But parapsychology holds out hope. If science has stranded us in an accidental world, perhaps rigorous empirical investigation will also show us a way out. J. B. Rhine, who moved the quest for miracles into the laboratory, believed his work could help religion to "expand and strengthen its foundations of accepted truth." Parapsychology could show us that

> there is something operative in man that transcends the laws of matter and, therefore, by definition, a nonphysical or spiritual law is made manifest. The universe, therefore, does not conform to the prevailing materialistic concept. It is one about which it is possible to be religious; possible, at least, if the minimal requirement of religion be a philosophy of man's place in the universe based on the operation of spiritual forces.[10]

This is particularly important today, when not only miracles meet with doubt, but even our minds are becoming less mysterious. We speak of Prozac, of evolutionary psychology, of brain development and artificial neural networks. Spirits who stand apart from matter and freely act upon the material world have almost vanished from serious intellectual life. Yet our religions still insist otherwise. No pile of mere matter—be it organized into brains or maddeningly complicated circuitry—should on its own be enough for consciousness, intentionality, intelligence.

There is a lot of common sense behind this intuition. After all, material objects seem very different than persons. People have beliefs, sensations, and purposes, but objects just sit there with a bunch of dull physical properties. So historically, philosophers usually thought that mind was wholly different from matter. We were souls housed in bodies; we might survive death. The mystery was why we were so intimately connected to bodies in the first place. And if miracles were not essential to reveal the spiritual aspect of the universe, this was because our very minds were already part of a higher reality.

Some philosophers continue to defend the soul, whether they adopt an old-fashioned substance dualism or a vague "emergent dualism."[11] And they find a formidable array of philosophical reasons to say it is impossible that minds arise from matter. They charge that there is too large a gap between mind and matter, that materialism is self-refuting, that the very argument presupposes minds that can respond to reason rather than be determined by physical causes. Some of these conceptual arguments raise good questions. But since dualism keeps the relation of spirit and body an impenetrable mystery, it faces a host of philosophical objections of its own. So many philosophers have become materialists, and others, while they still doubt that consciousness, reference, or subjectivity can be captured by physical processes, hesitate to talk of spirit.

If the soul has become unfashionable, this owes more to science than philosophy. Folk psychology is very basic to how we think, and our conceptual arguments tend to affirm its framework, finding that mind is radically different from matter. But though they are vital for navigating the social world, these intuitions may severely mislead us when we try to understand minds in detail. After all, the best thinkers were once intuitively certain that life was beyond physics, that organisms required "animal souls" to be alive. With modern biology, animal souls and vital essences vanished. So as we learn more about how our brains produce conscious experience, we might expect vestiges of personal souls to fade away. Of course, the task is formidable. We have but begun to understand how our brains put together our seemingly seamless perception of the world, how machines can exhibit creativity, how individually mindless operations in the brain might combine to produce intelligent behavior. Yet we have made significant progress—without quantum mysticism, without magic, without inventing souls under a new name. And those philosophers engaged with artificial intelligence or neuroscience are the ones least likely to have any truck with souls.

In this climate, defending the spirit calls for more than philosophy. Our souls would be safer if they were championed by a spiritual science—by a productive research program competing with materialism. Again, finding psychic wonders would work very well. As parapsychologist John Beloff puts it, "The significance of the paranormal is precisely that it signals the boundary of the scientific world-view. Beyond that boundary lies the domain of the mind liberated from its dependence on the brain. On this view, parapsychology, using the methods of science, becomes a vindication of the essentially spiritual nature of man which must forever defy strict scientific analysis."[12] It is particularly important that we find evidence that this immaterial mind *acts* on matter, that it is not just a mystery of the brain. If so, religious philosophers can repackage good-old-fashioned magic, calling it "agent-causation" or "noncontiguous causation," and reaffirm the supernatural realm.[13]

Bringing a God into the picture, though, requires some more work. If parapsychology succeeds, miracle stories about saints and prophets will become easier to believe. However, we will also come to see apparitions of the dead, predicting the future, or stilling storms as particularly impressive psychic phenomena.[14] In that case, someone like Jesus may seem just a superpsychic. Catholic miracles might be caused by a psychic force generated by intense belief. We may not need any God beside our own miracle-working spirits.

So the paranormal can also be at home in nontheistic spiritualities. Buddhists, for example, have historically used miracle stories for missionary and apologetic purposes.[15] Buddhist philosopher Gunapala Dharmasiri argues against God very much like a Western philosophical atheist, but also claims the Buddha had psychic powers: "He said that he could remember his past births and he maintained that anybody who developed the powers of paranormal vision could verify these truths. Normally these powers are achieved when one progresses on the meditational path."[16] Closer to home, many British Spiritualists of the late nineteenth century were distinctly anti-Christian. Many arrived at their beliefs after a phase of outright atheism; they were "unable to sustain the endless negativity required of the atheist,"[17] but their substitute religion focused on spirit and immortality—a personal God was optional.

Parapsychology will not directly support any one sect. The feats of a Hindu holy man do not look any less real—or any less illusory—than those of a Catholic saint. What might work, however, is a more ecumenical spirituality, with a God more like a cosmic mind than an authority dictating the sex lives of the faithful. Theologians impressed with parapsychology, such as David Ray Griffin, imagine a world pervaded by magic, where we can reaffirm commonsense perceptions of reality. Moreover, in rescuing the soul from oblivion, we need not crudely separate matter and mind. In Griffin's world, the supernatural and natural continuously shade into one another; God is not cut off from the world, and all religions appreciate something about divinity. Spirit infuses *everything*. Our religious journeys, extending beyond death, take us through a spiritual progression. And at the ultimate limit, we find a God who is the source of all.[18]

The Protestant and Catholic disputants of a few centuries back would not be happy with this New-Agey theology. But it helps us see how wonders might support a God. And it also gives us a new perspective on some of the other philosophical arguments we inherited from the Protestant critique of miracles.

Perhaps the best known is David Hume's variation on a Protestant theme—an argument he hoped would be "an everlasting check to all kinds of superstitious delusion." We propose natural laws to describe the world when we have repeated evidence that they hold true without exception. And we observe that while many people testify to wonders, people also regularly make mistakes, or even lie. We may have very solid testimony behind a miracle, but this can never overcome the repeated and further repeatable evidence backing up a natural law. Eventually the evidence for the anomalous event will become swamped by the cumulative evidence for lawful behavior, and the miracle report will be filed away as yet another mistake.

Some modern skeptics also make Humean arguments against both psychic powers and traditional miracles. Antony Flew points out that if we discover top secret information has leaked from a government department, we suspect a security problem—not a psychic employed by a hostile power. We assume that events take place within the boundaries of "basic limiting principles"; for example, we expect information cannot be transmitted without some physical means. These principles are very broad empirical generalizations. Though they do not tell us details such as whether information can be transmitted faster than light, whatever physical laws we discover must conform to the limiting principles. They pervade even our ways of interpreting evidence. For example, we may read testimony about Abu Said's miracles. But we can only say that ancient marks on paper are writing because we implicitly rely on regular natural laws in the present. We do not entertain the possibility that the writing was materialized by a psychic yesterday. So if we are to have miracles, we must at least have repeated evidence for them, evidence strong enough to overcome our repeated and pervasive evidence in favor of basic limiting principles.[19]

We certainly need extraordinarily strong evidence before accepting extraordinary claims. We cannot accept Abu Said's wonders just because someone swears they happened; it is far more likely that such reports are mistaken. Nevertheless, Flew sets the standards for accepting miracles too high.[20] Humean arguments overlook how parapsychology can accumulate evidence for its wonders. Paranormal reports do not present us with a haphazard collection of singular events; they fall in distinct categories such as apparitions, telepathy, or psychokinesis. So parapsychologists claim *repeated* evidence for consciousness-related anomalies which cannot be explained in any materialist fashion. A skeptical psychologist would try to explain the patterns of paranormal reports differently, but Humean considerations do not absolve skeptics from seeking nonparanormal explanations.

Even repeatability, while important, is not crucial. Say a miracle worker decides to silence the skeptics once and for all, by stilling a storm. She goes to Florida to confront the latest hurricane poised to sweep across the land. Just before the storm hits the shore, she raises her hand, and a flash of lightning leaps out from her palm into the sky. A wave of green light shimmers across the whole sky, and then it takes only one minute for all the menacing clouds to melt away, leaving a calm and sunny afternoon. The event is witnessed by thousands, broadcast by TV, recorded on videotape and with many scientific instruments.

Such an event should impress even the most Humean skeptic. Not because it would be absolute proof of a miracle; ruling out all naturalistic possibilities is a hopeless task. Perhaps our miracle worker stole a weather-control device from an alien flying saucer, so what everyone saw was really superadvanced technology which had nothing to do with spirit. But this is no less fantastic a possibility than magic. After the storm was stilled, any skeptic who came up with such a scenario would be scrambling to find an excuse, not proposing a serious explanation.

Parapsychologists do not have anything as spectacular as a stilled storm to offer, but they have a similar strategy. They try to accumulate anomalies which do not fit in a bottom-up picture of the world, and hope to force mainstream science into a cycle of excuse making. If they succeed, and especially if their spiritual science can provide an alternative theory to make sense of the anomalies, we will have a scientific revolution in the making. We will find that our limiting principles apply only to a limited domain of reality, and that we have to expand our horizons to include immaterial minds with strange powers.

So at least for now, we should set aside the philosophical issues raised by miracle claims. Psychical research, if it delivers on its promise, has a good prospect of showing that the world makes better sense from a spiritual perspective. The question now is what sort of case we have for thinking that miracles happen.

GREAT PERFORMANCES

Psychical researchers started out by investigating physical mediums, the wonder workers of late nineteenth-century Spiritualism. In darkened rooms, mediums produced communications from the dead, eerie spirit hands and materializations, self-playing musical instruments, and other miracles. Some were superstars whose performances still impress parapsychologists. D. D. Home, for example, was once reported to have levitated out one window, and back through another.

The debate over these reports followed a familiar script. Spiritualists claimed conventional science could never account for the events, while skeptics proposed natural explanations, very often trickery. D. D. Home appears to have fashioned his miracles out of his conjuring skills and the imagination of an audience he made sure could not observe the situation properly.[21] Never once did a medium levitate out a window in broad daylight in front of skeptical observers, but many were caught in the act of fraud. Apologists said that skeptics had not explained all the feats of mediums, and speculated that mediums filled in with trickery to satisfy their audience when their capricious psychic powers acted up.[22] But they were too obviously making up excuses, and physical mediumship eventually faded away.

Though the evidence of the mediums was disappointing, psychical researchers had the right idea. Enough psychic wonders could put an unbearable strain on a naturalistic view of the world. And it seems we have an abundance of great performances, from the endless testimony to the miracles of saints to the feats of psychic superstars. Something must produce this stream of miracle reports, and unless we have weighty reasons for doubt, the most straightforward explanation is that many such wondrous events do in fact happen.

In response, skeptics try to explain how a multitude of miracle reports can happen in an unmagical world. Sometimes the reason is simple ignorance. For example, some people walk barefooted over a bed of glowing coals, after being spiritually prepared by a religious ritual or a New Age seminar. While such invulnerability looks like an impressive demonstration of the mind's power over material forces, the reason behind it is largely that while very hot, the coals have a low heat capacity and conduct heat poorly.[23] Likewise, we can stick our hand in a hot oven, but we cannot touch a cookie sheet inside—the air is at the same temperature as the metal, but is a very poor conductor of heat.

Sometimes the human element is most important. Our normal mechanisms of belief and perception often help trigger strange experiences, and prompt us to explain them supernaturally.[24] Even multiple observers can make the same mistake when perceiving an ambiguous situation. In 1985, a figure of Mary began swaying in an Irish grotto. Hundreds repeatedly saw the effect, but a movie camera recording showed that no movement had taken place. In fact, *the observers* were swaying in unison. At dusk, they could not detect their own slight movement when they no longer had a clear reference point. Once the expectation of a moving Virgin was in place, and a few swaying observers to reinforce this expectation, others fell in line.[25]

Normal psychology is often sufficient to produce paranormal belief, but in many cases the process is helped along by a bit of deception. In a miracle like Abu Said's forty-day fast, we do not know if the event ever took place, or if it is a stereotypical saint tale someone added to make the Master look more impressive. And even if the challenge to go without food actually happened, it would have been easy for Abu Said to cheat. After all, we do know of a few other miracle workers who went without food, or rather preferred to take their nourishment under cover of night.[26] Miracles are very significant in our popular religions, and rightly so, but this results in an open invitation to self-deception and even outright fraud. Too much Higher Truth has a way of overwhelming garden-variety truth.

For the most interesting miracle reports, we need a complex interaction of many ingredients. Reacting to a secularizing world, Catholicism has produced many Marian miracles in the past two centuries, and one of the most spectacular took place in Fatima, Portugal, in 1917. Three children said they saw apparitions of Mary which were visible only to them. This invisibility did not deter the devout; after all, believing without seeing is virtuous. The Virgin made an appointment to show up every month, and at the time she was to make her final visit, tens of thousands gathered at the place she regularly appeared to the children. They were expecting a great miracle, and they were not disappointed. The invisible Mary came, granted the children some private visions, and then something happened which was visible to all. The leader of the children pointed skyward and lo! a "sun" appeared in the clouds. This sun danced, spun, moved around the sky, and generally behaved like a low-budget special effect. The Catholic hierarchy later investigated the events and certified the miracle.

One ingredient in Fatima seems to be a bright child bringing other children into a fantasy, and manipulating adults by using their shared religious beliefs. This happens often enough, and nothing too spectacular would have come about if not for the sun miracle. Even then, the sun perversely refused to put on its performance outside of Fatima; and in Fatima, the globe of light appeared in the wrong spot in the sky for the sun. It is hard to figure out what happened from the testimony we have, but a good possibility is that the public visions were triggered by a rare atmospheric phenomenon known as a "sundog" or "false sun." In any case, a crowd primed to witness a miracle confronted an ambiguous, unusual sight. This was interpreted—perceived—as a miraculous vision, and this perception spread among those present so that they literally saw a sun doing wondrous things.[27] We usually think multiple testimony is just about infallible—so many people cannot be all wrong at once. And *independent* testimony from many people does carry a lot of weight. But strangely enough, a crowd sometimes is the perfect setting for a contagious misperception.

So skeptics are not at a loss to explain where miracle reports come from. People make mistakes. And these mistakes are not arbitrary. We have some idea of the shape supernatural reports are likely to take; we can often understand even complicated cases by bringing our knowledge about physical science, psychology, and human cultures together in a narrative of what must have happened. We can observe how a religious community responds to reports of unusual experience by constructing a miracle story which fits their theological expectations. In contrast, supernatural explanations do not take us much further than the whims of the gods or the effects of capricious psychic powers. Proponents of the supernatural are very good at proclaiming various events unexplainable by natural causes, but they say very little about why God, a spirit, or a psychic ability would produce the particular bizarre phenomena which are supposed to be signs and wonders. As psychologist Nicholas Humphrey points out, there are too many "surprising limitations" on paranormal phenomena. We should ask "why messages from the dead should be so surprisingly vague, why poltergeists are particularly active around teenage children, why a statue of the Virgin Mary rocks her head at night but not during daylight, etcetera, etcetera."[28] Such arbitrary constraints tell us there is something wrong with a supernatural explanation, even when we have no ready-made naturalistic alternative.

Of course, this still leaves us with a large residue of unexplained mysteries. The most common argument for something paranormal is to produce a list of cases which have not been solved by conventional methods. But saying ours is a natural world is not the same as promising a full explanation for everything. In a complex world full of accidents, we often have to leave things unexplained. This is what we *expect*.

Consider the investigation of a murder. A detective brings many tools to the case, from experience with human behavior to sophisticated forensic methods. Yet large numbers of cases remain unsolved. Sometimes there is too little evidence; sometimes, more frustratingly, there is an abundance of information but all plausible suspects seem to be ruled out. Such unexplained cases, however, do not mean murderous space aliens are stalking us for their macabre medical experiments. For the work of an alien conspiracy, our unsolved murders have too many arbitrary constraints.

Besides the right tools, we also need a bit of luck to solve murders. There are many difficult cases, and sometimes the culprit gets caught only because someone stumbles on a clue which might easily have been overlooked. So there will also be cases where a clue is passed over, or where no one makes the crucial confession which would help everything fall into place. And these cases will not stand out from among other difficult murders except for the fact that they remain unsolved. We regretfully call them unexplained, but do not doubt there are normal causes for the crime. If we started finding corpses whose central nervous systems were removed by surgical techniques far beyond our technology, we might give the alien hypothesis a hearing, but not before.

Much the same applies to paranormal mysteries. The feats of mediums and psychics have always attracted investigators, some looking to debunk it all, some trying to find a true miracle among the prodigious nonsense. And since conjuring techniques are an excellent way to fake miracles, conjurers have often joined experimental psychologists in exposing psychics. Houdini was the bane of mediums; today James Randi is a merciless critic of psychics. Skeptics such as Randi—and Joe Nickell and Martin Gardner and others—manage to deflate an impressive number of popular miracle claims.[29]

It would be crazy, however, to promise a solution to every mystery—without supernatural powers of their own, no investigator can be that successful. Investigators of miracles need as much luck as police detectives. Skeptics, at their best, examine a sample from the best cases which proponents of the paranormal offer, solve them when possible, and draw some conclusions about how we might mistakenly think miracles have occurred. We can then conclude it is most likely the remaining unsolved cases were nothing supernatural, *if* these do not stand out among those difficult cases we were lucky enough to solve. And so far, this seems to be where we stand. Psychical researchers tell many puzzling stories, and with some of these, skeptics can do no better than speculate. But we never find outstanding performances like stilled storms, well-observed and recorded levitations in daylight, or a dancing sun seen by astronomers around the world.

As proponents of the paranormal like to say, science cannot explain everything. This is true, but not because there is magic in the world. In a complex, accidental world, we will often come across events we cannot fully explain. And if that is all we can say, so be it. Unexplained events are, after all, perfectly natural.

STATISTICAL MIRACLES

Psychical research began with case studies, trying to find performances beyond mere matter. One glaring problem with such studies, however, was the lack of control. It was all very well to frequent séances intending to act as an objective observer, but there was no escaping the fact that the medium ran the show. Psychical researchers found themselves acting not like experimental scientists but travelers to an unknown country. And even when they thought they were successful, the tales they brought back were mostly "here be dragons."

In the 1920s, J. B. Rhine tried to change this. Instead of watching mediums perform spirit materializations, he conducted experiments with ordinary people, asking them to perform unspectacular tasks like guessing which of five symbols was on a card. If the cards were drawn randomly, they would normally guess right one time in five. If in the long run they did better, they must have received information about the card by some nonphysical means.

Such procedures have many advantages. The researcher is now in control, so with some care, the experiment can be designed to preclude the more obvious forms of deception. Guessing at a rate above chance is a well-defined accomplishment, so no one need rely on fuzzy ideas of what counts as a miracle. Other scientists can repeat the experiment. If even ordinary people can perform better than chance, then psychic powers must tell us something important about all our minds. Best of all, by using statistics, parapsychologists can quantify their miracles. Controlled experiments which produce hard numbers—what more can a science ask for?

Rhine's experiments produced results. But critics soon found that they were poorly designed, allowing subjects information about the cards. For example, in some early experiments the subjects could see the back of the card they were guessing. All cards were not exactly identical from the back, and the subject could unconsciously pick up on which was which.

Even fraud did not disappear. By the 1950s, a series of experiments conducted by Samuel G. Soal appeared to be the best evidence for psychic powers. Soal had, under what seemed to be ironclad experimental conditions, found statistically significant results: the odds that a chance coincidence produced his data were less than one in many billions, often much less. Skeptics responded, but feebly. In a 1955 paper, G. R. Price brought up the possibility of cheating, and suggested a far-fetched scenario by which Soal could have perpetrated fraud. The heart of Price's argument, however, was a Humean conviction that good evidence for miracles was next to impossible. If fraud was at all possible, it was a preferable alternative to magic. Parapsychologists responded, with some justification, that Price's skepticism was dogmatic, and that suggesting fraud without evidence to back up the accusation was a personal attack which had no place in science.[30]

However, there were some small signs Soal might have cheated. A subject in Soal's experiments noticed that Soal could have altered the score sheet. Finally in the late 1970s a parapsychologist discovered a pattern of alterations in the old score sheets.[31] One of the best experiments of parapsychology was now dead, but such discoveries were much more significant than just shooting down old evidence. It became apparent that experimental parapsychology suffered from the same problem as the case studies. Fraud or small mistakes in the experiment were often found through a long process, with the help of some considerable luck. As with the great performances, skeptics could not discover a definite flaw in every experiment. But the unexplained experiments did not stand out from among those difficult cases which turned out to be due to well-concealed deception or a subtle mistake in experimental design.

Parapsychologists still forged ahead, designing new types of experiments. In the 1970s and 1980s, a kind of clairvoyance test known as "remote viewing" became popular, met with severe criticism, and receded from the limelight.[32] In the 1990s, the best statistical miracles showed up in ganzfeld studies and random event generator (REG) experiments.

In ganzfeld experiments, a subject tries to pick up impressions of a target—usually a picture or a film clip—which a "sender" in another location is concentrating upon. To reduce distractions the subject does her guessing with Ping-Pong balls taped over her eyes and headphones piping white noise into her ears. Later, a judge picks out which in a set of possible targets is closest to the subject's impressions. The experimenter then sees if the actual target matches the judge's choice. Many experiments find above-chance matching rates, and combining different studies by a technique called meta-analysis reveals a definite above-chance signal.[33]

REG experiments are more straightforward, eliminating the extra human element of senders and judges. An ideal REG experiment has a tamper-proof machine producing a random result—something like a perfect coin flipper. In each trial, the human subject tries to influence the machine to produce heads or tails as she chooses. And since each trial is simple and quick, subjects can make *long* series of trials, producing enough data to expose even very slight deviations from chance. Meta-analysis shows that REG experiments also tend to find nonchance behavior.[34]

Critics soon found imperfections in the ganzfeld and REG work. For example, in the best series of ganzfeld experiments there was a flaw. Most of the data were collected when there was an almost insignificant equipment fault: the sound from the film clips the sender was watching was leaking, almost imperceptibly, into the subject's headphones. The experimenters discounted this flaw, figuring it could not account for their results. Richard Wiseman reanalyzed the data, looking at those trials where this and a few other minor flaws were corrected. Performance dropped to chance level. Rounds of response and further criticism followed, establishing the familiar standoff between skeptics and believers.[35]

Though ganzfeld and REG studies promised clear and reproducible proof of psychic powers, they failed to change the nature of the debate. Historically, psychical researchers have regularly announced new and improved miracles, only to have critics start chipping away at their claims. Then yet another new type of experiment hits the stage, and the cycle starts over. This peculiar history is itself a reason for skepticism, as psychologist Ray Hyman observes:

> The most disturbing aspect of the case for parapsychology is the shifting basis for its claims. At any point in time there do seem to be a set of outstanding candidates for the repeatable experiment or the major breakthrough. But at a later time many of these candidates drop from the running because either they are subsequently found to be flawed, or it becomes difficult to replicate them, or suspicions of fraud have been raised.[36]

There seems to be no progress in spirit science, not even much continuity. Parapsychologists get new results with new techniques, skeptics find flaws. After more than a century of psychical research, skeptics like to say, we have gotten nowhere; most likely because there are no psychic powers.

Parapsychologists see the glass as half full. Critics suggest flaws, so parapsychologists keep improving their methods. And they always manage to find statistically significant effects. Surely there must then be something to psychic phenomena?

Parapsychologists certainly find significant effects; that is, effects which are very unlikely to be due to chance. But in another sense, statistical miracles are usually quite insignificant: they are very small deviations from chance. For example, say we are running an REG-style experiment using a coin-flipping machine. If a subject guesses right one hundred times in a row, it is a miracle. The probability of an exact match is $1/2^{100}$; it almost certainly did not happen by chance. A 100 percent success rate is also a very large effect compared to the 50 percent we expect. If our subject guesses 51 out of 100, there is nothing to be impressed with. But let us say she maintains a 51 percent success rate over many trials. When we do 1,000 coin flips, she matches them about 510 times. In 10,000, she gets about 5,100. As we increase the number of trials, it becomes very unlikely that this 51 percent success is a statistical fluke. We have, once again, found a statistically highly significant deviation from chance. However, 51 percent is much closer to 50 percent. It is a small, marginal effect, though it is very likely real.

Real REG experiments also produce marginal results, only more so. The Princeton Engineering Anomalies Research laboratory uses, as its coin flipper, binary pulses generated from electronic white noise. After many millions of guesses, they report highly significant deviations from chance, with likelihoods of being statistical flukes often as low as 10^{-7} or 10^{-10}. But their effect size is miniscule, roughly corresponding to 50.01 percent.[37]

Marginal effects coaxed into view by statistics are common in parapsychology, but also elsewhere at the fringes of science. Astrologers, for example, are fond of Michel Gauquelin's "Mars Effect," which allegedly demonstrates a connection between planetary positions at birth and a person's talents or personality. This is an extremely small effect, barely showing up among those "most eminent" in a profession, notably athletes. So the Mars Effect is only visible with the aid of statistics, and skeptics keep criticizing the statistical procedures and data-gathering process.[38] Then there is homeopathy, which claims to cure people with solutions so dilute that not one molecule of the original substance remains in the solvent. According to mainstream science, homeopathy is as crazy as astrology or psychic power; but of course, there are meta-analytic studies showing that homeopathy has very small but statistically significant benefits.[39] All this is a far cry from superpowers of the mind, stars as keys to our destinies, or homeopathic cure-alls; but today our occult sciences have to live with reduced expectations.

It may still seem a marginal effect means there is *something* paranormal happening, if ever so slightly. This is not true. If we find someone who can regularly guess how a coin comes up 80 percent of the time, we should be impressed. She either knows a very interesting trick, or our experiments have a major design fault, or she really is psychic. Since we are looking for a large information leak, there are a limited number of practical possibilities—we can expect to discover what is happening. If she guesses only 50.01 percent, life is much harder. Perhaps the machine does not produce a truly random sequence of heads and tails, and a nonrandom human guessing pattern exploits this.[40] A very subtle flaw in experimental design might allow a trick which is rarely useful, but is enough to give that 0.01 percent edge over the long run. Perhaps after having been used for a while in humid conditions, the machine emits a click barely above the threshold of hearing before it spits up a coin showing heads, but not before tails. And so on. As the effect size gets smaller, the number of alternatives to a miracle proliferates like mad. Though these alternatives are quite improbable individually, they add up—and there is no way to keep them under control.[41] There are just too many ways to get marginal effects, and no one can list them all, let alone find them after the fact.

If we were looking for a large effect which could stand out among the possible mistakes, we could ignore "nuisance" possibilities. Though the real world is a complicated mess in which the perfect experiment is impossible, our experiment could still be good enough. Eighty percent stands out against 50.01 percent as well as it does against 50.00 percent. But when investigating marginal effects, we can no longer say that 50.00 percent is our expected success rate in an unmagical world. In fact, there is little we can do in any specific circumstance to exactly determine the normal success rate, except to directly *measure* it. But parapsychologists cannot run a control experiment to measure this baseline, where everything is the same except that their subjects do not use psychic powers—that experiment would be identical to those they run to test for anomalies. So without outstanding effect sizes closer to 80 percent, they simply have nothing to report.

Marginal results in a complicated world are bad enough, but ambitions to radically change our picture of the world with such results make the case for psychic power even worse. Science is not a loosely connected list of facts, where psychic power would just be another fact to add to a textbook. Barely noticeable "facts" which contradict a tight-knit, highly successful picture of the world are always suspect. Physicist Philip W. Anderson emphasizes that parapsychology makes physical claims, and points out that "It is the nature of physics that its generalizations are continually tested for correctness and consistency not only by careful experiments aimed directly at them but, usually much more severely, by the total consistency of the overall structure of physics."[42] Fitting a result into a theoretical picture is crucial to avoid persistent mistakes. Without theory, experiment falters; in the history of physics, we often find experiments discounted, even when no specific errors were identified, as later theories and experiments converged on a stable picture.[43] Marginal experimental results, presented as bare anomalies within a threadbare theoretical framework, are never very impressive.

So parapsychologists look for patterns in their results. Gertrude Schmeidler, for example, finds that psychic success depends on psychological factors. When experimenters believe in magic, they tend to find psychic powers: the "experimenter effect." Believers score better than skeptics as subjects.[44] Skeptics, not to mention cynics, are not surprised. In any case, this is no theory; it is a series of excuses covering for peculiar constraints on psychic performance. And though some parapsychologists speculate that the results of quantum measurements are determined by conscious choices, or claim mind is unified by quantum fields,[45] efforts to tie psychic phenomena to modern physics are no more convincing. In the end, parapsychology has no solid theories, and no strong experimental results.

Parapsychologists protest that this critique is unfairly conservative.

Experiments are unreliable without a theoretical framework, and theories are castles in the sky without experiments to explain. How, then, can any experiment discover something radically new?

In the history of physics, some very small effects were discovered even though they violated strong expectations and had no theory to back them up. For example, modern physics is afflicted with some strange symmetry violations. Our ordinary physical laws are symmetric, and nothing in our normal experience suggests that in certain exotic interactions asymmetry reigns. Plus, theorists love symmetries because a symmetric world is simpler, with less that needs explaining. A world which is the mirror image of ours should have the same physics, as should a world we get by replacing each of our particles with its antiparticle. Perversely, these symmetries are violated by tiny amounts in our world. And we still have no solid theory to help us understand why. Yet experimentalists have convinced their skeptical colleagues that these tiny effects really happen. They did not do this by collecting zillions of ordinary events, exposing very small but statistically significant deviations from absolute symmetry. Though particles are very simple objects, incapable of deceiving themselves or physicists, the experimental environment is never simple. So physicists soon learn that experiment is dirty work, and that persistent mistakes can and do happen. The physics community would not accept a symmetry violation on the evidence of a marginal effect of varying magnitude, overwhelmed by uncontrollable nuisance possibilities.

We believe some symmetries are violated because we have found clear, large signals which stand out in a dirty experimental setting. "CP," the symmetry corresponding to replacing every particle with its antiparticle in a mirror-image world, is violated.[46] We first saw this in the decay patterns of an exotic particle called a neutral kaon. This is a highly sensitive system which magnifies even a ridiculously small CP-violation to produce a large effect. We can make very sharp predictions about how kaons will decay if CP-symmetry holds, and test them without having to sift through an immense amount of noise in hope of a marginal signal.

None of this is true with the statistical miracles of parapsychology. We have marginal effects set against an ambiguous "normal" background and vague hypotheses which do not add up to a real theory, all explored within a research tradition no more immune to mistakes than any other and which has no settled findings. As always, psychic powers have too many peculiar constraints, and they do not stand out from among the errors. In the laboratory, we have better control than with traditional miracles. And as the controls tighten, the wonders fade away to marginal effects. Perhaps parapsychology and theology really do belong together, for it appears both study what does not exist.

SOARING SPIRITS

If we are more than matter, we might have more direct signs of our true nature. Public miracles would be nice, but too many of them are pointless even if they really did happen—no decent spirit would be caught dead guessing cards and bending spoons. So perhaps we should ignore the parlor games and look within. After all, many of us have some very strange experiences. In the right circumstances, our spirits appear to detach from our bodies. Some go to death's door and reluctantly return to this world at the last moment. These are private experiences; no one still bound to this earth accompanies the soaring spirit on its journeys. But so many people independently tell similar tales that there is no doubt these experiences happen.

Someone having an out-of-body experience (OBE) typically finds that she has risen out of her body, and that she is viewing herself and her surroundings from above. Her spirit may then explore strange realms, or just float around and observe the mundane world by paranormal means. In one famous case, a woman hospitalized for a heart attack apparently noticed not only what was happening around her when she had lost ordinary consciousness, but even an otherwise invisible shoe caught on a window ledge outside.

Then there are near-death experiences, or NDEs, which are becoming more common now that medical technology helps many people recover from heart failure or other shocks which would have killed them a century ago. The dying person leaves her body, often hovering above for a while as in an OBE. However, the spirit soon leaves this world and approaches a tunnel of light. She passes through and often encounters a loving, comforting presence. Sometimes spirits of dead friends and relatives are there to greet her. She reviews her life. But the spirit does not, in the end, cross the boundary between life and death, returning instead to our world. Many who have had an NDE find it difficult to describe by words, and they are often led to a profound personal transformation.

In fact, there is an array of phenomena which appear to directly testify to spirits not bound to the brain. Instances of spirit possession look like an external entity taking over a body. Some people remember lives they lived long ago and in places far away. They may need the help of a hypnotherapist to regress them to past lives, but their memories come with appropriate emotional responses; they are not like stories told about someone else. Many people see apparitions. So it is no surprise that people in all traditional cultures believe in disembodied minds. Though our cultures vary wildly in how they picture the fate awaiting the dead, we find similar experiences backing up the basic idea of a surviving spirit.[47] Christian patients having an NDE meet Jesus while an Indian may be greeted by a Hindu god, but the overall structure of their stories is similar. Common experiences lead to similarities in our folk religions; even the way we picture our heavens and hells is alike.[48]

Private experience is weaker evidence than a public demonstration; it is easier to suspect it all happens in the head. However, spiritual experiences compensate by their intensity. An OBE does not feel dreamlike, but utterly real. NDEs can also feel more real than everyday reality; a patient who has gone through an NDE might say she no longer believes there is an afterlife, she *knows* there is. Memories of past lives can feel as real as any ordinary memory. Such experiences are personally much more compelling than guessing cards—it seems they give us firsthand knowledge that we are spirits.

Organized religions have room for such experiences, especially within their mystical traditions. But today, individualist spiritualities lay the strongest claim to this territory. So experiencers often resort to occult language in order to make sense of what has happened to them. P. M. H. Atwater, for example, tries to explain her NDE and her radically changed outlook on life in the context of a spiritual hierarchy:

> . . . to equalize pressure differences between latent spiritual potentiality and mundane personality development, descending currents of force (perhaps from the soul level, Higher Self, God) and ascending currents of force (perhaps from time/space ego states, lower self, personality level) meet to create a powerful lightflash. . . . [49]

As the muddled language indicates, the interpretation of the experience is up for grabs. Conservative Christians dislike occult speculation, since the Abrahamic religions have always been suspicious of freelance contact with the supernatural. Maurice S. Rawlings emphasizes that some NDEs are hellishly negative episodes, and warns that the "being of light" at the end of the tunnel is Lucifer, the angel of light. Apparently Satan is trying to deceive us into thinking there is no Hell and that salvation is possible without Christ.[50] But however interpreted, NDEs seem to be direct experiences of a supernatural realm, just as walking down a supermarket aisle is an immediate experience of the mundane world.

No spiritual belief is complete without a philosophical defense. C. B. Becker points out that a variety of events testified to by many people indicate an immortal spirit, so that "the survivalist hypothesis makes sense of all these phenomena: claimed memories of past lives, apparitions and OBEs, and NDEs with paranormal visions."[51] In fact, for Becker and other philosophers impressed with parapsychology, this fact is so clear that the really interesting question is why mainstream science is so reluctant to accept that we survive death.

Skeptics, then, once again have the burden of outlining a naturalistic approach. It does not mean much that spiritual experiences occur in all cultures; after all, people have similar brains. And a deeper understanding of the brain can reveal that though a variety of phenomena appear to support a spirit, this commonality is spurious. Whales and fish are not closely related, though many a folk taxonomy throws them together; we also may find that an OBE and a past-life memory are unrelated phenomena with a superficial similarity. To see if this is so, we must look to psychology—the boring kind which does not overthrow physics in every experiment.

A clue to the psychology of NDEs comes from comparing them with hallucinations.[52] Hallucinated images often include tunnels, so that a tunnel is one of the hallucinatory "form constants." Some hallucinations give us the impression we are off visiting strange realms, plus they certainly feel frighteningly real. Of course, this is only a starting point. OBEs and NDEs are obviously more than generic hallucinations; psychologists have to explain some of the details of the experiences. Why is there a tunnel in the first place, form constant or not? Why does someone having an OBE feel like she floats up around the ceiling, instead of, say, her astral body stepping out from her mortal shell and walking around the floor? Why does everything seem so real, instead of like a dream?

Susan Blackmore's work illustrates how we can start answering some of these questions. When working as a parapsychologist, Blackmore came to believe that psychic power was a nonexplanation for strange experiences. So she tries to do justice to the compelling nature of an OBE without detachable spirits soaring beyond the brain. She points out that OBEs typically take place when information from the external world is not available to our brain, but our awareness has not been switched off. Under conditions conducive to an OBE, she argues, our brain will focus attention on the most stable representation of the world available. As it happens, our memory of location is organized according to a bird's-eye view. In the absence of external input, this memory model is most stable, and so it becomes our working picture of "reality." While having an OBE we survey a scene including ourselves from above, and this feels no less real than our ordinary models based on continuing sensory input.[53]

NDEs usually begin with OBEs, since near death, the brain is cut off from outside information. Blackmore also explores how the NDE progresses beyond an OBE.[54] NDEs have also occurred in people who were not close to death; so while something in the dying process triggers a tunnel-image and so forth, it is not something unique about death. A good candidate is oxygen deprivation. Oxygen-starved neurons in the visual cortex fire in abnormal patterns, producing effects seen as concentric rings or spirals. Even gradually increasing electrical noise in the visual cortex can produce the basic elements of an approaching tunnel of light.

It is more difficult to understand the more complex aspects of an NDE, especially as not everyone responds the same way. Still, we are not completely in the dark. A flood of memories often accompanies oxygen deprivation in the brain's temporal lobe and other subsystems responsible for memory and time perception. Similarly, with random firings, seizures, and release of endorphins in the brain, mystical experiences such as feelings of timelessness are not surprising, though much remains to be worked out about the details.

Events in the dying brain, however, are just the beginning. A full-blown NDE is culturally constructed, and is comparable to other experiences of crisis, of separation from others.[55] Interpreting an NDE is a long process which begins before the experience, with the beliefs the patient brings to the NDE, and continues after the patient emerges back into the ordinary world. Supernatural theories are readily available and straightforward to understand, plus they accord NDEs a cosmic significance in proportion to their personal impact. Hence they take a large part in how patients respond to the experience. Though it is hard to disentangle what is a result of religious beliefs in the culture and what is due to events happening in any dying brain, it is fair to say that supernatural beliefs help patients construct the personality-altering aspects of the experience. This does *not* mean an NDE is a cultural artifact, even less that it does not have a profound effect. Blackmore suggests that as the brain gets closer to death, our model of self also falls apart, leaving us in a state where there is experience but no coherent self to experience it. It would be hard not to be changed after that, but how this change will unfold is not determined by events in the dying brain alone.

Phenomena like past-life memories, in contrast, arise from very different psychological processes which do not need a disruption to the ordinary functioning of the brain to be set in motion. When a hypnotized subject relates memories of past lives, she performs the role of someone who has lived before, recounting what she thinks are appropriate memories. She fashions this new identity out of common knowledge and stereotypes about the past, sometimes even fiction she had read long ago and had seemingly forgotten. As psychologist Nicholas Spanos points out, remembering past lives under hypnosis is closely related to experiences of spirit possession, channeling, alien abductions, multiple personalities, and therapist-induced false memories of ritual abuse. There is nothing paranormal about such experiences.[56]

This is not to say we fully understand extraordinary experiences. Something like an NDE is a messy event, not always triggered by any single, clearly identifiable cause like oxygen-deprivation.[57] However, mainstream psychology makes progress in tying our seemingly paranormal experiences to our other knowledge. In contrast, supernatural ideas are dead ends. We know enough to see that by lumping phenomena like OBEs and past-life memories together, we misclassify them on the basis of crude similarities. While psychologists can explain why OBEs take a bird's-eye perspective, to parapsychologists such features are yet more surprising constraints on how spirit manifests itself. Once again, we have no real theory to guide the quest for the paranormal, and no results that stand out against a complicated "normal" background. No one with memories of past lives comes up with information useful to historians or archaeologists. When skeptics investigate the story about the hospitalized woman who saw a shoe on an outside ledge while hovering in an OBE, they find that she easily could have obtained that information normally.[58] Of course, there still are reports which do not neatly fit any current psychological theory of unusual experience. But this is what we expect, precisely because consciousness is not some elementary property of the universe, but a very complicated process. We cannot hope to capture the detailed experience of a complex brain, like an NDE, with a short list of clear-cut causes. We have gaps in our knowledge, psychologists keep learning more, and that is all there is to say.

Still, there is something unsettling about trusting psychology rather than a compelling firsthand experience. We may be tempted to say that experiencers are directly acquainted with a spiritual reality, and if our theories call this a mistake, this only means the theories are wrong. That would, however, be another mistake. We rarely realize how much theory and interpretation is woven into our perceptions—even our biology. For example, we recognize faces with ease. We *see* faces, without being aware of the layers of computing our brain goes through to achieve that effortless identification. Different neural networks in our brain simultaneously process the information from our eyes, to deal with edges, color, motion and so on. Parts of our brain are also devoted to identifying features of faces, and recognizing the overall pattern.[59] This hardwiring is very useful in everyday life, but it also leads us to see Jesus' face on tortilla burns. Most of us are willing to admit that is a mistake, firsthand experience or not. NDEs and OBEs are also permeated by "theories" wired into our brain structure, and by the folk psychology through which we interpret our everyday experience. These theories are deeply ingrained, so much that many of us have endless trouble imagining how they could be false. Yet they are not infallible; the evidence of psychology is that our folk theories concerning spirits and firsthand experience are in fact wrong.

THINKING MEAT

Debating psychic powers can be frustrating. Skeptics often lose patience with a continuing lack of solid results, and go on to list the ways psychical research qualifies as a pseudoscience. Parapsychologists get exasperated at the blindness of orthodoxy, and portray mainstream science as a prisoner of its materialist assumptions. Following Thomas Kuhn's philosophy of science, they claim to herald a "paradigm shift." The present order blinds scientists to psychic realities; all parapsychologists can do is to accumulate the anomalies which will eventually break the old paradigm. When the revolution comes, science will go through a radical discontinuity, forcing us to rethink how we do science. When the paradigm shifts, we shall see the world anew, as if after a "conversion experience."[60]

Skeptics and believers are not, however, mirror images. Mainstream science *might* be limited, but its results are secure. Parapsychologists make extraordinary claims which they have not convincingly supported, and then respond to criticism by adopting even more sweepingly radical positions. Healthy intellectual enterprises do not need such desperate defenses. For all the overheated rhetoric, the skeptics are correct: parapsychology, like homeopathy or astrology, survives not on the strength of its results but because of its appeal outside of the scientific community.

Without psychic wonders, defenders of spirit have to fall back on philosophical tradition. But though we will never lack for philosophers saying our minds are magical, they do little but proclaim an invincible mystery wrapped in dubious conceptual certainties. As long as our sciences produce results, it is easy to ignore them. Artificial intelligence researchers, for example, tend to think that mind is what the brain, a highly sophisticated computing device, *does*. We are thinking meat, much the same way we are living meat without an animal soul. A properly designed, massively complicated machine could in principle think and feel; mind is in the function, not the hardware.[61] This may sound strange, since our folk theories and our intellectual tradition both have trouble imagining how mere matter can be animate, never mind alive or even conscious. Still, scientists can keep occupied indefinitely exploring the complexities of machine intelligence, and as long as they make progress, they will not be greatly concerned with this lack of imagination.

Neuroscience focuses our attention on how our particular hardware produces our kind of mind. Mental events are invariably accompanied by changes in the brain: drugs change our brain chemistry and alter our conscious experience; localized brain damage causes bizarre effects like destroying our ability to recognize faces. This does not close the case against spirit; after all, every change in the sound a radio receiver emits also corresponds to a mechanical change within the radio. The brain might be changing in response to an occult mental field, so that all interesting acts of perception, feeling, recall, and so forth are initiated nonphysically. Our cognitive and neurosciences make such scenarios implausible not just by correlating mental events with brain changes, but by learning how the brain goes about perceiving, remembering, and so forth. Studying PET scans and the effects of brain damage not only help us map the brain, but let us examine our neural architecture to see what sort of processing might be going on. We can test our ideas with technological analogues, however crude, like massively parallel computing and pattern recognition in neural nets. And the more we do this, the clearer it becomes that the brain is processing information, learning, and initiating action—not concentrating occult forces. There is less and less these days for a spirit to do.

In fact, philosophers Patricia and Paul Churchland argue that as we learn more about the brain, we may replace our folk psychology—our language of sensations, feelings, intentions—with a more accurate neuroscientific description. This seems crazy; after all, we think that if there is anything we can be certain about, it is that we see red or feel pain. But this language is part of a folk *theory* which can be wrong—just like an NDE is not an infallible taste of an afterlife. Folk psychology works fine in everyday life, but often radically misrepresents our inner states and how we actually process information. So the Churchlands expect that the concepts of a mature neuroscience will not smoothly translate to the language of folk psychology. When we discarded the demon-possession theory of disease, we did not translate demons into the new language of medical science; we eliminated demons altogether. Similarly, the Churchlands argue that folk psychology may be a good starting point, at least in mapping out the phenomena we need to explain, but it will not fare well as we make progress. We will have a better theory for what goes on inside our heads.[62]

We are far from settling such issues; scientists and philosophers still hotly dispute which are the best ways to construct a theory of mind. Explaining folk psychology rather than eliminating it, for example, is a more likely prospect. One thing, however, is clear. Now, more than at any other time in our history, the prospects look good for a naturalistic explanation of how our minds work. Spirits were always more excuses than explanations, but today we have hopes of penetrating the mysteries which supernatural hypotheses paper over. We may not yet be sure exactly how beliefs and feelings will figure in a fuller description of the world, but we can already tell that spirit will not show up at all.

If even in our minds we cannot find magic, psychic powers are completely out of place in our world. Signs and wonders are no argument for

God, because there are no miracles in the first place. It is bad enough to lose a traditional argument for the supernatural. But if we really are thinking meat, the very idea of God is in danger. To make sense of God, we need a religious theory, a picture of the world in which God has the leading role. We can start with analogies like the Great Programmer, asking what our world would be like if an unimaginably great power ran the show. But we also need an idea of how the Great Programmer could become more than an inflated version of a human designer. The Programmer must become more ethereal, mystical, more like an ultimate mind which is the creative foundation of all existence. We should be able to find creativity and personality at the bottom of everything, including our physics. But if our minds firmly belong in the material world, we no longer have anything real to hang such a picture of God on. If spirit is a deeply mistaken description of ourselves, how do we even begin thinking of ultimate spiritual realities?

So godly philosophers will continue to insist that our minds are mysterious, and that something spiritual lies behind the mystery. Perhaps the qualities of our inner experiences are beyond science, so that no interaction of a brain with light can begin to explain the redness of the color red. We may need a supernatural power to bind redness to physical changes in the brain.[63] Moreover, there are ineffable mystical experiences. Mystics do not only perform magic, but claim acquaintance with a reality so glorious it is beyond all description. Sophisticated believers talk of a mystical God, of firsthand knowledge freed from the conceptual shackles of science.

Popular religion needs miracles; most of us are likely to believe in paranormal wonders no matter what. The God of the intellectuals can survive without the Low Magic of psychic parlor tricks, taking refuge in the High Magic of mysticism. Those of us who are stubborn skeptics, well, we get along without magic. And late at night we sometimes wish we could still storms and read minds.

NOTES

1. Mark A. Shibley, *Resurgent Evangelicalism in the United States: Mapping Cultural Change since 1970* (Columbia: University of South Carolina Press, 1996), p. 72.

2. Bernard Ramm, *Special Revelation and the Word of God* (Grand Rapids, Mich.: William B. Eerdmans, 1961), pp. 44–48; Gerald A. Larue, *The Supernatural, the Occult, and the Bible* (Amherst, N.Y.: Prometheus Books, 1990).

3. From Abu Said, *The Secrets of Oneness*, quoted in F. E. Peters, *A Reader on Classical Islam* (Princeton, N.J.: Princeton University Press, 1994), pp. 318–19.

4. James McClenon, *Wondrous Events: Foundations of Religious Belief* (Philadelphia: University of Pennsylvania Press, 1994). Universality is socially modified; paranormal beliefs *increase* among those not rooted in communal religion. William Sims Bainbridge, "Wandering Souls: Mobility and Unorthodoxy," in *Exploring the Paranormal: Perspectives on Belief and Experience*, ed. G. K. Zollschan, J. F. Schumaker, and G. F. Walsh (Bridport, UK: Prism, 1989).

5. Harvey Cox, *Fire from Heaven: The Rise of Pentecostal Spirituality and the Reshaping of Religion in the Twenty-first Century* (Reading, Mass.: Addison-Wesley, 1995).

6. Paranormal claims sometimes go against common sense. Our mourning at deaths, for example, reflects a commonsense belief that death is the end, even though we also profess belief in immortality. On the other hand, folk psychology is dualistic, which supports paranormal beliefs. See Nicholas Humphrey, *Leaps of Faith: Science, Miracles, and the Search for Supernatural Consolation* (New York: Basic, 1996).

7. Janet Oppenheim, *The Other World: Spiritualism and Psychical Research in England, 1850–1914* (Cambridge: Cambridge University Press, 1985), chaps. 3, 4.

8. R. M. Burns, *The Great Debate on Miracles: From Joseph Glanvill to David Hume* (Lewisburg, Pa.: Bucknell University Press, 1981).

9. *Catechism of the Catholic Church* (English translation, Boston: Libreria Editrice Vaticana, 1994), ¶156.

10. Joseph Banks Rhine, *New World of the Mind* (New York: William Sloane, 1953), p. 185.

11. Gary R. Habermas and J. P. Moreland, *Immortality: The Other Side of Death* (Nashville, Tenn.: Thomas Nelson, 1992), chaps. 2, 3; William Hasker, *The Emergent Self* (Ithaca, N.Y.: Cornell University Press, 1999). Dualism is not confined to religious philosophers; e.g., Karl R. Popper and John C. Eccles, *The Self and Its Brain: An Argument for Interactionism* (Berlin: Springer-Verlag, 1977), but is rare otherwise.

12. John Beloff, "Historical Overview," in *Handbook of Parapsychology*, ed. Benjamin B. Wolman (New York: Van Nostrand Reinhold, 1977), p. 21; John Beloff, *Parapsychology: A Concise History* (New York: St. Martin's Press, 1993), pp. 215–25.

13. See Michael Stoeber and Hugo Meynell, eds., *Critical Reflections on the Paranormal* (Albany: State University of New York Press, 1996); note particularly contributions by Donald Evans, David Ray Griffin, and Terence Penelhum.

14. John J. Hearney, *The Sacred and the Psychic: Parapsychology & Christian Theology* (New York: Paulist Press, 1984).

15. McClenon, *Wondrous Events*, pp. 151–63.

16. Gunapala Dharmasiri, *A Buddhist Critique of the Christian Concept of God* (Antioch, Calif.: Golden Leaves, 1988), p. 179.

17. Oppenheim, *The Other World*, p. 90.

18. David Ray Griffin, *Parapsychology, Philosophy, and Spirituality: A Postmodern Exploration* (Albany: State University of New York Press, 1997).

19. Antony Flew, *Atheistic Humanism* (Amherst, N.Y.: Prometheus Books, 1993), chap. 3; Antony Flew, "The Problem of Evidencing the Improbable, and

the Impossible," in Zollschan et al., *Exploring the Paranormal*; David Hume, "Of Miracles," reprinted in David Hume, *Writings on Religion*, ed. Antony Flew (La Salle, Ill.: Open Court, 1992). On "limiting principles," see C. D. Broad, "The Relevance of Psychical Research to Philosophy," reprinted in *Philosophy and Parapsychology*, ed. Jan Ludwig (Amherst, N.Y.: Prometheus Books, 1978). Note that Broad *defended* psychical research.

20. David Basinger and Randall Basinger, *Philosophy and Miracles: The Contemporary Debate* (Lewiston, N.Y.: Edwin Mellen, 1986), chap. 2.

21. Gordon Stein, *The Sorcerer of Kings: The Case of Daniel Dunglas Home and William Crookes* (Amherst, N.Y.: Prometheus Books, 1993).

22. Some parapsychologists still argue this way; e.g., John Beloff, "The Skeptical Position: Is It Tenable?" *Skeptical Inquirer* 19, no. 3 (1995): 19.

23. Leonard Zusne and Warren H. Jones, *Anomalistic Psychology: A Study of Magical Thinking*, 2d ed. (Hillsdale, N.J.: Lawrence Erlbaum, 1989), pp. 61–64; Victor J. Stenger, *Physics and Psychics: The Search for a World Beyond the Senses* (Amherst, N.Y.: Prometheus Books, 1990), p. 102.

24. Zusne and Jones, *Anomalistic Psychology*, chaps. 4–9; James E. Alcock, *Parapsychology: Science or Magic?* (Oxford: Pergamon, 1981), chaps. 2, 3.

25. Joe Nickell, *Looking for a Miracle: Weeping Icons, Relics, Stigmata, Visions & Healing Cures* (Amherst, N.Y.: Prometheus Books, 1993), pp. 64–65.

26. Ibid., pp. 227–29.

27. Ibid., pp. 176–81; Zusne and Jones, *Anomalistic Psychology*, pp. 117–18.

28. Humphrey, *Leaps of Faith*, p. 88.

29. Popular skepticism is much less easy to find than pro-paranormal material. Still, there are skeptical voices. James Randi has written a number of books, including *Flim-Flam! Psychics, ESP, Unicorns and Other Delusions* (Amherst, N.Y.: Prometheus Books, 1982). Joe Nickell investigates forensic and paranormal mysteries; his best-known effort is *Inquest on the Shroud of Turin* (Amherst, N.Y.: Prometheus Books, 1987). Martin Gardner penned a classic of modern skepticism, *Fads and Fallacies in the Name of Science* (New York: Dover, 1957), and has been a prolific critic of the paranormal ever since. Most prominent modern skeptics have been associated with the Committee for the Scientific Investigation of the Claims of the Paranormal (CSICOP).

30. George R. Price, "Science and the Supernatural," reprinted in Ludwig, *Philosophy and Parapsychology*, followed by responses by Soal and Rhine.

31. C. E. M. Hansel, *The Search for Psychic Power: ESP and Parapsychology Revisited* (Amherst, N.Y.: Prometheus Books, 1989), chap. 8; Paul Kurtz, *The Transcendental Temptation: A Critique of Religion and the Paranormal* (Amherst, N.Y.: Prometheus Books, 1991), pp. 373–88. On the "problem of fraud," see Richard Wiseman and Robert L. Morris, *Guidelines for Testing Psychic Claimants* (Amherst, N.Y.: Prometheus Books, 1995); Paul Kurtz, ed., *A Skeptic's Handbook of Parapsychology* (Amherst, N.Y.: Prometheus Books, 1985), pt. 2.

32. David Marks, *The Psychology of the Psychic*, 2d ed. (Amherst, N.Y.: Prometheus Books, 1980); James E. Alcock, *Science and Supernature: A Critical Appraisal of Parapsychology* (Amherst, N.Y.: Prometheus Books, 1990); Richard Wiseman and Julie Milton, "Experiment One of the SAIC Remote Viewing Program: A Critical Re-evaluation," *Journal of Parapsychology* 63, no. 1 (1999): 3.

33. Daryl J. Bem and Charles Honorton, "Does Psi Exist? Replicable Evidence for an Anomalous Process of Information Transfer," *Psychological Bulletin* 115, no. 1 (1994): 4; Ray Hyman, "Anomaly or Artifact? Comments on Bem and Honorton," *Psychological Bulletin* 115, no. 1 (1994): 19.

34. Dean I. Radin and Roger D. Nelson, "Evidence for Consciousness-Related Anomalies in Random Physical Systems," *Foundations of Physics* 19, no. 12 (1989): 1499; Dean I. Radin, *The Conscious Universe: The Scientific Truth of Psychic Phenomena* (San Francisco: HarperEdge, 1997), chap. 4.

35. Richard Wiseman, Matthew D. Smith, and Diana Kornbrot, "Exploring Possible Sender-to-Experimenter Acoustic Leakage in the PRL Autoganzfeld Experiments," *Journal of Parapsychology* 60, no. 2 (1996): 97; Dick J. Bierman, "The PRL Autoganzfeld Revisited: Refuting the Sound Leakage Hypothesis," *Journal of Parapsychology* 63, no. 3 (1999): 271; Susan Blackmore, "What Can the Paranormal Teach Us About Consciousness?" *Skeptical Inquirer* 25, no. 2 (2001): 22.

36. Ray Hyman, *The Elusive Quarry: A Scientific Appraisal of Psychical Research* (Amherst, N.Y.: Prometheus Books, 1989), p. 237.

37. Brenda J. Dunne and Robert G. Jahn, "Consciousness and Anomalous Physical Phenomena," *PEAR Technical Note* 95004 (1995); Robert G. Jahn et al., "Correlations of Random Binary Sequences with Pre-Stated Operator Intentions," *PEAR Technical Note* 96003 (1996).

38. Suitbert Ertel and Kenneth Irving, *The Tenacious Mars Effect* (London: Urania, 1996); Claude Benski et al., *The "Mars Effect": A French Test of Over 1000 Sports Champions* (Amherst, N.Y.: Prometheus Books, 1996); Jan Willem Nienhuys, "The Mars Effect in Retrospect," *Skeptical Inquirer* 21, no. 6 (1997): 24.

39. David Reilly et al., "Is Evidence for Homeopathy Reproducible?" *Lancet* 344 (1994): 1601. Meta-analysis is a problematic technique in sciences with complex subjects; e.g., J. LeLorier et al., "Discrepancies between Meta-analyses and Subsequent Large Randomized, Controlled Trials," *New England Journal of Medicine* 337, no. 8 (1997): 536; Douglas M. Stokes, "The Shrinking Filedrawer: On the Validity of Statistical Meta-analyses in Parapsychology," *Skeptical Inquirer* 25, no. 3 (2001): 22.

40. Slight deviations from randomness are always a concern with REG experiments; Zusne and Jones, *Anomalistic Psychology*, pp. 173–76. Critiques of the major REG research programs, by Helmut Schmidt and the PEAR group, in their work up to the late 1980s, can be found in Hansel, *The Search for Psychic Power*, chaps. 15–18; and Alcock, *Science and Supernature*.

41. Alcock, *Parapsychology*, chap. 7.

42. Philip W. Anderson, "On the Nature of Physical Laws," *Physics Today* 43, no. 12 (1990): 9. See also William H. Jefferys, "Bayesian Analysis of Random Event Generator Data," *Journal of Scientific Exploration* 4, no. 2 (1990): 153, and the exchange between Anderson and Robert G. Jahn in *Physics Today* 44, no. 10 (1991).

43. Alan Cromer, *Uncommon Sense: The Heretical Nature of Science* (Oxford: Oxford University Press, 1993), chap. 8; Allan Franklin, "Do Mutants Die of Natural Causes? The Case of Atomic Parity Violation," in *A House Built on Sand: Exposing Postmodernist Myths About Science*, ed. Noretta Koertge (New York: Oxford University Press, 1998).

44. Gertrude R. Schmeidler, *Parapsychology and Psychology: Matches and Mismatches* (Jefferson, N.C.: McFarland, 1988). Such effects are compatible with ordinary psychology affecting possible mistakes in the experiments; Alcock, *Parapsychology*, pp. 124–25. See also Paul M. Churchland, "How Parapsychology Could Become a Science," reprinted in Paul M. Churchland and Patricia S. Churchland, *On the Contrary: Critical Essays, 1987–1997* (Cambridge: MIT Press, 1998).

45. E.g., Euan Squires, *Conscious Mind in the Physical World* (New York: Adam Hilger, 1990), pp. 201–30; Dean I. Radin, *The Conscious Universe*, pp. 284–85. Robert G. Jahn and Brenda J. Dunne, "On the Quantum Mechanics of Consciousness, with Application to Anomalous Phenomena," *Foundations of Physics* 16, no. 8 (1986): 721, proposes, with the aid of analogies to Jungian occult psychology, that consciousness is like a wave function.

46. Gustavo Castelo Branco, Luís Lavoura, and João Paulo Silva, *CP Violation* (Oxford: Clarendon, 1999).

47. Alice B. Child and Irvin L. Child, *Religion and Magic in the Life of Traditional Peoples* (Englewood Cliffs, N.J.: Prentice Hall, 1993), pp. 170–74.

48. McClenon, *Wondrous Events*, chap. 9.

49. P. M. H. Atwater, *Beyond the Light: What Isn't Being Said About Near-Death Experience* (New York: Birch Lane, 1994), p. 158.

50. Maurice S. Rawlings, *To Hell and Back: Life After Death–Startling New Evidence* (Nashville, Tenn.: Thomas Nelson, 1993); Habermas and Moreland, *Immortality*, chap. 6.

51. Carl B. Becker, *Paranormal Experience and Survival of Death* (Albany: State University of New York Press, 1993), p. 119. Other religious philosophers also cite evidence like past-life regression for survival; e.g., Hugo Meynell, in Stoeber and Meynell, *Critical Reflections on the Paranormal*, p. 40; Griffin, *Parapsychology, Philosophy, and Spirituality*, chaps. 4–8.

52. Ronald K. Siegel, "Life After Death," in *Science and the Paranormal: Probing the Existence of the Supernatural*, ed. George O. Abell and Barry Singer (New York: Charles Scribner's Sons, 1981).

53. Susan J. Blackmore, *Beyond the Body: An Investigation of Out-of-the-Body Experiences* (Chicago: Academy Chicago, 1992); Susan J. Blackmore, *In Search of the Light: The Adventures of a Parapsychologist* (Amherst, N.Y.: Prometheus Books, 1996).

54. Susan J. Blackmore, *Dying to Live: Near-Death Experiences* (Amherst, N.Y.: Prometheus Books, 1993).

55. Allan Kellehear, *Experiences Near Death: Beyond Medicine and Religion* (New York: Oxford University Press, 1996).

56. Nicholas P. Spanos, *Multiple Identities and False Memories: A Sociocognitive Perspective* (Washington, D.C.: American Psychological Association, 1996), pp. 135–41; Robert A. Baker, *They Call It Hypnosis* (Amherst, N.Y.: Prometheus Books, 1990), pp. 225–37. Nonhypnotic memories of past lives, as promoted by parapsychologist Ian Stevenson, are also suspect; Paul Edwards, *Reincarnation: A Critical Examination* (Amherst, N.Y.: Prometheus Books, 1996).

57. Susan Blackmore, "Near-Death Experiences," in *The Encyclopedia of the Paranormal*, ed. Gordon Stein (Amherst, N.Y.: Prometheus Books, 1996). Defenders of a survivalist interpretation of NDEs (e.g., Heather Botting and

Stephen E. Braude in Stoeber and Meynell, *Critical Reflections on the Paranormal*) claim support from the variety in NDE experiences, saying, for example, that medical explanations cannot account for both positive and negative NDE experiences. However, a naturalistic account is not committed to a single cause leading to uniform NDEs in every circumstance.

58. Hayden Ebbern, Sean Mulligan, and Barry L. Beyerstein, "Maria's Near-Death Experience: Waiting for the Other Shoe to Drop," *Skeptical Inquirer* 20, no. 4 (1996): 27.

59. Owen Flanagan, *Consciousness Reconsidered* (Cambridge: MIT Press, 1992), pp. 54–55.

60. Hoyt L. Edge et al., *Foundations of Parapsychology: Exploring the Boundaries of Human Capability* (Boston: Routledge & Kegan Paul, 1986), pp. 302–10; Thomas S. Kuhn, *The Structure of Scientific Revolutions*, 2d ed. (Chicago: University of Chicago Press, 1970), p. 151. See also Charles T. Tart, "Hidden Shackles: Implicit Assumptions that Limit Freedom of Action and Inquiry," in Zollschan et al., *Exploring the Paranormal*.

61. E.g., Georges Rey, *Contemporary Philosophy of Mind* (Cambridge: Blackwell, 1997); James H. Fetzer, *Philosophy and Cognitive Science*, 2d ed. (New York: Paragon, 1996), chap. 5; Daniel C. Dennett, *Kinds of Minds: Toward an Understanding of Consciousness* (New York: Basic Books, 1996).

62. Paul M. Churchland, *A Neurocomputational Perspective: The Nature of Mind and the Structure of Science* (Cambridge: MIT Press, 1989), chaps. 1, 6; Churchland and Churchland, *On the Contrary*, chaps. 1–4. For a critique, see Stephen P. Stich, *Deconstructing the Mind* (New York: Oxford University Press, 1996).

63. Robert Merrihew Adams, "Flavors, Colors, and God," in Robert Merrihew Adams, *The Virtue of Faith and Other Essays in Philosophical Theology* (New York: Oxford University Press, 1987).

Chapter Seven

Of Mystics and Machines

The world you saw was real, Rumi, not an apparition, etc.
It is boundless and eternal, its painter is not the First Cause, etc.

And the best of the rubaiyat your burning flesh left us
is not the one that goes, "All forms are shadows," etc.

She kissed me: "These lips are real like the universe," she said.
"This fragrance isn't your invention, it's the spring in my hair,"
she said.

"Watch them in the sky or in my eyes:
the blind may not see them, but the stars are there," she said . . .

—Nazım Hikmet, *Rubaiyat, First Series 1 and 6* (1947)

Mystics, or at least some of them, claim a direct experience of God. They also produce exasperatingly verbose poetry about their indescribable God, but that is part of the mystique of mysticism. The more incomprehensible the verbiage, the surer the sign of an encounter with a God wholly other than us. However, when we study mystical experience rather than sit around being impressed, we learn about our brains—not some sort of ultimate reality. For all our sublime experiences or powers of reason, we are biological machines.

FEELING THE SPIRIT

God should have been a necessary being; declared by the stars above, revealed by prophets, heralded by miracles. Unfortunately, as we learn more about the world, God disappears from public view. We might still feel a hint of an infinite power in an inspiring church service, or lose ourselves in a Quran recitation. But such feelings are fragile, ready to evaporate when we step back into a world of speeding automobiles and noisy children.

Emotion is, of course, vital to religions; many strive to make even everyday life a religious experience. We may not always feel the spirit, but religions still weave all that is significant to us into a whole way of life. Our Gods are never just ways to explain the world. Moreover, an intense religious experience is psychologically compelling in a way no involved analysis can be.[1] The trouble is, while vital to spiritual life, religious experience is rather feeble as an argument. A prophet may feel an inner certainty that the Space Brothers are watching over us from their flying saucers, but this does not add to the credibility of her claims. She might reach the heights of rapture when contemplating the coming fleet of UFOs, but her hopes are still wildly unlikely. If anything, beliefs backed up by rapturous emotions are doubly suspicious—it is too tempting to accept them uncritically.

Still, private spiritual experience is where many philosophers and theologians make their stand. In the early twentieth century, Rudolf Otto argued that while scientific rationality is no help in reaching God, we can get there if we start with our religious emotions. We approach God when we feel we are mere creatures facing a tremendously powerful mystery. We feel dependent on this "wholly other" or "numinous" aspect of reality. Such feelings are not derived from everyday experience or rational arguments; they are the "nonrational" basis of religion. There is more to this than a warm feeling in the stomach when contemplating God: religious emotions come from an encounter, an insight into the basic nature of reality which cannot be analyzed in ordinary rational ways.[2]

Otto's strongest examples of the numinous awareness behind religion are the experiences of theistic mystics. Indeed, mystics claim no mere warm and fuzzy feeling, but direct experience of ultimate reality. A churchgoer may believe Jesus is God; a mystic *knows*. However, experiential insight is not just the province of saints. Religious experience gives us another way of knowing, of capturing a spiritual dimension underlying ordinary physical reality. When not distracted by cars and kids, even nonmystics can catch glimpses of the deeper reality.

If religious feelings transcend reason, then it is no wonder our sciences fail religion. Spirit cannot be measured by instruments or trapped in theories. A scientific naturalist is like a blind person denying the colors of a sunset because she cannot hear or touch them. Ordinary religious people catch glimpses of light, and try to follow a path which leads them into seeing more and more. Then there are the mystics, virtuosos of spiritual awareness whose eyes have been opened. Drunk with God, they break free even of the words and concepts religions use to part the veil—they are like painters who best communicate their experiences in their art. They can get through, if ever so slightly, to someone who can see some light, and lead them to see a sunset like they do. But the reports of mystics mean nothing to those who are blinded by reason. By touch, they only experience a chaotic drivel of paint textures, and conclude it is all nonsense.

In the modern world, this is an attractive defense of religion. Science has chased God out of our picture of the world, leaving only a range of feelings. But religious scholars can now say this is exactly as things should be. Far from collapsing onto mumblings about spiritual experience, religion is returning to its roots in revelation, following what the logic of divinity demands. Denise L. and John T. Carmody describe mysticism as an experience of ultimate reality, and say that:

> The core truth about ultimate reality, as one gathers it from the traditional reports, is that only ultimate reality itself can render ultimate reality adequately: ultimate reality must be its own revelation and exegesis. The only way to know God is to experience God directly.[3]

Using our sciences to understand God, then, is bringing the wrong tools to the job. We must instead immerse ourselves in spiritual experience, opening ourselves up to a mystical way of knowing. A seeker must surrender her self to God; a critical attitude reserves judgment to the self, and so impedes spiritual progress. We must believe in order to see.

Mysticism has other advantages besides protecting religion from the spiritually blind. It is hard to reserve exclusive truth for any *one* tradition when the world is awash with so many different varieties of intense religious conviction. Mysticism might yield an esoteric core common to all religions. Each culture expresses its imperfect glimpses of ultimate spiritual reality through its own metaphors, but the ultimate reality, and the mystical experience leading to it, may be the same underneath.[4]

Infidels, of course, are not impressed. After all, another way of denying that any religious tradition has exclusive truth is to say they are all mistaken. If mystics hint that skeptics are blind, skeptics suspect mysticism is a delusion. Indeed, in the culture of science, "mysticism" usually stands for superstition or woolly-minded flights of fancy.

Nevertheless, there is more to the debate than two incompatible cultures trading insults. We can examine mystics' claims about their experiences and what they become aware of, and try to fit them into a picture of our world. We need not be bullied into abandoning our reason in doing so. But neither can we assume mysticism is a delusion because some mystics attack rationality. An adequate account of mystical experience might take us beyond ordinary reason. We do not have to grant absolute authority to either reason or mysticism as we know them.

We can start with the mystical claim to direct, unmediated acquaintance with ultimate reality. This is an attractive idea: knowledge unfettered by our limited ways of thinking. After all, our philosophical traditions also emphasize how our perspectives condition us. We should then seek truth by stripping away interpretations and attending to the clear, unadorned facts. Philosophers with an empiricist turn of mind departed from sensory experience. Others thought more ethereal ideas were more certain, like the Platonists who sought a stable reality behind changing appearances. With the European Enlightenment, philosophers tried to break free of the corruptions of custom and tradition. Descartes took this to an extreme, and declared that the foundations for knowledge were to be found only in those "clear and distinct ideas" present in a properly reflective state of mind.[5] But wherever philosophers found certainty, the idea was similar: theorizing distances us from reality. Even today, the popular mythology of science is that we do best by ignoring theoretical preconceptions, and letting reality speak for itself in the laboratory.

Mystics typically take this suspicion of theory further, seeking a point where even distinctions between knower and known melt away. As Richard H. Jones explains,

> Mystical knowledge is not "dualistic" knowledge of objects or events by a subject; it is not a knowledge of facts. . . . mystical wisdom is not mere intellectual assent to propositions concerning states of affairs in the world, but is a way of looking at the world tied to how we live and, in the case of mysticism, a path to mystical enlightenment that transforms the experiencer in a radical way.[6]

Mystical knowledge is supposed to be direct, unmediated by theories. And this knowing by acquaintance is intimately connected to religious emotions and a personality change which affects the mystics' whole outlook.

So a mystic defense of religion has a more basic quality than arguments that life is a miraculous creation or that every statement in the Quran is true. Even if someone were convinced such arguments were correct, this would only remove an obstacle to the real point of religion: getting in a right relationship with God. Religious experience suggests we can dispense with secondary issues. If we feel the spirit and get on a path to God, that is both our true objective and all the evidence we need.

Maybe so. But mystics also make straightforward fact claims like a cycle of rebirth or a Christian sort of God. These claims are central to mystics' own understanding of their experience;[7] we can question them without denying that mystics have extraordinary experiences which they describe as contact with ultimate reality. Our ideas of science and reason are not etched in stone; we may have to modify them in the process of examining mysticism. Mystical experience may be the gateway to God. But we cannot accept this blindly, even if we have such an experience. Even our most intense feelings of certainty can be mistaken.

THIRTY-ONE FLAVORS OF ULTIMATE REALITY

Mysticism certainly generates a lot of intimidating rhetoric. But it is not clear that if we shave our heads, wear a hair shirt, and rein in our critical devils, we achieve anything real. After all, even some religious traditions are ambivalent about esoteric experiences. Protestants who think salvation is accessible to all through a plain reading of scripture do not have much use for mysticism. Muslims who emphasize the Quran and Islamic Law are often suspicious of Sufi mysticism, especially since it is so often accompanied by unorthodox magical beliefs. Private revelation can be dangerous; the official Word of God must prevail.[8]

However, even unmystical traditions turn out to have mystic episodes tied up with their foundations. Early Christians cultivated ecstatic experiences. In 2 Cor. 12:1–10 Paul tells us he was taken to the third heaven and returned with secrets he could not reveal. Muslims speak of Muhammad's Night Journey, where he ascended to the Throne of God. Jews and Muslims developed similar techniques of ascending to heaven, even up to the Throne, and they returned to describe the journey in paradoxical language. Anyone who wants to argue that mysticism lies at the heart of Abrahamic religions can at least point to some pervasive mystical practices.[9]

Religious suspicions of mysticism also have a point, however. Though many have life-changing spiritual experiences, unless constrained by an official script, they do not always testify to the same reality. Mystics used to encounter angels; many people today meet alien beings who abduct them in the night. Psychiatrist John Mack is much impressed with their stories, even though they have difficulty describing their experiences. From a scientific perspective, UFO abductions look unreal, but Mack says, "it may be wrong to expect that a phenomenon whose very nature is subtle, and one of whose purposes may be to stretch and expand our ways of knowing beyond the purely materialistic approaches of Western science, will yield its secrets to an epistemology or methodology that operates at a lower level of consciousness."[10] Classic themes of inexpressibility and different ways of knowing emerge, though crudely, when Mack writes about the spiritual significance of UFO abductions.

If the ultimate reality behind spiritual experiences is to be God, we need a common theme beyond thrones and alien presences. A traditional favorite is the ineffability of the experience. Even lesser visions like UFO abductions suggest rational incomprehensibility. If God is wholly Other, something beyond human reason may be exactly what we are looking for.

So mystical theology usually takes the negative way, using paradoxical language highlighting how God is beyond all description. Classical metaphysics already ties itself into knots trying to convey the unsurpassable omni-everything greatness of God. Negative theology proclaims a God so great that all descriptions are inadequate. Concepts can only limit God. We cannot even say "God exists," as that might mean God is merely one real object among others. The logic of perfection, of absolute transcendence, drives negative theology to state only what God is not: not material, not limited in power, not nonexistent, and so on.

Since metaphysicians regularly ignore reality in search of Reality, and since the perfections of classical theism easily lead to paradoxical claims, the God of the philosophers and the God of the mystics should fit together very well. Indeed, ancient Greek philosophy culminated in the mystical theism of the Neoplatonists. They produced endlessly obscure doctrines about deeper realities, but all that merely hinted at the true God. As Proclus put it, "the One transcends all discursive knowledge and intellection and all contact. And only [mystical] unification brings us near the One, since just because it is higher than any existence it is unknown."[11]

Christian theology, and later Islamic philosophy, recruited Neoplatonism to defend the faith. The negative way set God beyond all possible criticism. Even saying "not this, not that" was not good enough; Pseudo-Dionysius described the nature of his God by saying:

> It is beyond assertion and denial. We make assertions of what is next to it, but never of it, for it is both beyond every assertion, being the perfect and unique cause of all things, and, by virtue of its preeminently simple and absolute nature, free of every limitation, beyond every limitation; it is also beyond every denial.[12]

Even the concepts of metaphysics, in other words, are only tools to frame what is always beyond concepts.

Hyperventilating prose piling one negation on top of another is a good way to drive critics away—from exhaustion, if not exasperation. The trouble is, negative theology also sets God beyond creation, immortality, salvation, revelation—everything we expect a God to be good for. In popular religion God makes a difference; the negative God seems only to fry the brains of mystics and metaphysicians once in a while. Purely negative language gives us no context to understand what we are talking about. Ideas become vague, slippery, content-free labels as we detach them from the network of concepts in which they take meaning. Consider "democracy." We have no ten-point checklist to test whether a country qualifies as a democracy, only prototypical examples like France or Sweden, usual features like multiparty elections, and political theories about how people express consent. If we start disconnecting "democracy" from our various examples and theories, it quickly becomes only a label with a positive feeling attached to it. Any dictatorship can claim it in its propaganda. Similarly, "God" is hopeless outside a network of religious concepts and fact claims. The purely negative God needs one more denial: it does not make sense.

And in fact, our religions do not rely on mindless negation. They set their Gods in a network of concepts which, while preserving God's distance, also allow us to make some sense of the mystery. For example, the Catholic God is unknowable "in essence"—whatever that means—but not totally beyond comprehension. There are *mediators* which bring us closer to God. Jesus, Mary, angels, the saints, and Church personnel are personal intermediaries; plus the whole idea of God is mediated through a network of claims including miracles, the accuracy of scripture, the created nature of the world, and so forth. If a critic asks too many inconvenient questions, she can be reminded that God is unknowable. Otherwise, the God who in "his infinite simplicity" cannot be captured in language turns out to be just approachable enough to back up Church authority.[13]

Neoplatonism is also an elaborate conceptual scheme which makes an unknowable God intelligible. Platonists envision a hierarchically structured reality. Those things which are cognitively clear, less subject to change, morally superior, and more causally potent are also more real. God is ultimate perfection, the only fully real One. Our world, the "sensible realm," is full of suffering and change, being a lesser reflection of a more perfect "intelligible realm." The intelligible realm is apprehended by Reason, and contains realities like mathematical truths by virtue of their rational certainty. The intelligible forms which lie behind the imperfect copies we ordinarily see are in turn caused by even higher realities. Existence is stretched out between God and Nothingness, where each level is causally dependent on a higher reality, and where in the end God is the only True Cause. This is not a God who one day gets bored of moving over the waters and decides to create something; God emanates lesser realities by impassive Necessity, like the sun illuminating the void. God first emanates personified "intermediate hypostases" like Reason, Soul, Wisdom, and so on. These intermediates then do the dirty work involving change and creation.[14]

The One God remains, of course, beyond ordinary comprehension. But with a Platonic background, we can read the negative theology of a Proclus or Pseudo-Dionysius in context. God is now a limit of a series of higher realities increasingly remote from our everyday concepts. By detaching our Rational Souls from the corruptible world and attending to the intelligible realm, we approach the realities above our sphere of being. Philosophy itself is like a mystical ascent toward the more Real, toward what cannot be captured by the sort of knowledge enmeshed in the world of accidents. The great unwashed might take an ascent to the Throne of God literally, but this is only an allegory of the mystic journey of the Rational Soul.

This is not just speculative metaphysics gone wild. Knowledge of Platonic realms should help explain our less perfect world which is, after all, caused by the higher realities. Philosophical contemplation should uncover the deeper principles behind appearances. Ancient Platonists thought a circle was the perfect shape, and so heavenly bodies should move in circles. In fact, they thought the heavens should be arranged in a series of spheres in perfect circular motion, all animated by cosmic souls emanating from God. Quite often Neoplatonism went hand in hand with astrology, since the divine macrocosm of the heavens was intimately connected to our personal spiritual microcosm.

Western mystical theology has always hammered on how God was beyond comprehension. However, the negative way itself only makes sense within complex conceptual schemes. And today, these schemes appear to be as deeply mistaken as the occult sciences they generated.

It may be tempting to say that grandiose metaphysical scenarios like Neoplatonism are just interpretations tacked onto a simple experience of God. Even though the most impressive mystical writings are drenched in metaphysics, we should eliminate the middlemen and look at the testimony of raw experience. Let us take the holy men whom ordinary people acknowledge as spiritual masters, and ignore the metaphysicians.

In folk mysticism, spiritual experience is closely tied to magic. Holy men perform miracles; and though such Low Magic is not supposed to be the goal, miracles confirm that the mystic is on the right path and lead to "more ineffable revelations."[15] While the Neoplatonists' incomprehensible God is a limit in a hierarchical reality, folk mysticism finds God in a limit of progressively more powerful and deeper paranormal phenomena. And this God is no more beyond concepts than the metaphysical version, since folk theories permeate popular mysticism, especially folk psychology about "direct experience" and feelings of certainty.

If interpretation is so intimately involved in mysticism, we should wonder how much our spiritual experiences are culturally conditioned. UFO abductees, for example, do not just have experiences which they narrate as an abduction story. Like past-life regression memories, an abduction tale is usually fashioned in therapy. UFO mythology is a part of our cultural environment, absorbed by both the abductee and the therapist. They bring a story template to the sessions where they forge a narrative, and their joint effort feeds back into the UFO culture to reinforce or change the standard story.[16] There is no raw, interpretation-free abduction experience.

Similarly, recent scholars have emphasized how mystical experience is constructed within religious cultures. First of all, mystical traditions have histories. For example, ancient Hebrew religion was full of mystic episodes. Especially in the early parts of the Bible, we find ecstatic experiences of God, but these are not described in terminally obscure language. A prophet might say "my own eyes have beheld the King, Yahweh of Hosts" (Isa. 6:5); God could be seen, and his messages were fearfully clear. Later Judaism revolved around *texts* which needed convoluted interpretation. God now communicated in obscure language which might require divine inspiration even to understand.[17]

Medieval Jewish mysticism continued this trend. Alongside early themes like a heavenly ascent and entry into the realm of angels, holy writ was central. As Steven T. Katz describes, the very letters of Torah have a magical significance in Jewish mysticism. Many Sufis treat the Quran similarly, creating elaborate schemes of letter magic; with Hindus, Vedic sounds are the magical, creative words. Layers of mysterious meaning and metaphor in texts run together with the depths to be plumbed by progress in mystical experience. Jewish mystics use texts like the biblical erotic poem, the Song of Songs. Their tradition trains them to reflect on these, to experience an undescribably great supernatural person, and to speak of this in a conventional language laced with the obligatory unknowables. Christians use the Song to create encounters with a Christ-like love—not the Jewish experience in different theological garb, but a distinctly Christian encounter. Religious cultures direct mystics to experience a particular "ultimate reality." Their cultural template helps shape their experiences, which feed back into the religious culture to reinforce or change the standard pattern. Katz denies that there is any raw, theory-free mysticism. Culture precedes and shapes the experience, as well as providing the resources to interpret it afterward.[18]

In support of his views, Katz points out how mystics in different traditions report different, highly specific details of ultimate reality:

> All these mystical personalities intended and experienced, they had *knowledge by acquaintance*, what their communities taught as *knowledge by description*. They had existential knowledge of what their coreligionists knew only through propositions. This existential knowledge moreover is quite specific, even to confirming minutiae of theological doctrine. For example, Theresa of Avila not only encounters Jesus but also receives existential verification of the doctrine of the Trinity, while Jewish mystics always experience reality as conforming to the *halachah* (Jewish Law). This is to say, experiences are highly specific, more often than not reinforcing the structures of established religion down to its most technical doctrinal detail.[19]

These are not culture-free, self-interpreting experiences. Mystics give us thirty-one flavors of ultimate reality. And these are conflicting realities, not an undifferentiated rush of ecstatic feeling.

Even some sort of God is not constant in mysticism. Classical Buddhist ultimate reality, which has as much mystic testimony behind it as any other flavor, is not theistic. Many Buddhists suggest that down deep, the world is but change and accidents. Our lives are changed by mystical experience, but mainly because, if we push far enough, we find that even our self breaks apart. Buddhist philosopher Gunapala Dharmasiri argues that mystical truth lies *beyond* experiences of Unity and Love.[20] In fact, Buddhist mysticism often harbors flashes of rationality. While some Buddhist sages adopt a heavy-handed negative language to describe the "emptiness" they seek, others charge them with "mistaking purely negative discourse for profundity" and advocate a more balanced approach.[21] A modern skeptic who thinks we live in an accidental world and who is convinced that the self is but a construct in the brain could agree with much in early Buddhist doctrines, though parting company about the existence of a cycle of rebirths from which we strive to escape.

We need elaborate conceptual schemes to speak of an unknowable God beyond our concepts. And if we seek unmediated experience of divine reality, we end up with conflicting certainties. Mysticism may still be a path to God, but we cannot rely on experience alone, no matter how sublime and ineffable. Defenders of a mystic God have to find ways to criticize rival flavors of ultimate reality, or find a common core to mysticism which amounts to more than negative talk and feelings of certainty. Otherwise, it looks like mysticism is constrained not by any ultimate reality, but by the accidents of history and human psychology.

BEYOND THE BRAIN?

It is hard to distill a mystical essence out of particular religious cultures. But the similarities of mystical traditions are as real as their differences. Though they disagree on its nature, mystics at least agree they taste an ultimate reality which is spiritually significant.

Mystical testimony, then, might still be firsthand evidence supporting a God, even if it is fallible. Theists can claim rival mysticisms are mistaken. A crude example would be a conservative Christian who believes that some of us encounter angels of God, while others are misled by lying spirits. Truthful experiences are those which fit the Bible and Christian tradition. More sophisticated theologians claim that a theory including a God does a better job of accounting for mystical experience and tying it to an overall picture of the world. Steven Payne argues that theistic mysticism is more likely to involve true perception than a Hindu monism which identifies the self with a transcendent but impersonal One. Theism, he claims, makes sense of religious experience, while if "everything is already identical with the transcendent One, then it is not clear how there could be any sense of dread, fascination, and unworthiness in the face of the Wholly Other." Plus, if mystical experience totally identifies the self with the One, it is hard to see why the experience is so fleeting. Why return to the world of illusion? Theism might explain monistic experience; "an intense experience of union with God might be *felt* as a kind of monistic identification," though a Hindu is mistaken to stop there and call it ultimate reality.[22]

Nontheists in turn criticize theistic mysticism, like the Buddhists who insist meditators should go beyond union to reach emptiness. But now the debate is about how to explain mystical experiences. Anyone criticizing a rival mystic tradition is acknowledging that experience is fallible. Mystics can be mistaken, though they often *feel* they cannot be. In that case, we should also consider the prospects for a naturalistic explanation.

Common mystical themes most likely arise from our common psychology. For example, many traditions use sexual imagery to describe contact with ultimate reality, since the emotions and vocabulary of sexual love are a natural metaphor for a feeling of ecstatic unification. A mystic may use ordinary love as a starting point to eventually attain a relationship with her God for which "love" is a barely adequate word. However, sexual imagery also suggests mystical experience is based on a mundane psychological substrate. It is not difficult to imagine that some people might build their capacity for sexual love up to an intense spiritual feeling, all without an actual object of this transformed love. In fact, the suppression of normal sexual expression common to many mystical disciplines is probably just the sort of thing to prepare for such an experience.[23]

So psychology promises to help explain mystical experience without going beyond the brain. And in any case, psychology should help us find some order in the grab bag of experiences called mystical. A shaman ingests a hallucinogen and journeys into a deeper reality, a contemplative withdraws into herself. Some mystics flog themselves, some attend academic seminars. We need a better sense of how their experiences are similar and different.

Figuring out mysticism is difficult, since even ordinary experience is very complex and nowhere near adequately understood. But we can start with a rough map. Many scholars recognize two broad forms: extrovertive, visionary episodes; and introspective, "depth mystical" experiences. Visionary mysticism is better known in popular religion. Saints encounter Jesus; prophets are transformed by the presence of God or a revealing angel. Like hallucinations, visions take place when we have high levels of cognitive and physiological activity. Techniques to get a vision often induce a hyperaroused state but cut down on external stimulation, so a stable external reality no longer occupies our attention. Depth mystical experiences, in contrast, come about when our level of arousal is abnormally low.[24] Introspective mystics typically use techniques to "clear the mind" and withdraw into a cloud of unknowing. Meister Eckhart's mysticism seems to have been of this sort:

> Therefore, if God is to speak his word to the soul, it must be still and at peace, and then he *will* speak his word and give himself to the soul and not a mere idea, apart from himself. Dionysius says: "God has no idea of himself and no likeness, for he is intrinsic good, truth, and being." God does all that he does within himself and of himself in an instant. . . . God acts without instrumentality and without ideas. And the freer you are from ideas the more sensitive you are to his inward action. You are nearer to it in proportion as you are introverted and unself-conscious.[25]

In depth mysticism, the seeker meets God not in a fiery vision which rips her apart and puts her back together differently, but in a deep quietness. The experience appears to be centered on a "pure consciousness event" (PCE), where, as Stephen Bernhardt describes it, "The subject is awake, conscious, but without an object or content of consciousness—no thoughts, emotions, sensations, or awareness of any external phenomena. Nothing."[26]

Meister Eckhart probably was familiar with the PCE; it was likely an aim of his spiritual discipline, to be sought after and interpreted in Christian terms. Eckhart locates his experience in a Neoplatonism-influenced Christian metaphysics, even quoting Pseudo-Dionysius. He ties the content-free awareness of a PCE to the God beyond all concepts. But a PCE without concepts—without even visions of deep reality or feelings of an overwhelming presence—does not guarantee the truth of Eckhart's conceptual scheme. Though we might find a place for God in an explanation of PCEs, a PCE is by no means a direct experience of God.

And in fact, naturalistic psychology gives us a better handle on PCEs. Robert Forman describes how an introspective mystic uses a process of "forgetting." Through repetitive mental routines, she begins to turn off her cognitive and conceptual thinking, producing an effect similar to what we see in sensory deprivation experiments. "Meditators and mystics practice a technique in which they recycle a constant mental subroutine. This technique serves as a catalyst to enable the mind to come to 'forget' all thought and sensation."[27] It seems the neurological basis for attention can continue to function for a while even when those parts of the brain dealing with cognition and memory are swamped by a recycled subroutine. Because consciousness is not a unitary, magical thing, we can have awareness without content. Attention can sometimes go its own way even without anything to attend to.

Introspective mysticism is a bit too free of concepts to support a God.

Visionary mysticism is more promising. Here we have encounters with an overwhelming presence, the self melting away in divine love, hints of spiritual realms increasingly difficult to describe by ordinary language. It seems this is exactly what to expect from experiences that bring us closer to God.

Visionary episodes are similar to other extraordinary experiences that we think of as paranormal events; for example, near-death experiences. People who have had NDEs often report that they felt more real than everyday reality, and that they emerged transformed. They met a loving presence, time became unreal to them, their selves lost their boundaries and even broke apart completely. These are mystical themes. This suggests that mystical experiences, like NDEs, take place when the normal function of the brain is disrupted in certain ways. Mystical traditions cultivate techniques to induce these experiences, and fit their transforming personal significance into a picture of ultimate reality. The mysticisms of different traditions have much in common, since they rely on similar states in similar brains. But the experiences are also very much entangled with the traditions, since little in our brains is untouched by our cultural environment.

Another clue that mystical experiences are due to anomalous brain states comes from observing the effects of drugs. Traditional peoples often use psychoactive substances to enter the spirit realm, and the Western occult tradition is not averse to dabbling in psychedelic mysticism. This does not mean encountering Christ or Krishna is a matter of selecting the right drug; neither does it mean drug-induced visions are the same as what saints see. Crudely messing with the brain's chemistry is bound to have different effects than what native brain endorphins produce after we practice mystical techniques within a supportive religious environment.[28] Still, it is impressive how many weird and ineffable aspects of mystical experience show up in drug-induced states. In Walter Pahnke's famous Good Friday Experiment, seminary students were given either a placebo or a psychedelic drug. All immersed themselves in a religious environment of chapel, prayer, and meditation. The psychedelic group's experiences closely echoed classic mystical accounts; many reported transformed lives in a follow-up study.[29]

The psychedelic evidence is suggestive, but we would like to know more about how the brain behaves in mystical experiences. For example, we know that the temporal lobe of the brain has an important role in time perception, and that distorted time perception and feelings of timelessness are typical in mystical experiences. Indeed, by stimulating the temporal lobe, experimenters find they can reproduce not only time distortion, but often other mystical feelings like an overwhelming presence as well. The amygdala, a part of the brain involved with very basic affective emotions, is tightly connected to the temporal lobe along with the hippocampus, and they appear to act in concert during mystical experience. Naturally occurring abnormalities in temporal lobe function also give us clues about mystical experience. Sufferers from temporal lobe epilepsy, for example, have frequent visions, and are extremely religious and often hypersexual. Chronic hyperstimulation by mystical techniques produces similar results. In our modern, secular world we often see such conditions as pathologies, but in a different social environment they would be valued signs of the visionary.[30]

We still do not have a fully detailed theory telling us exactly what happens in visionary states. But we know enough to say that they belong together with those abnormal brain states accompanying phenomena like NDEs, drug highs, and temporal lobe seizures. We cannot expect a simple, straightforward theory of mysticism, since we are not studying a single, simple experience. But we have made progress, and have excellent prospects to learn more. What we have learned fits a naturalistic picture rather well. If we immerse ourselves in a mystical tradition and practice techniques to drive our brain into an anomalous state, we can enjoy a rush of feelings about ultimate reality. No Gods needed.

This does not yet end the argument. All experience involves the brain, but this does not mean it testifies to nothing beyond the brain. The early psychologist William James developed a philosophy which was both radically empiricist and thoroughly mystical.[31] He knew mystical episodes were accompanied by physical conditions creating abnormal brain states, including his own ineffable experience after inhaling nitrous oxide. But he argued that an unusual brain condition might be necessary to perceive ordinarily occluded realities. If sensory deprivation, fasting, or drugs help produce spiritual visions, perhaps such techniques are precisely what open the doors of perception. After all, ordinary perception works similarly. To see a cat and identify it as a cat, we need a properly prepared brain operating in the right conditions. And when we perceive the cat, something changes in our brain. If we had a full-blown neuroscientific theory telling us exactly what happens when we see a cat, this would not mean the cat is a figment of our imagination. Similarly, neuroscientists may one day find out exactly what happens in the brain during a mystical vision, but this will not decide whether our Gods are real.

At first it may be hard to take a mystic who believes that her experience opens a "third eye" seriously. But mystics seek a very extraordinary reality. Mysticism is supposed to break us out of our ordinary conceptual frameworks; giving the brain a good jolt might do just that. Psychologist Stanislav Grof claims psychedelic and "transpersonal experiences" provide mystic knowledge and transcend the "Cartesian/Newtonian conception of the world."[32] If so, we need something beyond the brain to explain mystical brain states. An everyday cat sighting is best explained by the independent existence of a cat. Perhaps the same goes for a vision of an angel.

This analogy does not quite work, however. According to the best of our knowledge, the cat is an external source of information. An angel is not. To see this, consider Daniel C. Dennett's argument that a truly strong hallucination is impossible, as is a virtual reality computer game which is as rich as real life.[33] Say we find a subject to don a virtual reality helmet, which shows her a lifelike picture of a sleeping cat. We need a lot of information to produce that picture, but this much is doable. However, when we allow her to move her head and see the cat from different angles, the amount of information we must feed her multiplies enormously. The situation gets completely out of hand when we allow the cat to move around independently. At some point, only the real world and a real cat can keep up with the massive amount of information we have to supply the subject. Better to short-circuit the whole process and just jolt her brain into feeling certain she saw a cat—a cat which will of course be difficult to describe clearly.

Hallucinations, then, should not be as rich as real life: the brain cannot produce so much information internally. The best we can hope for is to feed the brain some random noise under conditions of sensory deprivation and let the imagination elaborate. And even then, it is difficult to generate a coherent vision with stable facts which can be rechecked in detail.

Mystical visions are much more like hallucinations than cat sightings. A mystic is typically *passive*, responding to an outside spiritual force even when she initiates the encounter. If she sees an angel, she has little opportunity or desire to circle around it and get massive amounts of coherent information. She never returns with consistently accurate and surprising knowledge about the ordinary world which is supposed to depend on the supernatural reality she perceives. She may, in describing her experience, use vast amounts of obscure metaphysics but end up saying very little of substance. If she narrates a heavenly journey, it starts out as a tale imaginatively fashioned out of echoes of her training, probably in conditions of sensory deprivation. It becomes more concrete when retold and further connected to traditional story templates about mystical ascents.

Though visionaries are not much for producing information, they excel in emotion. Indeed, mystics often insist their experience does not result in information—mere "dualistic knowledge"—but in intense feelings and personal transformation. Maybe so. However, sublime feelings and even personal transformation can easily take place in a brain driven into an anomalous state. Nothing mysterious need participate.

We can criticize mystical claims, and ask how we might best explain mystical and paranormal experiences. And whether with UFO abductions, visions, or introspective experiences, we invariably find unusual psychological events which serve as ingredients for elaborate cultural constructions. We learn that our basic psychological capabilities like love, awareness, and person-recognition can be scrambled and disconnected from the outside world. We find changed lives, intense convictions, ambitious doctrines. We do not find any spiritual reality.

THE LIMITS OF LANGUAGE

Mystics discover too many flavors of ultimate reality. Mystical experience can be understood within the material world. These objections, in one form or another, are standard criticisms of mysticism. This is no surprise; after all, they *are* excellent reasons to remain skeptical. Still, we might be missing something. Though we can approach mysticism fairly well within a naturalistic framework, in their talk of paradox and ineffability, mystics gesture at something beyond all frameworks. Perhaps critics ignore the radical nature of mysticism, treating it as just another claim. The mystic challenges the arrogance of reason, denying that language can express everything.

This threatens to take us back to the negative way. But perhaps we should move beyond mystical theologies as well. Any conceptual scheme, it seems, is based on certain metaphysical presuppositions, or at least rules which tell us what fits and what does not. If we insist that any scheme—common sense, science, theology, whatever—is *the* correct way, we risk becoming ensnared in its rules. We end up like computers, rigidly following predetermined programs. Yet we live lives which are touched by creative novelty. Perhaps we should think of spirituality as being open to creativity, and see mysticism as a particularly impressive example of our ability to leap beyond rules.

In this way we can also understand why mystics talk in negatives and paradoxes. By "directing the listener away from the realm of objects," mystics try to break us out of our ordinary conceptual sets.[34] As a way of supporting a claim or conveying information, negative talk is useless. As a way of pointing to an unknowable God in a conceptual scheme like Neoplatonism, it is a failure. But paradoxical language might have a different purpose: breaking us free of the well-worn ruts of everyday thought. Mystics tell us we have reached the limits of language. They invite us to join the experience, but cannot say anything positive about what lies ahead. Theodore W. Jennings argues that everyday religious language often functions similarly. When we call out, "Oh God!" we are not doing theology, but instead using "God" as an expletive. Jennings believes this "signals the rupture of the structures of our experience and is therefore rather like swearing."[35] God hovers at the limits of language, as does mystical experience.

This is all very well—religion can make us see things differently, sometimes to the point of radical personal transformation. But as yet, none of this supports the notion that there is, in fact, a God. Mystics are jolted out of their conceptual sets, but they come back to earth and settle on a more intensely religious way of seeing the world. They then say their experience *validates* their new point of view. They claim that a direct contact with ultimate reality changed them; they do *not* say they went through an arbitrary rearrangement of conceptual furniture. And against such claims, the standard objections apply. If the experience itself is to count for God, we need something extra in it—something beyond language, yet not out of reach. Something like perception but different than acquiring information, which creatively changes our perspective.

To find what this something might be, consider an analogy used by Richard H. Jones. Say we have to describe our world to intelligent beings who live in a flat, two-dimensional environment. If they know geometry, we can bring up the concept of an extra spatial dimension. If they know some physics, we might even convince them they live in a thin slice of a larger three-dimensional universe. We can acquaint them with a different mode of existence by using familiar categories in "two-dimensional language." But the crucial experience is missing. No amount of talk can convey our effortless familiarity with 3-D life, our qualitative certainty that we move in 3-D. Perhaps language fails mystics for a similar reason.[36]

Indeed, many philosophers believe there is something ineffable about experiential qualities, or "qualia." Frank Jackson tells the story of Mary, a "color scientist" who knows everything about light and how the brain reacts to light. She knows what wavelengths will dominate in an eye looking at a red spot. She understands how the retinal cells respond, how the optic nerve conveys information to the brain, and the exact physical changes which take place in the visual cortex and other parts of the brain. But Mary has spent all her life in a room where everything is shades of grey. One day she steps out, and sees a tomato for the first time. And the experience, the redness, is something new. Since she already knew all *about* red, she cannot have acquired any new information. It seems, instead, that she has become directly acquainted with a quality which cannot be reduced to information.[37]

In that case, maybe mystics experience something qualitatively new, creative, beyond all conceptual sets—only something much more dazzling than a newfound color, more fundamentally important than discovering that one lives in a thin slice of a 3-D world. A spiritual quality which is the creative ground of our existence infuses reality. God might be simple beyond information, but qualitatively rich beyond imagination.

Before we get carried away, however, let us see if qualia are all they are advertised to be. This "redness," for example, seems to be a peculiarly useless addition to the physical world of light and brain states.[38] It just sits there looking pretty. And even that is dubious. After all, philosophers say our behavior would remain the same without the qualia, since the brain changes underlying an appreciation of red would not differ. Even in an aesthetic response, qualia come along for the ride without actually doing anything. Something so impotent is not a promising clue about an all-powerful God. Moreover, since ineffable qualia are so useless, we should suspect there is no ethereal redness apart from the physical brain.

Let us go back to Mary's adventures with color. There is an important *physical* difference between Mary's brain before she stepped outside, and after she saw red. Previously she knew how a brain would change when perceiving red light, but she herself had not undergone this change. Having a complete database about how brains respond to light does not automatically put her brain in the state of "seeing red." That can only be done firsthand, but it is still a physical event. As Owen Flanagan says,

> 'An experience of red' is not in the language of physics. But an experience of red is a physical event in a suitably hooked-up system. . . . Completed physics, chemistry, and neuroscience, along with a functional-role description, will explain what an experience of red is, in the sense that they will explain how red experiences are realized, what their functional role is, and so on. But no linguistic description will completely *capture* what an experience of red is like. That is only captured in the first person. You have to be there.[39]

It might seem that Flanagan promises too much, that it is still hard to see how anything physical can add up to the difference between red and blue. However, Austen Clark has shown that just with our current knowledge of psychophysics, we can go a long way toward bridging this gap and understanding how our "quality spaces" are constructed.[40] Our eyes have three kinds of color-detecting cell, so Paul Churchland suggests that "a visual *sensation* of any specific color is literally identical with a specific triplet of spiking frequencies in some triune brain system." This gives us "an effective means of expressing the allegedly inexpressible. The 'ineffable' pink of one's current visual sensation may be richly and precisely expressed as a 95Hz/80Hz/80Hz 'chord' in the relevant triune cortical system."[41] But of course, knowing this is not the same as placing ourselves in that brain state. Sometimes this will not even be possible. We might know exactly what a pigeon's color sensations are, but we cannot share them, since pigeons have four kinds of color-sensitive retinal cell. Though we might possess the same information as a pigeon, we cannot represent it the same way.

Such an approach promises to make sense of a lot which is otherwise mysterious. But we still might suspect that it conceals a huge jump from mechanical information processing to the rich qualities of conscious experience. So it might help if we can reduce this gulf to smaller gaps, and see how new qualities of experience might emerge in a physical world.

Imagine that a ship of colonists lands on a desert planet inhabited by life-forms which seek out mineral nuggets and "eat" them by a process releasing an intense blast of heat. The blast leaves behind a valuable liquid as a waste product. The liquid drains away quickly, so it has to be harvested immediately after the blast. The strange life-forms die when confined. So the colonists wander the desert, trying to detect rapid local temperature variations to avoid being killed, and to get to the precious liquid in time. They start out with hand-held optical temperature gauges, pointing them around and trying to catch surges in the readings. In other words, they do what we think of as mechanical data processing—work with sets of numbers. The temperature of the landscape, for them, has no "felt quality."

Carrying an instrument around is crude and inefficient. So the colonists devise a way to "see" temperature: a helmet with a small sensor mounted on top. The sensor rapidly scans the area in front, and then converts the data into a continually updated map projected on a transparent visor. So a temperature map is superimposed on the view of the terrain. Noticing that nothing on the planet is red, they decide to code temperature ranges in shades of red. "Red" is still a spiking pattern in the visual part of the brain, but for a trained forager, its meaning shifts. She now "sees" temperature without conscious attention, without having to crunch numbers. She deftly avoids blasts and quickly moves toward their centers once they pass. Like an expert driver "feels" the road and the condition of his car as an extension of his body, she has begun to "see" temperature.

The next generation uses biotechnology. Foragers acquire a permanent graft of an advanced temperature sensor, a "third eye." This projects signals directly into the optic nerve. Some neural rewiring connects "red" sensations to initiate reflex movements—to avoid very rapid blasts even without much training. The foragers' perception of red has become like a new sense for temperature; not only are we far from conscious number crunching, we should begin to think in terms of an entirely new experiential quality. Of course, this new sense still cannibalizes much of the old circuitry for red. But our bioengineers can keep working. They can give "red" back to normal vision, and add a new area to the brain, sort of like a monochromatic visual cortex. They can connect this directly to the third eye, and broadcast information from the temperature cortex to the rest of the brain, to be available to subroutines which handle attention and compete to assemble conscious experience.

At this point, the structure of the bioengineered brain will define a new quality space. The foragers will have a completely new sense, with new sensations linked to the temperature map of their environment.[42] But there is no point in the development process where qualia appear magically out of nowhere. Crude gauge reading and effortless experiential knowledge of a temperature map lie on a continuum—a continuum of increasingly sophisticated information processing.

If qualia are not as ineffably mysterious as they seem at first, this is bad news for mysticism. Once again, we are left with an experience which tells us about our brains, not a spiritual reality. It also means there is no such thing as knowledge beyond information, no change of perspective which can set us free of all theories. We are not mystic souls trapped in a world of dry, lifeless information. We are machines. In all their terror and magnificence, in all their rich depth, we have experiences *because* we are biological machines.

HOLY REASON

Robert Gimello observes, "Were one to subtract from mystical experience the beliefs which mystics hold to be therein confirmed and instantiated, all that would be left would be mere hedonic tone, a pattern of psychosomatic or neural impulse signifying nothing."[43] Theistic mysticism has plenty of hedonic tone, but few beliefs which can reasonably be called true. Mystics might insist finite human reason cannot grasp their truths anyway. But any fool can go beyond reason, and even a ridiculous idea can be stamped with sublime feelings of certainty. Experience alone simply does not amount to much.

So mystical arguments need a change of emphasis. Instead of seeking an experience which trumps reason, mystics can try and change our view of reason. Maybe mysticism goes beyond *mechanical* reason. Machine intelligence, if even possible, might be inferior to the best of human reason. And in what separates our intellect from mere computation, we might get a glimpse of what mystical insight may be.

Joining mysticism and reason is not a new idea. To the Platonic philosophers, reason was more akin to mystic insight than to plodding computation. Imagining a hierarchically structured reality, they also believed there were different levels of reason. The lowest faculties of the Soul were the sort of automatic tasks an animal might be capable of, keeping it alive but no more. Next came faculties like imagination and practical reason, the intelligence we use to negotiate the world of sensible objects. Still higher was the Reason philosophers cultivated, putting them in touch with the universal truths of the intelligible realm. And there was more. The great Muslim Neoplatonist Ibn Sina described a power of "holy Reason," granted by God only to a few prophets. Holy Reason apprehended universal truths directly, without need for learning or reflection. Perfected Reason was like inspiration or direct experience of Reality, not like computing or inquiry.[44]

Ibn Sina's version of holy Reason was also entangled with some peculiar metaphysical disputes, such as whether it could produce any knowledge of the sensible realm. But the idea that learning and computing inflicted only imperfect intermediate stages of Reason put deep roots into Western theism. Christian theology, for example, came to depict angels as intuiting facts directly and immediately. Even the Devil instantly knew all he needed about the universe and his own nature, and rebelled against God almost as soon as he was created.[45]

Modern philosophers can no longer get away with piling layer upon layer of Reality and Reason; certainly none claim, like some Neoplatonists once did, that Plato's philosophy was a divine revelation.[46] But Platonism runs deep in our intellectual culture. Many of us think our minds cannot be confined by machines. Many also suspect our minds are somehow involved with the deepest aspects of reality. And if we can no longer speak of holy Reason with ease, we can still find something mystical in how the inventions of mathematicians can describe our world so well. The inspiration of mathematicians and physicists might be something beyond a machine trying to blunder its way through life.

Physicist Roger Penrose, speaking for many mathematical Platonists, asks "how is it that the physical world indeed accords—and has accorded since the beginning of time—with such extraordinary precision with subtle and beautiful mathematical laws?" This mystery is a bit overblown, of course—our ideas of physical elegance are shaped by the interplay of theory and experiment; they do not spring from the brow of Platonists contemplating perfect circles. Elegant, "fundamental" physics is a framework for accidents, not a gateway to perfection. However, Penrose also sees that a serious Platonism requires a new theory of mind, one quite different from the ideas prevailing among neuroscientists and artificial intelligence researchers. He suspects that a solid classical world emerges from quantum physics not by a process of decoherence, but by some unknown feature of quantum gravity. He also proposes that consciousness is not a product of a system of classically interacting neurons, but is a quantum gravity phenomenon. If "consciousness could tie in with the behavior of the universe at its deepest levels, and be relevant even to the very geometrical structure of space-time," a Platonic view of the world might become more respectable.[47]

Penrose's physics and neuroscience is very speculative; neurophilosophers Rick Grush and Patricia Churchland describe his arguments as "merest possibility piled on merest possibility teetering on a tippy foundation of 'might-be-for-all-we-know's.' "[48] But Penrose has a deeper reason motivating his belief that mind is beyond the machine, so that some strange new physics and neurobiology must be involved in consciousness. In proper Platonic form, this reason comes from mathematics. Gödel's incompleteness theorem, according to Penrose, shows that human mathematical reasoning cannot be captured by any set of mechanical procedures. A computer executes an algorithm: a set of rules. Penrose thinks we have a nonalgorithmic mind—beyond even the most complex set of rules.[49]

No algorithm exists for certain functions. For example, no computer program can possibly implement Turing's halting function h, which is simply $h(p,n) = 1$ for programs p that halt when given input n, and $h(p,n) = 0$ for code which grinds on forever without producing any result. Of course, humans are also stumped by meaningful nonalgorithmic functions like h; indeed, there is no way to know $h(p,n)$ for all p and n without some sort of mathematical holy Reason. But Gödel's Theorem suggests we can achieve a less ambitious nonalgorithmicity. In any axiom system powerful enough to be interesting, Gödel tells us there are theorems which are true but which cannot be proven. For any theorem-proving algorithm we devise, a mathematician inspecting the algorithm can find a theorem which is true and yet which our algorithm cannot prove. All systems of rules have blind spots, yet it appears a mathematician is not locked into any formal system. She can always step outside of any system and examine it from the outside.[50]

Intuitively, we feel we are more than any set of rigid rules. A computer can only execute its program, but we should have free will—*we* make our choices, however much we may be influenced by our circumstances and our history. But if we are machines, this means our acts are ultimately physically determined. Like the computer bound by its instructions, we only do what the laws of physics drive us to do. But this position seems absurd, even self-refuting. If all we do is determined by physics, the very idea of a reasoned argument becomes a farce: the decision we make is shaped by the dynamics of some molecules, not the logic of the argument.[51]

This does not yet mean we can resurrect holy Reason. But if Penrose is correct, we may begin to think about it. If our special talent is breaking free of rules, mystics claim to break free of all conceptual sets. If Platonic reason mysteriously grasps the deep order of reality, Geoffrey Parrinder says, "Mysticism is not a plug for gaps as yet unfilled by science but, on the contrary, its conviction of the mystical unity at the heart of things may alone provide that order and continuity upon which all other studies depend."[52] Maybe science and mysticism can come together in a deeper harmony.

Artificial intelligence researchers are not impressed by Gödelian arguments. To say we are not machines is to make an interesting claim about the world, yet Penrose does not give us a test to tell us from machines. We do not know what in a human mathematician's performance could set her beyond any possible machine. Consider chess. A general algorithm for playing a perfect game exists, but the number of possible chess games to evaluate is so incredibly huge that no machine will ever implement it. But computers today still do a pretty good job playing chess. It is probably only a matter of time until computers will routinely defeat even the best human players. Algorithms need not be perfect to do a respectable job; indeed, we humans also seem to get by doing imperfect but good enough jobs.[53]

On the other hand, rule-bound machines *do* have blind spots, and it seems avoiding ruts should be an important aspect of intelligence. Something is missing when we compare human intelligence to a very complex computer program, something connected to the Gödelian intuition that we are not bound by rules.

This something, however, is quite unmystical: a touch of randomness turns out to be all we need to keep us—and machines, for that matter—away from blind spots.[54] At first this seems crazy. After all, randomness is the only thing more mindless than rigid rules. Platonists at heart, we think the essence of Reason is a lawful, transcendent order. In contrast, randomness is a total lack of pattern, as in a sequence of coin flips. Such chaos can only corrupt Reason. But if we ignore Plato for a moment, we find that randomness is rather interesting. A patternless function is maximally nonalgorithmic. It is a nonalgorithmic function which is meaningless, useless for everything but keeping us from following preset rules all the time. So if we want a nonalgorithmic intelligence, distinguished not by a magical knowledge of functions like Turing's h but by its ability to jump outside of any system of rules, randomness is just what we need. A nonalgorithmic machine uses randomness as a device to introduce novelty, as a way to break out of ruts.

Consider a very simple game. There are two houses; a cop starts the game in one and the robber in the other. In each turn, they independently decide either to stay where they are or move to the other house. If the cop ends up in the same house as the robber, she catches him and the game is over. Usually the cop will be successful within a few turns. But say she follows a search procedure described in the police manual. In that case, her search algorithm will have a blind spot: a robber who has managed to steal the manual will never be caught. All he has to do is follow the same rules, switching when the cop switches and staying put when she stays.

The police chief in charge of writing the manual is in the same Gödelian predicament as a mathematician trying to devise a perfect theorem-proving algorithm. The chief will notice that her search procedure has a blind spot. But every time she tries to patch up the police manual by working around the previous blind spot, she will introduce a new one. A robber can defeat any search algorithm by following those same rules.

If our chief knows some game theory, however, she will supply her officers with a coin to flip every now and then. Someone playing a game against an intelligent opponent has to guard against behaving too predictably, since her opponent might figure out her strategy and play accordingly. So sometimes it is best to act randomly.[55] In our simple game, all a cop has to do to guarantee she will eventually catch the robber is to decide whether to switch or stay by flipping a coin. In different games, randomness may be useful at different levels than just deciding moves. For example, a police chief might issue a new manual every week, but generate the search algorithm randomly. Mechanical computing includes not only algorithms but also arbitrarily complex combinations of rules and randomness. Whenever it needs to explore new ways of thinking or to avoid falling into ruts, a machine can start flipping coins.

Novelty only brings us part of the way to creativity. But we already know how to get there. Biological evolution creates wondrously complex life-forms, using random variation as a source of novelty. Like Darwin, we can start with an analogy to artificial selection. Our computers are very good at calculating 1 + 1 = 2, but they could just as easily be programmed to conclude 1 + 1 = 3. However, any program which did its sums so blatantly incorrectly could not survive long in the software market. Programs with an improved performance, on the other hand, have a selective advantage. Now, dispose of the artificial selection, and imagine a population of machines which have to survive and reproduce in a hostile environment. To various degrees, our machines will be able to represent their environment and communicate with one another. In a world where one machine next to another makes two, our machines would evolve toward consistently concluding 1 + 1 = 2. In fact, as Daniel C. Dennett argues, here we can see the beginnings of abstract reasoning, of machines which evolve the structures enabling them to respond to the logic of the arguments they present one another.[56] Perhaps machines following fixed rules cannot genuinely respond to arguments. Darwinian machines, however, can.

In the real world, evolution has an extra twist. Genes do not carry that much information. Yet brains which let machines represent their environment in detail, imagine possible future scenarios, and communicate with similar brains would be quite useful in certain ways of life. It would be best if these brains were not so rigidly rule-bound as to abound with blind spots which their competitors can exploit. So evolution adds another layer of natural selection, now in the brain. In our minds just as in biological evolution, random novelties plus selection produce creativity. Hence Darwinian thinking has become increasingly common in neuroscience and artificial intelligence. Scientists will debate the details for long years to come, but a good picture of our minds is already emerging. We are nonalgorithmic *machines*. The physical world is awash with noise, which is a source of random variation in the brain. We feed this randomness into variation-and-selection mechanisms and intricate learning strategies. The result is a machine which runs a complex, unpredictable "program" combining algorithms and randomness. Intelligence, like life, walks a knife edge between order and chaos.

Machines, then, can avoid the Gödelian trap. Even so, talk of Darwinian machines might seem too vague, however much this promises further progress. It turns out we can make a much stronger statement: the *only* way for our minds to reach beyond machines is through holy Reason, and we simply do not possess that. Next to Gödel's incompleteness theorem, we can also prove a completeness theorem showing that *every* function, no matter how nastily nonalgorithmic, can be expressed as a finite combination of algorithms and randomness. All nonalgorithmicity, in other words, comes from randomness. Mechanical nonalgorithmic strategies like a cop flipping coins to catch a robber depend only on the fact the coin is random. Which sequence of heads and tails actually turns up is immaterial. A Platonic intelligence can only improve on the machine by performing a task which requires a particular random sequence of heads and tails. In other words, it must compute a meaningful nonalgorithmic function like Turing's h, and the only way to do that is to somehow intuit the exact random sequence defining h. There is no middle ground between machine intelligence and holy Reason for Penrose or anyone else to occupy.[57] And holy Reason is only a fantasy. Nothing and nobody we know of has such godlike talents.

Before Darwin, most biologists were Platonists. Each individual reflected, more or less perfectly, the essence of its species. After Darwin, we learned to give accidents their due. Platonic essences melted away. And as we learn more about our minds, Darwin is once again eclipsing Plato. Human intelligence is a product of an accidental world, relying on those very accidents to work properly. If our minds have anything to do with fundamental physics, it is only that quantum mechanics is a source of randomness. So the debate between science and mysticism does, in fact, change our views of human reason—but not in a way useful to religion. Reason, we have found, is not a transcendent force governing the world, nor a mystic power of our Souls, but a way to muddle through in an accidental world.

ACCIDENTAL REASON

God has not changed much for the past few millennia. We still look around, ask where it all comes from, and conclude there must be a creator. Many of us still hope to get closer to God through miracles and spiritual experience. And metaphysicians still come up with Gods a Greek philosopher could have recognized without too much difficulty. Perhaps more of us today are sophisticated believers who speak gently of an infinite unknowable, suggesting our religions only symbolically hint at an inexpressibly deep Reality. But even these views echo ancient mystical theologies.

Modern science changes the ancient debates, though not because it opens a new front in an ancient war between Reason and Revelation. That war is more myth than reality. True, philosophers have always pulled out their hair trying to reconcile God with evil, and worried about revealed religions which do not even bother to be internally consistent. Even so, premodern philosophers rarely undermined faith. They occasionally thought orthodoxy depended too much on arbitrary authority, and often considered popular religion vulgar. But if philosophers disdained the aid of the priests, their transcendent Reason led to God anyway. Even this conflict of styles was no great issue; many Christians and Muslims brought Athens and Jerusalem together quite effectively.

Science changes the debate because it forces us out of our traditional conceptual sets. It has roots in previous ways of thinking, of course. Unlike Near Eastern mystical sects devoted to preserving the wisdom of a founder, Greek philosophy encouraged criticism; a student could disagree with the master. And unlike Eastern Christianity, the Western Church was never sufficiently mystical, so it tolerated interest in the material world. Chinese and Muslim civilizations, though advanced, never nurtured a critical tradition; Europe inadvertently started science off and failed to strangle it. Scientists took the critical attitude and the interest in the world, and made something new and bizarre. Modern science emerged as a historical accident—and a precarious and disruptive accident at that.[58]

Like scientists, mystics and metaphysicians realize common sense is not enough, and they pursue the reality behind appearances. But they desire certainty, and only the certainty which comes from a direct, theory-free apprehension of ultimate reality is good enough. As Ben-Ami Scharfstein points out, this is no way to understand the world:

> The non-mystic may be naïve in taking the world as it looks, yet if he does, he may begin to explain himself, the mystic, and the world. In ordinary experience, the world is more or less one; in mystical experience it is split, it lacks the possibility of at least intellectual coherence. Mystics try to sweep the world under the rug. Then they try to ignore the lumps in the rug.[59]

Science *starts off* with a naive view—our commonsense, folk-theoretical description of the world. Not because common sense has any special virtue, but because we can only start from where we happen to be. Science then goes further, often undermining common sense in the process. Though we test reality by experiments, this is not like a quest for pure experience unfettered by theory. Every measurement, every bit of "raw data," depends on a theoretical background, whether folk theories or sophisticated mathematical models. When we improve our theories, we not only fit the facts better, but we also go back and rethink what the facts are. "Mary saw a red object" is not the same fact in folk psychology and in a mature neuroscience. In the light of our best theories, "Mary encountered God" is not a raw fact but a mistake.

So science harbors a curious tension. On one hand, we hope that in refining our theories and experiments, we fall into a positive feedback loop.[60] In that case, we converge onto some very powerful and solidly confirmed theories. The planets really do behave as described by our celestial mechanics. Evolution is a theory, but also solid enough to call a fact without apologies. On the other hand, nothing is absolutely certain. Even our most basic assumptions, like an external world independent of our whims, are only starting points. If parapsychology had succeeded, we would have had to rethink the notion of a mind-independent external world. Not only would we know some new facts, but we would have had to reconsider how we do science.

This lack of certainty can be unnerving. After all, transcendent Reason was supposed to be automatically trustworthy. The accidental reason of a fully naturalistic science is a bit tarnished in comparison. We might wonder if we have pushed reason so far as to undermine itself—whether without some transcendental guarantee, reason is just an intellectual game with no real claim to produce truth. After all, as Paul Churchland remarks,

> Human reason is a hierarchy of heuristics for seeking, recognizing, storing, and exploiting information. But those heuristics were invented at random, and they were selected for within a very narrow evolutionary environment, cosmologically speaking. It would be miraculous if human reason were completely free of false strategies and fundamental cognitive limitations, and doubly miraculous if the theories we accept failed to reflect these defects.[61]

Churchland is correct—and also overly pessimistic. Yes, our brains are terrible at intuitively grasping things like quantum mechanics. Our intelligence is mainly a tool to negotiate a *social* environment, so we tend to see purposes everywhere in nature, inventing gods to supply them if necessary. But our brains and social enterprises are also quite flexible. Our history may predispose us toward certain conceptual sets, but we are not locked into them. In well-trained physicists' brains, and more importantly in the community of physicists, we have created a "virtual machine" which sees the world in terms of quantum physics.

Some of us live without gods. Accidental reason is never perfect, but it is probably free of *permanent* limitations. We learn.

Still, accidental reason is too fragile for many of us. Science might oppose mystical visions of reality, but without transcendental assurances, why accept science? Perhaps every way of looking at the world, mysticism or science, depends on fundamental metaphysical presuppositions. And without transcendent Reason, there may be no unbiased, truly rational way to decide between these. Mystical systems and science all come as a package deal, where assumptions about ultimate reality include the very criteria used to judge whether any claim is acceptable.[62] So no one can criticize a faith from the outside, they can only convert from one all-encompassing point of view to another.

Such arguments seem to misunderstand science, particularly how its presuppositions are only starting points. But their distrust of reason resonates in these postmodern times, so we still have to see whether accidental reason self-destructs. Meanwhile, if we accept science, the verdict on mysticism is clear. Certain, theory-free experience of ultimate reality is an illusion. There is nothing which remains when the surface reality of accidents is boiled away—all we can find behind accidents are another layer of accidents. Against our intuitions, facts become clearer, more solid, as we wrap them in ever-more sophisticated theories. Paradoxically, refusing certainty leads to the most reliable knowledge.

NOTES

1. See Thomas V. Morris, ed., *God and the Philosophers: The Reconciliation of Faith and Reason* (New York: Oxford University Press, 1994). A frequent reason for belief is a paranormal or mystical experience.

2. Rudolf Otto, *The Idea of the Holy: An Inquiry into the Non-Rational Factor in the Idea of the Divine and Its Relation to the Rational*, 2d ed., trans. John W. Harvey (New York: Oxford University Press, 1950).

3. Denise Lardner Carmody and John Tully Carmody, *Mysticism: Holiness East and West* (New York: Oxford University Press, 1997), p. 21.

4. Frithjof Schuon, *The Transcendental Unity of Religions*, trans. Peter Townsend (New York: Harper & Row, 1975). This is unfashionable today, but otherwise, attempts to compare *and* affirm diverse traditions produce inconclusive hand waving; e.g., Robert Cummings Neville, ed., *Ultimate Realities* (Albany: State University of New York Press, 2001).

5. Ernest Gellner, *Reason and Culture: The Historic Role of Rationality and Rationalism* (Oxford: Blackwell, 1992), chap. 1.

6. Richard H. Jones, *Mysticism Examined: Philosophical Inquiries into Mysticism* (Albany: State University of New York Press, 1993), p. 5.

7. Ibid., pp. 32–39.

8. Bernard McGinn, *Foundations of Mysticism* (New York: Crossroad, 1992), pp. 267–91; John B. Henderson, *The Construction of Orthodoxy and Heresy: Neo-Confucian, Islamic, Jewish, and Early Christian Patterns* (Albany: State University of New York Press, 1998).

9. E.g., Karen Armstrong, *A History of God: The 4000-Year Quest of Judaism, Christianity and Islam* (New York: Alfred E. Knopf, 1994).

10. John E. Mack, *Abduction: Human Encounters with Aliens* (New York: Charles Scribner's Sons, 1994), p. 43.

11. Proclus, *Commentary on the Parmenides* 7, quoted in McGinn, *The Foundations of Mysticism*, p. 61.

12. Pseudo-Dionysius, *The Mystical Theology*, quoted in ibid., p. 176. One of my favorites is the Gnostic *Apocryphon of John*, in James M. Robinson, ed., *The Nag Hammadi Library*, rev. ed. (New York: Harper & Row, 1988), which starts out with "The Monad is a monarchy with nothing above it," and proceeds to a negative theology gone mad; about two pages of immeasurability, incomprehensibility, and ineffability.

13. *Catechism of the Catholic Church* (English translation, Boston: Libreria Editrice Vaticana, 1994), ¶¶39–43.

14. R. T. Wallis, *Neoplatonism* (London: Duckworth, 1972); John Peter Kenney, *Mystical Monotheism: A Study in Ancient Platonic Theology* (Hanover, N.H.: Brown University Press, 1991).

15. Geoffrey Parrinder, *Mysticism in the World's Religions* (New York: Oxford University Press, 1976), pp. 165–67.

16. Nicholas P. Spanos, *Multiple Identities and False Memories: A Sociocognitive Perspective* (Washington, D.C.: American Psychological Association, 1996); Elaine Showalter, *Hystories: Hysterical Epidemics and Modern Culture* (New York: Columbia University Press, 1997), chap. 13. This is not to say the mythic explanation must be the whole story. See Thomas E. Bullard, "UFOs: Lost in the Myths," in *UFOs & Abductions: Challenging the Borders of Knowledge*, ed. David M. Jacobs (Lawrence: University Press of Kansas, 2000).

17. James L. Kugel, in James L. Kugel and Rowan A. Greer, *Early Biblical Interpretation* (Philadelphia: Westminster, 1986), p. 59.

18. Steven T. Katz, "The 'Conservative' Character of Mystical Experience," in *Mysticism and Religious Traditions*, ed. Steven T. Katz (New York: Oxford University Press, 1983).

19. Ibid., p. 21. See also Bernard McGinn, *The Flowering of Mysticism: Men and Women in the New Mysticism—1200–1350* (New York: Crossroad, 1998), describing late medieval vision narratives as imaginative creations meditating on ever-present religious imagery; and Bernard C. Farr, "Becoming Spiritual: Learning from Marijuana Users," in Laurence Brown, Bernard C. Farr, and R. Joseph Hoffmann, eds., *Modern Spiritualities: An Inquiry* (Amherst, N.Y.: Prometheus Books, 1997).

20. Gunapala Dharmasiri, *A Buddhist Critique of the Christian Concept of God* (Antioch, Calif.: Golden Leaves, 1988), chap. 6. See also Ray Billington, *Religion Without God* (London: Routledge, 2002).

21. Robert Gimello, "Mysticism in Its Contexts," in Katz, *Mysticism and Religious Traditions*. Quotation from Tsung-Mi, pp. 77, 80. Other religious traditions also moderate negative theology; Steven T. Katz, "Mystical Speech and

Mystical Meaning," in *Mysticism and Language*, ed. Steven T. Katz (New York: Oxford University Press, 1992).

22. Steven Payne, *John of the Cross and the Cognitive Value of Mysticism: An Analysis of Sanjuanist Teaching and Its Philosophical Implications for Contemporary Discussions of Mystical Experience* (Dordrecht, The Netherlands: Kluwer, 1990), p. 143.

23. Paul Kurtz, *The Transcendental Temptation: A Critique of Religion and the Paranormal* (Amherst, N.Y.: Prometheus Books, 1991), pp. 99–102; Robert Charles Zaehner, *Mysticism: Sacred and Profane* (Oxford: Clarendon, 1957), p. 151f.

24. Roland Fischer, "A Cartography of Understanding Mysticism," *Science* 174 (1971): 897. Suggestibility and hyperarousal are important in quasi-mystical episodes like evangelical conversion experiences as well; Leonard Zusne and Warren H. Jones, *Anomalistic Psychology: A Study of Magical Thinking*, 2d ed. (Hillsdale, N.J.: Lawrence Erlbaum, 1989), pp. 91–94.

25. Raymond Bernard Blakney, *Meister Eckhart: A Modern Translation* (New York: Harper & Row, 1941), pp. 99–100.

26. Stephen Bernhardt, in Robert K. C. Forman, ed., *The Problem of Pure Consciousness: Mysticism and Philosophy* (New York: Oxford University Press, 1990), p. 220. A PCE is no more a cultural artifact than an orgasm, but Steven T. Katz's thesis that mystical experience is culturally constructed remains broadly correct.

27. Robert K. C. Forman, ibid., p. 38.

28. Payne, *John of the Cross and the Cognitive Value of Mysticism*, pp. 204–206, lists some differences and concludes that no good physiological account of mysticism is likely to become available. However, comparison with drug experiences is not meant to produce a single simple account for the phenomenology of all mysticism, especially the depth mysticism Payne observes to be most different from a psychedelic experience. Mystics themselves do not always have similar experiences, relying on a folk taxonomy to lump physiologically distinct events together. Much the same can be said for visionary experiences. Phillip H. Wiebe, *Visions of Jesus: Direct Encounters from the New Testament to Today* (New York: Oxford University Press, 1997) argues that while traditional supernaturalism is inadequate, neuroscience has not yet produced a comprehensive explanation, hence there is hope for a transcendent explanation linked to parapsychology. Nevertheless, we have reason to expect neuroscience will continue to make progress explaining visions, while religious approaches take us nowhere.

29. Walter N. Pahnke and William A. Richards, "Implications of LSD and Experimental Mysticism," in *Altered States of Consciousness*, ed. Charles T. Tart (New York: Wiley, 1969); Kurtz, *The Transcendental Temptation*, pp. 102–104.

30. Rhawn Joseph, *Neuropsychiatry, Neuropsychology, and Clinical Neuroscience*, 2d ed. (Baltimore: Williams & Wilkins, 1996); Michael A. Persinger, *Neuropsychological Bases of God Belief* (New York: Praeger, 1987). Ben-Ami Scharfstein, *Mystical Experience* (Baltimore: Penguin, 1974), pp. 164–66, considers the most important difference between mystical and pathological cases to be that psychotics lack *control* over their experience.

31. G. William Barnard, *Exploring Unseen Worlds: William James and the Philosophy of Mysticism* (Albany: State University of New York Press, 1997).

32. Stanislav Grof, "Beyond the Brain: New Dimensions in Psychology and

Psychotherapy," in Christian Rätsch, ed., *Gateway to Inner Space: Sacred Plants, Mysticism and Psychotherapy*, trans. John Baker (Bridport, UK: Prism, 1989). Grof's argument is based mainly on occult physics and alleged paranormal feats in psychedelic states.

33. Daniel C. Dennett, *Consciousness Explained* (Boston: Little, Brown, 1991), chap. 1.

34. Jones, *Mysticism Examined*, p. 112.

35. Theodore W. Jennings, *Beyond Theism: A Grammar of God-language* (New York: Oxford University Press, 1985), p. 135.

36. Jones, *Mysticism Examined*, pp. 109–11; Edwin A. Abbott, *Flatland: A Romance of Many Dimensions* (New York: Barnes & Noble, 1963). A degree of ineffability in experience and language can be defended without accepting the claims of mystics; e.g., Ben-Ami Scharfstein, *Ineffability: The Failure of Words in Philosophy and Religion* (Albany: State University of New York Press, 1993).

37. Frank Jackson, "Epiphenomenal Qualia," *Philosophical Quarterly* 32 (1982): 127. William G. Lycan, *Consciousness and Experience* (Cambridge: MIT Press, 1996), argues that Mary does obtain "phenomenal information." But what Mary gets out of her experience is a reorganization and rerepresentation of available knowledge—calling this "acquiring" information is misleading.

38. In other words, such qualia are epiphenomenal. Philosophers who accept nonphysical qualia then try to bind qualia to the brain, usually by some very hypothetical laws connecting mental and physical realms. E.g., David J. Chalmers, *The Conscious Mind: In Search of a Fundamental Theory* (Oxford: Oxford University Press, 1996).

39. Owen Flanagan, *Consciousness Reconsidered* (Cambridge: MIT Press, 1992), p. 100.

40. Austen Clark, *Sensory Qualities* (Oxford: Clarendon, 1993).

41. Paul M. Churchland, *A Neurocomputational Perspective: The Nature of Mind and the Structure of Science* (Cambridge: MIT Press, 1989), pp. 104–106. Churchland also expresses dissatisfaction with the limits of language; however, this is due to the practical impossibility of conveying full brain-state information by ordinary language. See also Paul M. Churchland and Patricia S. Churchland, *On the Contrary: Critical Essays, 1987–1997* (Cambridge: MIT Press, 1998), chaps. 5–11.

42. Some philosophers seriously raise the possibility of zombies: creatures identical to humans in every physical way, except that they're not really conscious. So the availability of a new sense is not assured. But then, zombies are a modern version of transubstantiation, only worse. Dennett, *Consciousness Explained*, pp. 404–406.

43. Robert Gimello, in Katz, *Mysticism and Religious Traditions*, p. 62.

44. Majid Fakhry, *A History of Islamic Philosophy*, 2d ed. (New York: Columbia University Press, 1983), p. 142.

45. Jeffrey Burton Russell, *Lucifer: The Devil in the Middle Ages* (Ithaca, N.Y.: Cornell University Press, 1984), p. 175.

46. Wallis, *Neoplatonism*, p. 17.

47. Roger Penrose and Stuart Hameroff, "What 'Gaps'? Reply to Grush and Churchland," *Journal of Consciousness Studies* 2, no. 2 (1995): 98; Roger Penrose, *The Emperor's New Mind* (Oxford: Oxford University Press, 1989); Roger Penrose, *Shadows of the Mind* (Oxford: Oxford University Press, 1994). Pen-

rose's Platonism is less hierarchical than the classical version; see figure in *Shadows*, p. 414.

48. Rick Grush and Patricia Smith Churchland, "Gaps in Penrose's Toilings," *Journal of Consciousness Studies* 2, no. 1 (1995): 10.

49. Roger Penrose, "The Emperor's New Mind—Concerning Computers, Minds, and the Laws of Physics," *Behavioral and Brain Sciences* 13 (1990): 692; and accompanying critical comments.

50. J. R. Lucas, "Minds, Machines, and Gödel," *Philosophy* 36 (1961): 112. Many of Penrose's arguments are a variation on the same theme.

51. Gary R. Habermas and J. P. Moreland, *Immortality: The Other Side of Death* (Nashville, Tenn.: Thomas Nelson, 1992), pp. 39–41; Antony Flew, *Atheistic Humanism* (Amherst, N.Y.: Prometheus Books, 1993), pp. 119–23. For a materialist response, see Patricia Smith Churchland, *Neurophilosophy: Toward a Unified Science of the Mind/Brain* (Cambridge: MIT Press, 1986), pp. 335–46.

52. Parrinder, *Mysticism in the World's Religions*, p. 6.

53. Daniel C. Dennett, *Darwin's Dangerous Idea: Evolution and the Meanings of Life* (New York: Simon and Schuster, 1995), chap. 15.

54. Taner Edis, "How Gödel's Theorem Supports the Possibility of Machine Intelligence," *Minds and Machines* 8, no. 2 (1998): 251.

55. James O. Berger, *Statistical Decision Theory: Foundations, Concepts, and Methods* (New York: Springer-Verlag, 1980), pp. 10–13.

56. Dennett, *Darwin's Dangerous Idea*, chap. 14; Daniel C. Dennett, *Elbow Room: The Varieties of Free Will Worth Wanting* (Cambridge: MIT Press, 1984); Daniel C. Dennett, *Brainchildren: Essays on Designing Minds* (Cambridge: MIT Press, 1998), chap. 3.

57. Edis, "How Gödel's Theorem Supports the Possibility of Machine Intelligence."

58. Toby E. Huff, *The Rise of Early Modern Science: Islam, China, and the West* (New York: Cambridge University Press, 1993); Gellner, *Reason and Culture*, chap. 7.

59. Scharfstein, *Mystical Experience*, p. 59.

60. Ronald N. Giere, *Explaining Science: A Cognitive Approach* (Chicago: University of Chicago Press, 1988).

61. Churchland, *A Neurocomputational Perspective*, p. 141. For criticism see Jerry Fodor, *In Critical Condition: Polemical Essays on Cognitive Science and the Philosophy of Mind* (Cambridge: MIT Press, 1998), chap. 16.

62. E.g., Jones, *Mysticism Examined*, chap. 3; Donald Rothberg, "Contemporary Epistemology and the Study of Mysticism," in Forman, *The Problem of Pure Consciousness*. The one constant in most philosophy of mysticism seems to be obscurantism. Building metaphysical castles in the sky used to dominate (McGinn, *The Foundations of Mysticism*, pp. 291–326); as philosophers lost confidence in grandiose metaphysics, equally obscure postmodern hand-wringing replaced it.

Chapter Eight

Leaps of Faith

Reason is the greatest enemy that faith has: it never comes to the aid of spiritual things, but—more often than not—struggles against the divine Word, treating with contempt all that emanates from God.

—Martin Luther, *Table Talk* (1569)

Faith always finds a way around any difficulty. If modern knowledge casts doubt on spiritual things, the obvious remedy is to denounce science and despair about objective knowledge. Then God can come galloping in to save the day, now as the foundation of Reason. We do not, however, have to enlist the gods to defend science; neither do we need much from traditional philosophy. If we begin to explain knowledge from within science, we can also see that we do not need any transcendent foundation to lend security to what we know.

EXORCIZING DOUBT

Doubt need not destroy faith. A vital belief which overcomes challenges grows stronger, tempered by doubt. The faithful can seek understanding, questioning as well as worshiping God. After all, reason and religion have no essential conflict; a religious person can hope to emerge from honest doubt with a trusting relationship with a personal God.[1] The best of our knowledge could have confirmed a religious picture of our world. This would have been all the more impressive for being independent confirmation.

Honest doubt, however, means taking genuine risks. The beliefs we start with might not survive criticism. And in the last few centuries, this appears to have happened. After the European Enlightenment, skepticism about religion came into its own. Philosophers began to suspect that traditional metaphysics was smoke and mirrors, which meant God was no longer safe as a necessary being. This would have been no more than an irritation if our real-world knowledge delivered what speculative metaphysics could only promise. Unfortunately, our sciences continually erased the signs of God. Today, God is invisible in the heavens and silent in history. Little remains but magic and mystic feelings, also steadily eroded by science. Western intellectual culture put its faith at risk, and found multiplying doubts instead of a more secure belief.

So today, defenders of God need to exorcize the doubt. For many of us, God remains at the center of our morality and identity, no matter how invisible as a fact. When our doubts threaten to spin out of control, it is tempting to just make a blind commitment. If God offends reason, all the better—true faith demands we crucify our intellect and submit to God.

Such stark irrationalism, while invincible, is rather unappealing. A more popular way to dispel doubt is to say religion has its own brand of truth. Science is supposed to describe the material world but keep silent about spiritual matters. In turn, proper religion does not trespass on science but reveals the ultimate purpose behind our world. Reason cannot decide ultimate matters; when it comes to fundamental metaphysical principles like God, all we can do is make a commitment of faith.

Unfortunately, like most half-measures, granting religion a special dispensation does not really work. Our sciences have something to say about claims like "God created the universe" or "God revealed the Quran," even when they are stamped with ultimacy. Moreover, acknowledging science as objective, but declaring ultimate matters arbitrary, devalues religion. Even without positivists to say that only verifiable claims are meaningful, religious "truths" begin to look inconsequential when compared to the accomplishments of science.

Still, the idea that God must be a fundamental commitment captures something important. For the religious person, *everything* must depend on God. Of course, blindly accepting what looks false does not help, and neither does making God optional. But what if every view of the world, scientific naturalism no less than theistic religion, is based on arbitrary presuppositions? Enlightenment Rationalists cause their mischief by demanding that every claim be judged by reason, objectively and independent of all faith. What if this universal reason is an illusion—if reason is only a way of working out the consequences of our foundational beliefs?

If so, even the most blatant conflict with science need not cast doubt on a religious belief. Imagine an argument with a young-earth creationist. He asks why the earth is billions and not thousands of years old, and so we point to evidence like radiometric dating. But now he observes that radiometric dating *assumes* many things, like constant decay rates, and that there was no outside interference with the process.[2] We answer that physical theory and experiments indicate decay rates are constant under the relevant range of conditions. Our results fit together nicely with other, independent evidence for an old earth. It is barely possible that an external force intervened to contaminate all our evidence, but such cosmic conspiracies are no more likely than the earthly variety. Still, this does not put an end to the questions. Granted, the earth is old according to scientific notions of method and evidence. But why, the creationist can ask, should we accept the presuppositions of modern science? God's revealed word is just as good a foundation, if not better.

It might seem the creationist is being bloody-minded. Actually, he is being philosophical—one needs an expensive education to tell the difference. We often think arguments proceed from basic presuppositions, upon which all else is grounded. These foundations, which are of course the most important principles, are uncovered through philosophical reflection. From this perspective, the conflict over creation is not between reason and stubborn faith, but between different presuppositions. A creationist can reason capably enough, drawing valid conclusions from his basic premise of inerrant scripture. A young earth follows pretty directly from a commonsense reading of scripture; it requires very little independent evidential support. If science concludes otherwise, this reveals a disagreement at the level of bedrock principles.[3]

When a disagreement cuts so deep, it becomes difficult to see who is right. Rationalists usually think science is an impartial method; we plug in the empirical data, crank the machinery, and find out whether the earth is young or old. The trouble is, creationists can also dispute the principles defining the method. If rationalists try to justify their method by appealing to more basic principles, creationists can always deny this new foundation as well. And if our principles somehow justify themselves, this is merely circular reasoning. Reason, it appears, is not self-sufficient, because there can be no rational choice between foundational principles which define reason and evidence differently.

So apologists can use doubt to exorcise doubt. If rationality is severely limited, if we all depend on presuppositions which we cannot criticize, we might as well take God as a basic principle. An Enlightenment Rationalist is in no position to object to this basic God; evidence that ours is a godless world is irrelevant because our very notion of evidence is at stake. In fact, theists are better off, since they realize reason cannot stand alone without a transcendent guarantee. God is just the sort of power which is beyond reason but which can save reason from runaway doubt.

Those criticizing God in the name of reason, then, ignore their own arbitrary commitments. The Abrahamic traditions, always suspicious of self-sufficiency apart from God, often see their critics as in the grip of the sin of pride as well. So religions often appeal to faith against criticism.

The Counter-Reformation, for example, used runaway doubt against Protestants who charged Catholicism with irrationality, offering Church authority as the only security against an all-devouring skepticism. Protestant rationalism later ran into trouble of its own, especially as the quest for the historical Jesus turned up disturbing results. Protestant theology also began to proclaim a need for commitment.[4] Today, postmodern philosophers have picked up the torch. Skeptical of all sweeping "metanarratives," of the theories of modern science as well as of classical metaphysical schemes, postmodernists tend to see all views of the world as ideologies rooted in the way of life of a community. There is no neutral standpoint outside of all communities and traditions from which we can judge differing ways of life. The best we can do is be aware of this, and try not to privilege any community and its metaphysics over another. Any claim to objective knowledge, even by a conspicuously successful science, is to be unmasked as an exercise of power.[5]

With its claims to objectivity exposed as a sham, Enlightenment Rationalism becomes vulnerable to moral criticism. As James H. Olthius charges,

> Control through reason and science has left wide swaths of destruction in its wake: systematic violence, marginalization, oppression, suffering, domination of the "other." It is that sorry history that both lies at the root of the postmodern attack on the totalizing power of reason and gives shape to the postmodern ethical imperative to include the "other" and to make room for the "different."[6]

Old-time religion harbors similar sentiments. Cutting reason off from God is a supreme display of human arrogance which inevitably leads to totalitarian oppression.[7] Both traditional and postmodern religion roots our fundamental commitments in our identity and our moral convictions. In the end, we are more certain of our loyalties than our science.

This is all very frustrating to the Enlightenment Rationalist. After all, what more can an honest critic do, other than sympathetically consider how a claim might be true, examine the relevant evidence and arguments, and try to find the best explanation? Now it appears anything can be protected by declaring it a fundamental commitment. This has to be cheating.

Indeed, it is. But worries about the limits of reason have some substance as well. We too often give science and reason transcendent attributes, which makes them vulnerable to runaway doubt. Fortunately, a fully natural, accidental reason works very well. We do not need invulnerable foundations for our knowledge. When we lay all our principles open to criticism, we will no longer have certainty. We will, however, still have some confidence in our knowledge, and that is good enough.

REASON REFORMED

At first, it may seem strange that a postmodern mood could help God. After all, postmodernists dabble in relativism, while religious leaders usually proclaim God-given Absolutes. The Abrahamic God is jealous; there is but one saving truth, best revealed in one tradition. Other beliefs are more or less in error. In contrast, relativists just say there are many perspectives on the world, and give up trying to sort out true from false. Enlightenment Rationalists at least agree that there is an objective reality, possibly including a God. In a postmodern world, nothing can be the One True Faith.[8]

Relativism, in other words, threatens religion as well as rationalism. In our diverse societies, we avoid criticizing each other's religion. We want a plumber to fix our pipes correctly; whatever faith he chooses to practice in his spare time is irrelevant. So we might be tempted to say that every faith is correct by its own standards, and those are the only standards that matter. But some plumbers believe prayer cures cancer, or that Jesus is about to return, or any number of things that do not affect the pipes but would matter very much if they were true. We need some way to tell not only if our pipes work correctly, but if it is a good idea to sell everything and gather on hilltops. When there is anything real at stake, it does not help to talk about many perspectives and shrug off objective knowledge; the result is not a benign tolerance but cognitive paralysis.

Nevertheless, religion can use an ally against the Enlightenment. As Stanley J. Grenz says, "because of our faith in Christ, we cannot totally affirm the central tenet of postmodernism as defined by Lyotard—the rejection of the metanarrative. We may welcome Lyotard's conclusion when applied to the chief concern of his analysis—namely, the scientific enterprise."[9] By treating science as a kind of mental imperialism, postmodernists create intellectual space for religion. Religion scholar Bruce B. Lawrence, for example, observes that the creation-evolution wars are more than a dispute about facts:

> Creationists fail to realize the extent that they have not invoked true scientific principles because they are not trying to identify new, testable hypotheses but rather to cast doubt on those hypotheses that already exist. Yet at the same time positivists falter in not recognizing the ideological and nonuniversal character of their claims on behalf of science. . . . Even as a public spectacle, the debate on creation is less an objective inquiry into knowable facts from value-free perspectives than it is a testing of constituencies who advocate two variant worldviews.[10]

Lawrence portrays Enlightenment science as one view among many, which stands out only in the power it commands to exclude other visions. It illegitimately claims privileged access to reality. So he says, "Even contemporary exponents of evolution must admit that the arguments for its validity rely as much on ideology as facts. The pretense to scientific objectivity needs to be unmasked, the claim to universal validity scaled down."[11]

For those of us who trust science, these are disturbing words. If even as solid a fact as evolution is tainted with ideology, how can we have confidence in anything? Paralysis looms. However, attacks on science need not explode into an anything-goes relativism. Postmodernists can admit we know a bit about plumbing, or even that science is a good way to solve some problems. They only snarl at more ambitious projects, like using science to criticize whole worldviews. Reason can still be useful, as long as it is limited in scope. For example, many milder postmodernists take a pragmatic view. They deny that science is objective, that it makes progress by formulating theories which increasingly correspond to reality. Instead, they portray science as a way of coping. Plumbing is a practical skill; similarly, science is useful in dealing with many puzzles we face. But then so is faith. Many of us cope better with life when we believe there is a deeper purpose behind everything.

This pragmatic attitude is most at home in a mushy therapeutic spirituality. Old-time religion, however, can also take advantage of the limits on reason. Stanley J. Grenz says, "postmodern thinkers rightly alert us to the naïveté of the Enlightenment attempt to discover universal truth by appeal to reason alone. Ultimately, the metanarrative we proclaim lies beyond the pale of reason either to discover or evaluate." If reason cannot do the job, we need another way to decide between rival worldviews, and "this criterion is the story of God's action in Jesus of Nazareth."[12] Our sin-corrupted intellect can lead us away from God, but if we open ourselves to experiencing the Holy Spirit, we find the truth.

This talk about experiencing fundamental truths beyond reason sounds quite mystical. Indeed, postmodern religion of all stripes often adopts a mystical tone, even reclaiming negative theology.[13] This is no surprise. Like many mystics, postmodernists distrust reason. The problem then is cognitive paralysis, and the most straightforward way to avoid that is to ground ourselves in personal experience. In the end, our everyday sense of reality overcomes runaway skepticism. After a long hard day denouncing objective reality, the philosopher has to go out and buy groceries. Such common sense, however, is not science. At this point, postmodernists resolutely defend "other ways of knowing"—intuition, feeling, mystical illumination, faith in the basic beliefs of one's community—all which impress themselves strongly on our commonsense reality. Privileging science over these other ways, they say, is nothing but an arbitrary commitment, since reason cannot justify itself. Only spirituality, which is openness to and trust in the Other, can lead us out of this impasse.[14] This trust is not an embrace of irrationality, but an opportunity to reconstruct reason. So Mohammed Arkoun asks us to reject

> the dualist framework of knowledge that pits reason against imagination, history against myth, true against false, good against evil, and reason against faith. We must postulate a plural, changing, welcoming sort of rationality, one consistent with the psychological operation that the Qur'an locates in the heart. . . .[15]

Rationalists often dismiss postmodernism as aggressive irrationality. It is much more interesting—the postmodern mood becomes a diffuse mysticism, not of the self melting into God, but of experiencing a truth beyond reason even in the most mundane life. The result is not a mindless relativism, but a hope of remaking reason in the image of God.

Hope is fine, but we need to get more specific. The Calvinist or Reformed Protestant tradition has a long history of trying to reform reason. If anyone can fit an old-fashioned God into a newfangled philosophy, they can.

Reformed thinkers make a career of denigrating bare human reason. Some theologians, like Karl Barth, go so far as to reject all evidential apologetics, all of natural theology. Reason can never reach the true God; even theistic philosophy is ultimately an idol of human devising. Faith is a gift, a miracle, there is no reasoning which can attain it. We accept the Word of God in reverent obedience, and that is all there is to say. Of course, with this degree of irrationalism we might as well hire a street-corner loony to proclaim that the moon is made of green cheese. It even goes against the Bible, which, however ineptly, makes use of reasoned arguments.[16]

The better idea is to reconstruct reason from the ground up. Some Calvinists propose a "presuppositional apologetics," arguing that the biblical God is the unprovable premise of all coherent thought. In a world of mere accidents, we cannot trust in reason. We cannot even be confident the sun will rise tomorrow; since a day with no dawn is perfectly consistent with all we know, how can we say, without an act of faith, that things will continue the way we are accustomed to? The Bible is the bedrock which grounds all else, not a collection of claims which needs to be defended. If Enlightenment science says life evolved, or casts doubt on salvation history, this only means we must reconstruct science on the foundation of the Word. The alternative is mindless chaos.[17]

This is promising, but overambitious. It is also overly sectarian; even if only something transcendent can prevent runaway doubt, this need not be the God of the Bible. We need a more respectable way to build on

Reformed intuitions about God and reason. In other words, the recent "Reformed Epistemology" which declares God is a "basic belief."

Religious philosophers lay the groundwork by a "parity argument."[18] God, they say, is a philosophical problem; and no, the old proofs of God do not work. But this is not the same as disproving God, so we have a standoff. God is a groundless belief, but so are many other beliefs we rely upon. We cannot decisively prove that the sun will come up tomorrow, that an external world exists, or that other people have the same sort of inner life as we do, but they still seem to be eminently rational beliefs.

The question, now, is why certain groundless beliefs are acceptable while "the moon is made of green cheese" is not. The usual solution is to seek beliefs which force themselves on us. Classical foundationalists found bedrock in logic, which was inescapable, and sensory experience, which was simply a given. But while *perhaps* there are such things like "self-presenting properties" at the bottom of our observations,[19] they do not imply that other minds or an external world exists. They are certainly very remote from scientific practice. This, Alvin Plantinga argues, means that properly basic beliefs should extend beyond the self-evident truths of the classical foundationalists. When the circumstances are right, beliefs like the reality of other minds come to us naturally, and it is reasonable to believe them even if the evidence leaves us short. At gut level, we *know* other people have an inner life, no matter what skeptics might say. For Christians, God may be just such a belief. We have a natural tendency to believe in God, though this is often obscured by our sinful nature. When encountering the glory of the heavens, or contemplating the Bible, many of us become convinced there is a God. Christians need not offer further evidence for this God, any more than they have to defend other perceptual beliefs arising in the appropriate circumstances.[20]

This is a nicely postmodern God, supported by common sense and spiritual experience, further constrained by the standards of the Christian community. Still, since the Reformed God is not self-evident to everyone, it remains a somewhat arbitrary belief.[21] It would be nice to recover a sense of God as an objective reality. So once theists install God in their foundations, they can set out to build a more solid view of the world. For example, some Christians concerned about the corrosive effects of evolution envision a "theistic science" which would restore intelligent design to the center of nature; many Muslims hope for an "Islamic science."[22] Enlightenment science prefers naturalistic explanations, but a theistic science would be based on different presuppositions. It would not only be open to correction by experiment, but by scripture; it would fit a firm moral vision together with its description of nature; and it would affirm spiritual truths manifested in human needs.

If all this works, Reformed reason will be a viable alternative to

Enlightenment reason. This does not mean reason demands that God exists, of course, only that the existence of God does not go against reason. In other words, Reformed philosophers only remove an obstacle to belief. Reason is not enough; Protestant tradition often deems a conversion experience necessary to truly know God. We must believe in order to see.

To motivate a conversion, a Protestant apologist can once more turn to runaway doubt. There is, she can say, no deeper principle of rationality to help us choose between Reformed and Enlightenment reason. Any principle we propose will fall into yet another, wider circle of presuppositions. And this circle will once again have no justification outside of itself. Again, cognitive paralysis creeps upon us. However, this crisis of reason can be resolved by a leap of faith. Protestants often convert because they find themselves convicted of sin, without forgiveness in this world. They then accept a promise of otherworldly salvation, even though this appears too good to be true. After that, they *know* they are saved. Runaway doubt brings us to an intellectual version of being convicted of sin. Reason alone self-destructs, so from our paralysis we reach out to a God beyond reason but who is the foundation for reason. Rationality emerges from relativism, but under the authority of the Word of God, which is its only sanction. We do not reach God without an act of faith. But once we take that step, our reasoning faculty reaches salvation as well as our souls.

Postmodern attitudes and old-time religion go well together for a good reason. As its promoters claim, postmodernity *is* a revolution—but in the old sense of the word, meaning a return to a previous state. It brings us back to a time when reason had not yet broken loose from community and authority, when we had not learned that the universe was an unpleasantly impersonal place. Even its insistence on a plurality of views is not new; it can happen any time an empire throws together many different peoples. Nevertheless, postmodern thought poses a serious challenge to the Enlightenment. To the discomfort of rationalists, pure reason appears to be a myth, and common sense today stands on the side of a diffuse postmodern mysticism.

PROGRESS? WHAT PROGRESS?

Most natural scientists do not take postmodern moods too seriously. Philosophers keep complaining that knowledge is impossible, but science makes progress. We have a better picture of our world than a generation ago, not to mention a more impressive array of gadgetry. We expand our knowledge, improving our approximation to reality while opening up new horizons to investigate. Philosophy rehashes Plato; science *advances*.

To postmodern ears, though, talk of science marching forever forward sounds like a myth legitimating a soulless technological civilization. Science pretends to rise above culture and deliver objective knowledge valid for all. But what if scientific progress is an illusion rooted in Enlightenment culture?

We usually think scientists gradually add to existing knowledge. For example, about a century ago physicists discovered the electron, and have been making increasingly accurate measurements of its charge and mass ever since. Along the way, experimentalists also found unexpected properties like an intrinsic "spin." Theorists calculated how conduction electrons behave in metals and semiconductors, and they contributed to the emergence of a new electronic technology. They developed relativistic quantum mechanics, which explained spin. But though physicists achieved a more solid picture of electrons, this was not just a matter of adding to what previous generations knew. Nineteenth-century physics included the ether, a medium pervading space in which light traveled as mechanical waves; physicists toyed with ideas like the electron being a disturbance in the ether. Relativity got rid of the ether, and altered the very meaning of space, time, and observation. With quantum mechanics, our basic conception of a particle and even measurement changed once again, so electrons were no longer tiny billiard balls with a negative charge. Learning about the electron included discarding or radically changing some very fundamental concepts of classical physics.

The history of science, in fact, seems full of revolutions ushering in heliocentric astronomy, plate tectonics, evolutionary biology, and more. Scientists used to depend on concepts like the ether or a designing Intelligence, but such things do not appear at all in our present theories—not even approximately. Physics was once steadily adding to our knowledge of the ether, and then, in a few decades, the ether vanished altogether. Science, it appears, does not progress just by adding to established knowledge; even the most central ideas in our theories can be revised.

Postmodernists, however, think correcting our idea of progress is not enough. Conceptual revolutions change *everything*: explanations of accepted facts, but also basic notions of observation and evidence, even the standards of merit for theories. After Darwin, for example, scientists not only adopted a naturalistic explanation of life, but became less forgiving toward supernatural hypotheses. So instead of judging accumulating data in the light of a universal method, science would seem to shift between rival circles of presuppositions or "paradigms." Indeed, scientific change comes on like a "conversion experience," solidified only when the older generation dies off.[23] Not only do scientists not simply add to what was known before, but there is no fixed yardstick of progress independent of the paradigms scientists operate within. Without an external standard of rationality to judge theories by, science cannot claim to approach truth.

This is runaway doubt again, now dressed up in some history of science. Postmodernists further strengthen their case by seeking social explanations for scientific change. Rationalists claim that proper science weighs the merits of a theory, recognizing, for example, how evolution is better than creation. In contrast, creationism persists only because it enjoys social support. As conservative religious people join the professional classes, they feel a need to defend their view of the world. They cannot reject technical expertise out of hand, so they promote a populist vision of science affirming religious beliefs.[24] Postmodernists agree, but apply such explanations to the scientific community as well. They insist on a "symmetry principle," saying all beliefs have similar social causes. And if, as rationalists imply, creationism is determined by social causes and not biological reality, then scientific theories can also be seen as social artifacts. Talk about objective reality begins to look like a community's way of congratulating itself.[25]

Sociologists of knowledge can now dig up dirt on how proponents of different theories struggle to win over their community. They can show how science is closely tied to the wider society, and how scientists act as an interest group suppressing other ways of knowing. Sordid politics invades the scientific process; many, for example, have suspected that Darwin's emphasis on competition reflected Victorian capitalism, or that evolution was an antireligious ideology. Soon the Darwinian revolution starts to look not like progress, but a result of a power struggle. Bruce B. Lawrence claims that

> Evolution prevailed not on its intrinsic merits but because Darwin benefited from the general prestige conferred on science as an independent inquiry that heralded progress but lacked firm criteria for correlating truth with success.
>
> Once evolution has been understood as dubious science seeking acceptance as a universal ideology, it becomes possible to make sense of the entanglement of Darwin with religious issues. The natural selection of the species precluded any divine agent or ulterior purpose in the genesis of human life. A battle of "orthodoxies" was inevitable, not only because of Darwin's truculent disposition but also because of the expectations of his scientific colleagues.[26]

Even today, belief in evolution is sustained by an ideology which invests the authority to define knowledge in an elite, while creationism goes together with religious populism.

So in postmodern eyes, debates over the nature of our world are but surface manifestations of a contest for power. Whether creation science succeeds depends on whether religious conservatives can exploit dissatisfaction with the Enlightenment, promoting creationism as part of a culture providing a stable community, moral guidance, and a perception of clear-cut meaning in history and the natural world. Truth does not come into this picture; neither is it relevant to explain belief in evolution. If creationists can sustain institutions more sophisticated than Bible colleges, practicing a Baconian science devoted to collecting facts and avoiding "speculations" like evolution, it will no longer be easy to dismiss them as intellectually marginal crackpots. Evolutionary biology might be immensely successful *on its own terms*, but creation science will also succeed by its own criteria. And once again, we will have no place to stand outside these paradigms to decide which is correct.

This cannot be right. People who talk of paradigms and ideologies spin some elaborate theories about science. If there are no good reasons to prefer one theory over another, why should we believe the theory which says there are no such reasons?[27] Of course, we might exempt social science from the rule, so only natural science fails to deliver objective knowledge. But this would be a strange claim. After all, social thinkers are notoriously unable to reach the kind of consensus on solid results which natural scientists achieve.[28] It is too easy to suspect that postmodern social scientists project their frustrations on natural science.

On the other hand, we cannot embrace the myths of additive progress and a scientific community isolated from the social world. Conceptual revolutions *do* happen. The symmetry principle *is* correct; scientists debate and reach consensus by social means. And we should ask whether science really advances. "Science is a good way to evaluate fact claims" is itself a fact claim, and if our sciences do not support this, then there is something rotten with Enlightenment science.

The symmetry principle is troubling because in a realm of social causes and power struggles, there seems to be no room to acknowledge the reasons one theory might be better than another. But this worry is as misguided as saying all that happens in our heads is neurons firing, and therefore we cannot truly think. Our brains are structured in such a way as to *enable* us to reason. Of course, we are best at negotiating our social world, not at the peculiarly impersonal reasoning which has turned out to be so successful in natural science. Nevertheless, we press our brains into scientific, mathematical, and other uses which have little to do with our biological history. As so often happens in evolution, we coopt existing structures for new functions.

Similarly, social mechanisms enable science.[29] A close look at science will find rampant careerism, big-name scientists throwing their weight around, research driven by funding opportunities, and all sorts of unsavory stuff which does not fit the idealized image of dispassionate truth seekers. Still, such social behavior, though not immediately directed toward learning, makes up a community which does a good job learning about our world. As with sausages and laws, the process of forming knowledge is not always pretty. So Philip Kitcher admonishes,

> . . . do not think that you can identify very general features of scientific life—reliance on authority, competition, desire for credit—as epistemically good or bad. Much thinking about the growth of science is permeated by the thought that once scientists are shown to be motivated by various types of social concerns, something epistemically dreadful has been established. On the contrary, . . . particular kinds of social arrangements make good epistemic use of the grubbiest motives.[30]

For example, sociologist James McClenon studied attitudes toward parapsychology among scientists, finding that those in senior "gatekeeper" positions, especially, treat psychical research as a form of deviance. He argues that science has a "megaparadigm"—a guiding ideology excluding psychic phenomena.[31] But this is mistaken. Since science does not command infinite resources, it makes sense to discourage unpromising avenues of research. To accomplish this, the scientific community coopts common social mechanisms, like labeling a practice deviant. Though crude, such measures work fairly well. Parapsychologists shoulder a heavy burden of proof, but they are not so severely excluded as to have no opportunity to engage interested critics. If they could produce more than marginal results, they *would* get the attention of other scientists.

The fact that science is a social enterprise means only that we use social mechanisms to learn about the world. A more serious worry is whether science is overly influenced by society at large. Sometimes we fall into a false consensus. Soviet biology and agriculture, for example, were devastated when the state endorsed Lysenko's genetics. Closer to home, Western anthropology was long infected with scientific racism. Even with little solid evidence, intrinsic European superiority was obvious to European scientists—a fact to be explained rather than a claim to be examined critically. In such cases, however, it is too easy to see how ideology distorted the process of inquiry. This history teaches us to keep science independent, not that science is unreliable. For that, we would need to see if theories which seem to have been arrived at freely, with mountains of evidence backing them up, also reflect social myths. If Darwinian evolution, for example, turned out to be a secular, capitalist myth, science would be in trouble.

It is not difficult to find ideology in the debate over evolution—especially in the wider public debate. Before Darwin, ideas of descent with modification were closely associated with social radicals. So historian Adrian Desmond suggests that acceptance of evolution was delayed until

Darwin produced a theory congenial to Victorian elites.[32] No doubt this is partly correct. But there is much more to the Darwinian revolution. Victorian biologists shared not only a social background, but a similar view of the relevant facts and the problems which needed work. Their community could engage in *biological* debates which were not locked into ideologically determined outcomes. And in fact, far from just reflecting a competitive ethos, Darwin's theory was quite original.[33] We find much the same in other conceptual revolutions. Scientists consistently come up with new ideas and technologies which affect society in unforeseen ways.

Of course, even if scientific change is not a social artifact, revolutions still challenge the notion of progress. Like creationists denying the continuity of macro- and microevolution, postmodernists distinguish between paradigm shifts where change is swift and radical, and periods of "normal science" where scientists ploddingly work within the reigning paradigm. Since paradigms are radically discontinuous, we cannot say whether replacing one with another is progress.

This is, however, a misleading description of science. To begin with, normal science is not quite so tranquil. For example, high-temperature superconductivity is a tough puzzle for physicists. They try out ideas similar to the theory of low-temperature superconductivity, but also explore new possibilities. A few suspect that even the old, successful theory has to be rethought. Experiment is no bedrock of certainty either. Even basic issues like exactly where superconductivity occurs in the new materials are disputed. Physicists will be satisfied only when their theories and experiments converge on an explanation which fits the many unusual properties of superconductors. They have a fair idea of what kind of research is most promising, but much is up for grabs, and many smell exciting changes in the air.[34]

A mini-revolution in superconductivity would not overturn our view of the world. But the process would be similar to the more exciting revolutions. Revising existing theories while looking for a better overall explanation is everyday science; it is not confined to times of paradigm shift. Scientists exploring the frontiers of their field put much up for debate, even much that had been thought settled. But they do this while holding vast areas of our background knowledge constant. Physicists might reconsider an accepted theory of superconductivity, but they will *not* overthrow quantum mechanics to explain high-temperature superconductivity. They will develop new laboratory techniques, but they will still practice what is recognizably physics. The larger revolutions proceed similarly, with scientists constructing new ideas against a stable background of other theories and practices. Darwin did not invent long geological ages or decide fossils were remains of once-living things. Though evolution helped solidify these ideas, they were already largely settled.

Biologists collected specimens and pondered anatomy in similar ways, as creationists and afterward as evolutionists. And of course, evolution is not just radically different from creation. It is an alternative constructed to contrast with creation and improve upon it. Darwin did not lose his ability to think as a creationist; he argued that evolution explained life better than creation. Biologists did not convert overnight; they first learned to use evolution, to solve problems or even just to critique it.[35]

No one can guarantee what will survive unchanged—not the basic objects of the old theory, nor the methods scientists use. Yet without simple additive progress, or even fixed principles to judge theories by, there is a thread of continuity in science. Scientists have to convince their community that the change they propose is an improvement. Typically, the new idea must incorporate the strengths of the old view, while doing better in other areas. Evolution did not abolish all creationist classifications of life; it explained why a nested hierarchical structure had worked best in comparison to other schemes of divine order. Electrons are no longer Newtonian particles moving in the ether, but physicists still train students in classical mechanics, and require relativity and quantum mechanics to reproduce classical physics as a limiting case. Science does not jump from paradigm to paradigm, forgetting old ideas. Change must fit a narrative of progress—scientists have to be able to understand old theories and explain how the new ideas work better.

Science is messy. It is entangled with all sorts of social debates, and change comes at all scales. Still, it builds on and improves upon previous descriptions of the world. Evolution and quantum mechanics simply allow us to do a lot more; it is perverse to say they are no better than earlier theories. We have learned more about how we learn, and science, with all its ups and downs, looks like learning.

ROUND IN CIRCLES

History and sociology let us explain science from within science. Studying the scientific enterprise, we can see how scientists engage in critical debate, genuinely advancing our knowledge. But there is a problem here. Any ideology with a self-justifying story, it seems, can also claim this kind of self-consistency. For example, someone who believes the Quran can find confirmation in 10 Yunus 37:

> This Qur'an is not such as could be composed by anyone but God. It confirms what has been revealed before and is an exposition of (Heaven's) law. Without any doubt it's from the Lord of all the worlds.

Standing alone, this seems no more convincing than a used-car dealer's assurances of honesty. But if so, science certifying itself also looks like a circular argument.

To break the circle, rationalists usually seek an external foundation for science. Science, in this view, embodies a method which ensures objectivity and progress.[36] Justifying a theory is to show that it is the correct inference from the proper set of data. Since method transcends the facts it helps reveal, it can ground science directly, without risking circularity. Unfortunately, this Platonic, hierarchical view of knowledge is more the problem than the solution. Method comes first, shaping what we make of incoming data. So a change in method radically alters the whole structure of our knowledge—it upsets the foundations, collapsing all that depends on them. And without Platonic intuition to make them self-evident, it is all too easy to doubt our foundations. Seeking security in transcendent principles, we end up with a painfully fragile conception of knowledge instead.

To get around this problem, defenders of science often make a crude connection back to the real world. As history indicates, science *works*, so even if our method lacks a transcendent justification, we must be doing something right. This is an improvement, but it still retains the troublesome gap between the methods and results of science. After all, we flesh out the notion that science works from within science. We validate the claim that science succeeds by using method, and use this result as evidence that the method is correct. This awkward loop of mutual support does not seem quite enough.

Instead of patching up traditional rationalism, we should get rid of the Platonism in our thinking about science. There is no Scientific Method residing in a realm separate from the results of science; our methods *are* results. Consider, for example, how psychology advances. As always, we start with our present knowledge, which may be no more than a folk theory. Testing and revising this, we also learn about how we make mistakes, and reconstruct our methods to avoid these. In medical experiments, we use double-blind protocols to prevent researchers' and subjects' expectations from influencing the results. Examining paranormal claims, we learn not to just accept the testimony of gentlemen, but to tighten up our methods. And of course, we apply these lessons to psychology itself. With double-blind procedures, rigorous statistics, and automated experiments, we go back and revise the knowledge which suggested that such measures were necessary in the first place. Nothing in this process acts as a psychological method, directing our progress while remaining free from revision. We change psychology against a stable background, but this is our other knowledge, like the physics which enters into our theories of observation.

The same goes for our more general ideas of progress and method in science. We start with the commonsense observation that science succeeds, and proceed to refine it; discovering, for example, that progress is not a simply additive affair. Reflecting on the history of science, we come up with theories about what promotes progress.[37] Cognitive norms are not ethereal philosophical pronouncements; they arise from science no less than physical theories. "If you want an explosion, mix chemicals *X* and *Y*" is a scientific result. So is "if you want reliable experiments, use clean test tubes"; or, for that matter, more ambitious claims about proper practice. The philosophy of science is not an outside judge, justifying or condemning science. It is itself part of the broader scientific enterprise.

We have too often thought that learning is like adding bricks to a building in progress. If we remove a lower brick, those on top fall. Tamper with foundations, and everything collapses. Instead, we can compare knowledge to a web. The theory of evolution, for example, is a network of interconnected ideas about fossils, genetics, natural selection, and so forth. Some sections of this tangled web are fairly stable, like the overall sequence of fossil forms; some are under active construction, like the space occupied by rival subtheories about the relationship of birds and dinosaurs. And all this is richly connected to the larger web of our knowledge: thermodynamics, geology, methods of historical reasoning, and much more.

This picture of a web helps us see, for example, what is wrong when a young-earth creationist charges that evolution rests on an arbitrary "uniformitarian presupposition." First of all, a uniformitarian picture is not a first principle but a result of science, one much modified from the crude starting assumption of the early days of geology. Secondly, even if geology turned out to be a big mistake, leaving us with no geological evidence concerning the age of the earth, this would only slightly weaken evolution. Evolution does not rest on a few key presuppositions; it is bound to our background web by a multitude of independent threads. If some of the geological threads were severed, what remains would be more than enough to support evolution. Of course, a creationist can propose to simply rip evolution from the web. But this would only produce a massively disrupted structure of knowledge, with loose ends of severed threads now terminating in a series of creationist excuses. This impoverished network cannot explain much in biology, let alone promote progress.

If our knowledge is structured like a web, and if our methods and norms of judgment are themselves part of the web, there are no isolated foundations upon which everything depends. Everything is open to criticism: we can always hold part of our web constant, using it as a base while trying to reconstruct another section. There is no great difference between criticizing a claim because it is not supported by what we think are reliable methods, and rethinking a cognitive norm because it fails to capture successful scientific practice. And in a web, criticism works best when we weave an alternative to the theory we replace, achieving an overall structure that hangs together better.[38]

Treating knowledge as a web is only part of the solution. After all, there is still a subtle hint of circularity in this picture. If everything hinges on the self-consistent coherence of our knowledge, we face the danger of being left with nothing but words referring to other words.[39] We might have a tight-fitting, coherent network of beliefs which is pure fiction. In other words, we need to anchor our web on reality. Normally, we think science succeeds when it captures reality, when "the earth is round" is true because the earth *is* round. Without such correspondence, our web might be impressive but it will not deserve to be called knowledge.

The job of anchoring our theories obviously falls to our observations. Rationalists often take an empiricist view, emphasizing the steady stream of data from our senses. This data is simply given, a brute fact of life, while theories are a way to organize the data and predict what might come next. And in fact, this is not a bad starting point for theorizing about how we learn. Our sciences always depend on some facts more basic than the models we construct in order to explain them. Physicists do not agree on theories of superconductivity, but none doubt that some materials superconduct. Biologists say evolution fits the facts best, and not even the most perverse creationist will deny facts about anatomy or act as if no fossils exist. There seems to be something coercive about facts: skeletal shapes in a rock are simply there, no matter what we come to think they signify.

The trouble, once again, is lurking Platonism. Empiricists introduce a large gap between theory and observation, with no clear way to bridge this gap. Soon we end up with a picture where theories are constrained by the method from above and by the data from below. Between the intuitive certainties of method and the experienced certainty of data, our fallible theories face a second-class existence. This situation positively invites runaway doubt. Moreover, this picture fails to capture the way theory and observation are entangled with one another in science. An experiment to measure the properties of an electron, for example, depends on some sophisticated background theories to obtain reliable, meaningful results. Our physics and our theories of measurement are strongly interconnected; without the theories of modern physics, we could neither devise the extremely complex equipment in our labs nor make sense of what they produce. A reading on an instrument is just a number; it could not further constrain our theories without the extensive network of background knowledge we depend upon.

So we need to see how observations become part of our web. A fact is very difficult to deny, which means it is bound tightly in place by many independent supportive threads. We build this support as our experiments and theories converge onto a stable, richly connected result, producing facts which are much more solid than fleeting sensations. Our theories are vital to anchoring our knowledge on reality; they are not just models erected on the certainties of sense-data.

Consider, once again, how our knowledge about the electron developed. Early on, physicists did not have much to go on but some puzzling experiments, plus intuitions about how to relate these to the rest of physics. They learned more by taking these intuitions as starting assumptions and bouncing them off experiments. Their theories suggested ways to perform better experiments, which in turn prompted them to revise their theories. The most convincing results came when this back-and-forth interaction converged onto a stable network integrated with the rest of our knowledge. For physicists, this sort of convergence is coercive. When we can make incredibly precise measurements of an electron, when we have a detailed theory telling us what an electron is in the context of the rest of our physics, and when we can make accurate predictions and build powerful devices based on this knowledge, then we *have* locked onto something about reality.[40]

Such convergence is real learning, because theory and observation *independently* correct one another. Physicists seek a broad theory of the electron and its place in the world, and such a theory would affect how we do experiments. But meanwhile, we already have a well-developed network of knowledge about how to probe nature, which allows us to perform measurements which can stand as reliable results even in the absence of a broad theory. As we revise and draw connections between theoretical proposals and the background which allows experimentation, both aspects of our knowledge can get stronger. But this mutual support is not foreordained—nature can surprise us. At every step of the way, things can go wrong. We achieve a stable picture when our web of knowledge resists revision except in the direction of further precision, even though things could have gone otherwise.

We anchor ourselves in reality, in other words, by going round in circles—but these are critical feedback loops, not vicious circles. Our starting assumptions set the stage for inquiry but do not predetermine the outcome. Sometimes we bind our starting knowledge more tightly in place, but sometimes we converge on a theory very different from our starting point.

Science is always working in such critical loops; not only learning about electrons and evolution, but also confirming that science advances, and that stable, surprising results indicate we are constrained by an external reality.[41] There is more to this self-consistency than science patting itself on the back, since this result is also not predetermined. Explaining science brings in everything from physics and neurobiology to sociology and history. None of these disciplines rubber-stamp science; in every case, we have an opportunity to critically revise our knowledge. Of course, in this picture, science is only the best we can do; it does not enjoy the certainty which philosophies of transcendent Reason promised. Our points of local stability may always be undermined later. Nevertheless, pulling off the kind of revolution which would uproot what we thought was an anchor in reality is very difficult. Without certainty, we can still have plenty of confidence in our knowledge.

What, then, of "other ways of knowing"? Quite simply, nothing compares to our sciences in the depth of their critical self-analyses and the range of their productivity. Religion can attempt a similar achievement; after all, we can read 10 Yunus 37 sympathetically, observing that it says that the Quran confirms Jewish and Christian revelations, and that the Quran is perfect beyond human ability. It does not just remind Muslims of a basic belief, but suggests ways to support its claim to divine inspiration. If this had succeeded, we could have begun to see the Quran's testimony for itself as a trustworthy statement within a book of other trustworthy claims. Unfortunately, this fails—we converge on something quite different. In fact, religions fail consistently. So they avoid criticism, either by writing God into their foundations, or panicking about the very idea of truth and becoming therapeutic mush. In contrast, science happens to work, even without a transcendent guarantee that it must work. That is all we can say, and it is enough.

THE SUN ALSO RISES

If all elephants are pink, and Mortimer is an elephant, we can deduce that Mortimer is pink. There is an unbreakable link between our premises and Mortimer's color. If science is a web always under construction, however, we must depend on weaker links. For example, geologists can get a good idea of a rock layer's age from the fossils it contains. But an index fossil is only evidence, not a certain indicator of age. Perhaps the fossil was embedded in the rock as an elaborate hoax. Or the theories of geology might be all wrong. Or perhaps this particular layer is an exception. Science is fallible.

This awakens ancient philosophical worries. A mere web spun of theories and memories of past sunrises cannot guarantee the sun will rise tomorrow. What can we say if a paranoid prophet shows up, proclaiming we are deceived by an all-powerful Evil Demon? The Demon has been controlling what we observe, covering up all traces of itself to lull us into complacency. Any day now, we will wake up to a reality which will start to unravel as in a Philip K. Dick novel.[42]

David Hume described the problem this way:

> We are determined by *custom* alone to suppose the future conformable to the past. When I see a billiard ball moving towards another, my mind is immediately carried by habit to the usual effect, and anticipates my sight by conceiving the second ball in motion. There is nothing in these objects, abstractly considered, and independent of experience, which leads me to form any such conclusion; and even after I have had experience of many repeated effects of this kind, there is no argument which determines me to suppose that the effect will be conformable to past experience.[43]

We usually generalize: the sun comes out every morning, so we expect it will tomorrow. But a paranoid prophet trembles: the longer the sun keeps coming out, the more tempting for the Demon to strike. And to Hume, the world looks like the result of a coin toss. Just as observing ten heads in a row is no reason to expect heads next time, *even after experience* we have no warrant to expect the next sunrise.

So perhaps science rests on something arbitrary after all: the way we link our knowledge in a weak, nondeductive fashion; the way we jump between known and unknown. Of course, deductive proofs are not immune to mistakes—a real concern for mathematicians when proofs go hundreds of pages long. We have to infer, from our fallible knowledge, that the proof was performed correctly. So without some weak links, we cannot reason at all. Unfortunately, exactly *how* we link known and unknown may be quite arbitrary.

This is a vexing problem. And of course, some say a rational God ensures the continuity of nature presupposed by our reasoning. So our very ability to do science indicates a God.[44] However, this is a close cousin of the cosmological argument, with similar flaws. If we knew a God would impose the kind of continuity we hope for, we might get somewhere. To some, divine perfection demands rational order, but aside from the usual problems with appeals to perfection, theological intuitions do not agree here. Some traditions, like classical Muslim philosophy, emphasize divine inscrutability and freedom to the point of arbitrariness. We can, of course, require God to produce our kind of order. But then, this is like inventing a fire demon which causes fires—such a God does not actually explain anything; it is just a generic label for mystery.

Setting the gods aside, treating knowledge as a web will help once again. Our confidence in tomorrow's sunrise is not due to a simple generalization. Science works by weaving many weak but independent links together, developing industrial-strength connections. Our knowledge about the sun comes embedded in broad, very well anchored theories; we cannot rip them apart on a whim.

Observing how we build a coherent web of knowledge also helps us understand why the Demon is not credible. A paranoid prophet gets nowhere; she cannot even support her own paranoia. In contrast, trust in our cognitive abilities is a self-supporting belief.[45] We can start there, and see what happens—including, possibly, converging on paranoia. Instead, we have learned how ordinary deceptions and mistakes work, and gained a measure of trust in our abilities. Our present background theories leave little scope for comprehensive paranoia. We can imagine being kidnapped by a gang of mad engineers and being subjected to a massive virtual-reality prank. But to believe this we would need strong evidence, more so than with everyday deception. The Evil Demon directs a total conspiracy, hence it would disrupt our web even more violently, for even less reason. We have, of course, no evidence against the Demon, since it makes sure all our experience is misleading. But this is still a conspiracy theory; after all, the Illuminati or the Elders of Zion are also supposed to be extremely powerful and able to cover up all evidence of their foul schemes.

Still, all is not well. Even an extensive, well-supported knowledge base cannot cover everything. And we want to be able to learn in entirely new circumstances. We might, for example, know very little physics and be unfamiliar with billiard balls. In that case, we have to start with crude generalizations, and Hume's problem comes back to haunt us. At a very basic level, we have a Darwinian, trial-and-error sort of learning in which an organism identifies significant features of the local environment, and implicitly generalizes to deal with similar conditions.[46] But we also want to do better than an insect, and extend our knowledge beyond a local environment. Broad background theories help us do just this: they constrain our expectations, letting us make reasonable guesses without a long and costly process of trial and error. But then, if our background knowledge is inadequate, how do we guess about the next sunrise or billiard ball? And if we have a good theory, this must have been developed from a state of less knowledge, using evidence. How, if Hume is right and experience tells us nothing about the unknown, is all this possible?

Now, we know a few things about generating plausible hypotheses to account for new observations.[47] We also have the mathematical apparatus of statistics, which is all about testing hypotheses given some evidence. Scientists use an extensive toolkit of formal and informal reasoning to explore new territory. We would like to know how much these tools depend on existing background theories, and how close they are to being good advice for any new situation. Since even without an absolute Method we hope to discover some reliable methods, we should see if our ways of learning stand up to scrutiny.

A good way to address this question is to examine Bayesian statistical inference. We often speak of probabilities; we might say, for example, that an index fossil makes a particular age for a rock layer most likely. We start out with a fuzzy, indistinct idea about the rock's age—it could just as well be twenty million years old as two hundred million. After we find the fossil, two hundred million years looks much more likely than twenty. In fact, Bayesian philosophers often treat probability as *the* basic weak link, taking the effect of evidence on our theories as a matter of adjusting their probabilities. Indeed, arguments from betting behavior, requirements of consistent reasoning, and other considerations regularly converge on the probability calculus. Though Bayesians overstate their case, at least a modest form of probabilistic reasoning is a good candidate for a general principle independent of particular background theories.[48] At the least, probability concepts give us a rigorous way to state Hume's problem, and explore some possible ways around it.

Imagine a spaceship exploring a planet in a distant galaxy. It sends down a robot to make a visual survey under the thick cloud cover. The robot soon notices some green, boulderlike objects which emit radio noise. The survey team dub these "norks," and observe about a thousand large, green norks before calling it a day.

The team then maps radio signals from the whole planet and finds millions of norks. They remark what a strange planet it is, covered with green radio noise emitters. But Mission Control has included a philosopher in the crew; he corrects the survey team, saying they have not found that norks are green, only that norks in the area they explored were green. Unobserved norks are wholly distinct and separate; the survey learned nothing about *them*. One of the engineers replies that after the first twenty or so norks they started to expect others would be green. And as other norks kept turning up green, their confidence increased. After all, these norks could have been blue instead. They were not.

The philosopher agrees. Call the hypothesis "all norks are green" G, and the evidence of a thousand green norks e. The survey starts out with an initial probability for G, which is $P(G)$. As green norks accumulate, the probability that norks are green increases, so that $P(G|e) > P(G)$. But this comes about only because e eliminates some rival hypotheses such as "all norks are blue." Consider W, "norks on the Western hemisphere, where the robot landed, are all green, and those on the Eastern hemisphere are all blue." The evidence e is equally compatible with G and W; in fact, it confirms both to the same degree.[49] In other words, the survey has learned nothing about unobserved norks—the probabilities concerning norks in the Eastern hemisphere remain the same as they were prior to all experience.[50]

Normally, the philosopher continues, our background knowledge helps set prior probabilities. Animal species, for example, usually share traits like color, so someone observing only black crows could rightly conclude that most crows are black. But norks are entirely new and strange, so our background knowledge does not help. The engineer presupposes uniformity of color, starting with $P(G) > P(W)$. Someone else may figure that all color distributions are equally likely, so their $P(G) = P(W)$. Who is to say otherwise?

Now thoroughly confused, the survey team sends the robot down to the Eastern side, and finds green norks. But this does not quiet their worries. There are many more areas which they have not explored, and their philosopher can repeat the same speech for the unseen norks in any of these areas.

Clearly, Bayesian reasoning is not enough; like any pure empiricism, it sets up a large gap between data and theory and provides no way to link the unknown to what we know. In Bayesian inference, this problem manifests itself as arbitrary prior probabilities.[51] "Objective Bayesians" try to get around this by determining prior probabilities which reflect a state of ignorance. This is an attractive idea, since learning in new circumstances is exactly what we are trying to capture. And ignorance-maximizing principles find use in both Bayesian and non-Bayesian statistics.[52] However, a solid background theory is necessary to determine the appropriate statistical model and hence the prior probabilities. This trick will not work with the norks.

In fact, the trouble runs deeper. Philosophers often conceive of the relationship between evidence and theory very narrowly. We have a range of possible worlds, and all we do by observing a piece of evidence is to eliminate those worlds inconsistent with that information. Evidence does not help us construct theories, nor does it hint at anything beyond itself. For Bayesians, this means that everything interesting is contained in the prior probabilities, independent of experience. For an empiricist approach, this is a curiously Platonic picture.

So we need to go beyond using evidence to passively eliminate options. And we might be able to use probabilistic reasoning, and a modest version of ignorance maximizing, to do just this. Say our philosopher takes two balls, blue and green, juggles them in a bag, and asks the engineer to pick one blindly. Most everyone would agree that her probability of selecting the green ball, $P(g)$, is 1/2. Of course, we do not know the engineer's propensity to pick either ball. Perhaps the green ball has a more pleasant texture, which influences her blind choice. Or she may have an occult affinity to green, so she will pick green over blue ninety-nine times out of a hundred. We do not know. But we do not know if the engineer has an affinity to blue rather than green either. We have two alternatives which are similar in all relevant ways—in what we are ignorant of as well as what we know. Setting $P(g)$ to anything but 1/2 would be to assume knowledge we do not have.

The engineer now can reason similarly about the norks. The survey observed a thousand norks out of a few million. And as any pollster knows, almost all samples of this size are *representative* of the whole population.[53] It is possible the robot had the bad luck to land in the only pocket of green norks on the planet. But this is very unlikely. Precisely because norks are entirely new and strange, the philosopher has no reason to claim the survey took a biased sample. Without presupposing a principle of color uniformity, the engineer can say, quite objectively, that norks on the other side of the mountains are more likely to be green.

Of course, it would be a mistake to calculate a precise number for $P(G)$ and go back to Bayesian thinking.[54] Taking a first sample does not fix the best statistical model; a better way of thinking about what the survey learned is that they now know how to guess. They can take their evidence as being representative of the true state of affairs, and start forming theories about norks accordingly.

In this picture of learning, we do not follow a set rule of inference or just eliminate possibilities. We start with trial and error, and basic pattern recognition, and extend it by crude generalizations based on sampling. We use our ignorance to gain a foothold in the unknown. As we obtain more complex, interlocking generalizations, we also develop more sophisticated models, including statistical models. We then not only have larger, more structured samples, but multiple, independent ways of extrapolating from our evidence. If these converge, we can construct broad background theories which frame our ignorance. This then allows us to sample new information and generalize again, in ways not entirely controlled by our background theories. We operate in feedback loops, not chains of deduction.

The sun is very likely to rise tomorrow. We do not know this because of an ironclad proof which we can wave in the face of paranoid prophets. Nevertheless, we do not just expect the sun to rise because this is a habit of mind. We can defuse the challenge of complete skepticism, and make progress in understanding how we learn. We actually *know* something about tomorrow. And we achieve this knowledge without Gods or any other transcendent guarantees.

UNIVERSAL REASON

Too much in postmodern thought is grandiose theorizing laced with anti-Enlightenment moral fervor and ignorance of the science being denounced.[55] Still, postmodernists are not wholly wrong. The Enlightenment tradition has too often treated reason as a Platonic ideal. Philosophers hand down standards of Reason from on high, and many friends of science think it is too obvious that science produces truth to subject scientific institutions to criticism. No wonder such Rationalism looks like an organized metaphysical faith.

We know better now. Platonic Reason was a starting point; we revised it in response to criticism. We thought there was a hierarchy of knowledge, with the eternal principles of philosophy at the top, theoretical science next, and downward as we got our hands more dirty. And since philosophy was terribly important, uncertainty about first principles became a reason to panic. But we do very well without certainty. We preferred our science ethereal as well, so we worried if the imperfections of the scientific community tainted their results. But jury-rigged social arrangements still help us learn about the world.

So we no longer need appeal to philosophy alone as a justifying authority for our knowledge. This does not, however, mean that philosophy is over and, as some philosophical pragmatists have said, that we can dance off into the sunset while enjoying a series of edifying conversations. C. G. Prado points out that instead, philosophy and science become part of one another:

> As there is growth in scientific knowledge, our philosophical narratives and metanarratives may prove more and more durable and become less and less amenable to narrational restructuring, not because of "conceptual" necessity or a privileged methodology, but because they approach being empirically *right*—that is, not as narratives and metanarratives of a special "conceptual" sort, but as very high level scientific hypotheses in the broadest sense of "scientific."[56]

In our sciences, broadly construed, we recover a very real sense of universal reason. This is not like the old version, where rational people could come together on a royal road to Truth. We do not acquire all our knowledge by the methods of the natural sciences, or by any single Method. Yet, contrary to what postmodernists say, knowledge is *not* fragmented between many self-contained ways of life. All our knowledge claims are connected, and all can be criticized from within our broadly scientific web. There is no such thing as a religious way of knowing; at least, not in the sense that religion has self-sufficient internal standards of knowledge immune to scientific criticism.[57] Furthermore, especially when probing the natural world, science very clearly achieves objective knowledge. It is idle to deny this in the name of culture. In Ernest Gellner's words, "The ability of cognition to reach beyond the bounds of any one cultural cocoon, and attain forms of knowledge valid for *all*—and, incidentally, an understanding of nature leading to an exceedingly powerful technology—constitutes *the* central fact about our shared social conditions."[58]

This universality is a source of tension, precisely because the power of science is such a clear social fact. For while our sciences produce results valid for all, our ability to do science depends on a distinct cultural style. The scientific community cannot be value-free; it must embody norms which promote learning.[59] Doing science means taking on, at least in some compartment of our lives, an Enlightenment identity. And science cannot flourish within just any overall social arrangement. Most scientists hesitate to get entangled in issues of identity and social morality. Nevertheless, questions like, for example, how reliable environmental science can be in a corporate-dominated world belong squarely within science. Universal reason has a way of breaking down the walls we erect between science, religion, or politics.

Here, perhaps, lies the greatest weakness of Enlightenment Rationalism. Producing knowledge is all very well, but being engaged with science draws us into a rationalist way of life. And though they are wrong in dismissing science's claims to knowledge, the postmodernists' dissatisfaction with Enlightenment culture still stands. It seems that universal reason falters in matters of community and identity, and yet cannot escape them. Religion, in contrast, easily claims morality as its domain. In our social lives, commitment reigns. Furthermore, religions promise to unite our lives, which modernity fragments into different roles and loyalties. In an accidental world, there may still be pragmatic reasons to *believe* in God, regardless of the truth.

In 1943, Sidney Hook published a well-known essay denouncing "the new failure of nerve."[60] He attacked the antiscience philosophies of his time and the liberal theists who used them to carve out a special sphere for God. As a good rationalist, Hook pointed out that "the chief causes of our maladjustments are to be found precisely in these areas of social life *in which the rationale of scientific method has not been employed*." We could more fully realize the hopes of the Enlightenment with a scientific approach to our social problems.

Today we have yet another failure of nerve, but for rationalists, optimism seems harder to come by. Social life is too complicated for clean answers. Worse, extending science to social life is not always welcome—knowledge about us is also a means to manipulate us. And science, it seems, has been great at building bombs but useless in producing the moral wisdom to stop us blowing up one another. So, many claim that we need religion for morality. The argument is still alive.

NOTES

1. Ian Barbour, *Religion in an Age of Science: The Gifford Lectures 1989–1991,* vol. 1 (San Francisco: Harper, 1990), pp. 62–65.

2. John D. Morris, *The Young Earth* (Green Forest: Master Books, 1994).

3. Taner Edis, "Relativist Apologetics: The Future of Creationism," *Reports of the National Center for Science Education* 17, no. 1 (1997): 17.

4. William Warren Bartley III, *The Retreat to Commitment*, 2d ed. (La Salle, Ill.: Open Court, 1984).

5. E.g., Stanley Aronowitz, *Science as Power: Discourse and Ideology in Modern Society* (Minneapolis: University of Minnesota Press, 1988). See Gerald Holton, *Science and Anti-Science* (Cambridge: Harvard University Press, 1993), chap. 6.

6. James H. Olthius, in James H. Olthius, ed., *Knowing* Other-*wise: Philosophy at the Threshold of Spirituality* (New York: Fordham University Press, 1997), p. 235.

7. Giles Kepel, *The Revenge of God: The Resurgence of Islam, Christianity and Judaism in the Modern World*, trans. Alan Braley (University Park: Pennsylvania State University Press: 1994), pp. 55–59, 143–48.

8. Don Cupitt, *After God: The Future of Religion* (New York: Basic Books, 1997); Paul Heelas, David Martin, and Paul Morris, eds., *Religion, Modernity and Postmodernity* (Oxford: Blackwell, 1998).

9. Stanley J. Grenz, *A Primer on Postmodernism* (Grand Rapids, Mich.: Wm. B. Eerdmans, 1996), p. 164.

10. Bruce B. Lawrence, *Defenders of God: The Fundamentalist Revolt Against the Modern Age* (San Francisco: Harper & Row, 1989), p. 188.

11. Ibid., p. 184; Aronowitz, *Science as Power*, p. 12.

12. Grenz, *A Primer on Postmodernism*, pp. 164–65.

13. James H. Olthius, "Crossing the Threshold: Sojourning Together in the Wild Spaces of Love," in Olthius, *Knowing* Other-*wise.*

14. Hendrik Hart, "Conceptual Understanding and Knowing *Other*-wise: Reflections on Rationality and Spirituality in Philosophy," in Olthius, *Knowing* Other-*wise*.

15. Mohammed Arkoun, *Rethinking Islam: Common Questions, Uncommon Answers*, trans. and ed. Robert D. Lee (Boulder, Colo.: Westview, 1994), p. 37. Contrast Arkoun to a modernist Muslim, Bassam Tibi, who is puzzled to find even fundamentalists resorting to cognitive relativist rhetoric; *The Challenge of Fundamentalism: Political Islam and the New World Disorder* (Berkeley, Calif.: University of California Press, 1998).

16. James Barr, *Biblical Faith and Natural Theology: The Gifford Lectures for 1991 Delivered in the University of Edinburgh* (Oxford: Clarendon, 1993).

17. Cornelius Van Til, *The Defense of the Faith*, 3d ed. (Philadelphia: Presbyterian and Reformed, 1967).

18. Terence Penelhum, "Do Religious Beliefs Need Grounds?" reprinted in *Contemporary Classics in Philosophy of Religion*, ed. Ann Loades and Loyal D. Rue (La Salle, Ill.: Open Court, 1991); Terence Penelhum, *God and Skepticism: A Study in Skepticism and Fideism* (Dordrecht, The Netherlands: D. Reidel, 1983).

19. Roderick M. Chisholm, *Theory of Knowledge*, 3d ed. (Englewood Cliffs, N.J.: Prentice-Hall, 1989).

20. Alvin Plantinga, "Is Belief in God Properly Basic?" reprinted in Loades

and Rue, *Contemporary Classics in Philosophy of Religion*; Kelly James Clark, *Return to Reason: A Critique of Enlightenment Evidentialism and a Defense of Reason and Belief in God* (Grand Rapids, Mich.: Wm. B. Eerdmans, 1990). See Mark S. McLeod, *Rationality and Theistic Belief: An Essay on Reformed Epistemology* (Ithaca, N.Y.: Cornell University Press, 1993), for a sympathetic critique.

21. Michael Martin, *Atheism: A Philosophical Justification* (Philadelphia: Temple University Press, 1990), pp. 268–77; Anthony Kenny, *Faith and Reason* (New York: Columbia University Press, 1983), pp. 12–16.

22. J. P. Moreland, *Christianity and the Nature of Science: A Philosophical Investigation* (Grand Rapids, Mich.: Baker Book House, 1989); J. P. Moreland, ed., *The Creation Hypothesis: Scientific Evidence for an Intelligent Designer* (Downers Grove, Ill.: InterVarsity, 1994); Osman Bakar, *The History and Philosophy of Islamic Science* (Cambridge, UK: Islamic Text Society, 1999).

23. Thomas S. Kuhn, *The Structure of Scientific Revolutions*, 2d ed. (Chicago: University of Chicago Press, 1970). In Thomas S. Kuhn, *The Essential Tension: Selected Studies in Scientific Tradition and Change* (Chicago: University of Chicago Press, 1977), he moderates his views to recover a kind of objectivity. Postmodernists often dismiss science with an offhand reference to the early Kuhn; e.g., Arkoun, *Rethinking Islam*, p. 4.

24. Raymond A. Eve and Francis B. Harrold, *The Creationist Movement in Modern America* (Boston: Twayne, 1991). James Gilbert, *Redeeming Culture: American Religion in an Age of Science* (Chicago: University of Chicago Press, 1997).

25. Introduction to Diederick Raven, Lieteke van Vucht Tijssen, and Jan de Wolf, eds., *Cognitive Relativism and Social Science* (New Brunswick, N.J.: Transaction, 1992), p. xxii–xxiii; Bruno Latour, *Science in Action: How to Follow Scientists and Engineers through Society* (Cambridge: Harvard University Press, 1987), p. 192. Even more careful sociologists often write in a way suggesting that scientists' interactions with nature are negligible; e.g., Thomas F. Gieryn, *Cultural Boundaries of Science: Credibility on the Line* (Chicago: University of Chicago Press, 1999).

26. Lawrence, *Defenders of God*, pp. 175–76.

27. James F. Harris, *Against Relativism: A Philosophical Defense of Method* (La Salle, Ill.: Open Court, 1992).

28. Ernest Gellner, *Relativism and the Social Sciences* (New York: Cambridge University Press, 1985), chaps. 1, 4; Mario Bunge, "Counter-Enlightenment in Contemporary Social Studies," in *Challenges to the Enlightenment: In Defense of Reason and Science*, ed. Paul Kurtz and Timothy J. Madigan (Amherst, N.Y.: Prometheus Books, 1994).

29. Steve Fuller, *Philosophy, Rhetoric, and the End of Knowledge: The Coming of Science and Technology Studies* (Madison: University of Wisconsin Press, 1993), pp. 335–39; Susan Haack, *Manifesto of a Passionate Moderate: Unfashionable Essays* (Chicago: University of Chicago Press, 1998), chaps. 5, 6.

30. Philip Kitcher, *The Advancement of Science: Science without Legend, Objectivity without Illusions* (New York: Oxford University Press, 1993), p. 305. This is not to say grubby motives are the best—it depends. Just as thought should be possible in various physical systems, science can be implemented by various social arrangements. For example, the American particle physics community

tends to be hypercompetitive, while Japanese physicists are more consensus-oriented; Sharon Traweek, *Beamtimes and Lifetimes: The World of High Energy Physicists* (Cambridge: Harvard University Press, 1988). Still, they do substantially the same physics—their communities embody very similar cognitive norms.

31. James McClenon, *Deviant Science: The Case of Parapsychology* (Philadelphia: University of Pennsylvania Press, 1984). Sociologist Charles F. Emmons, *At the Threshold: UFOs, Science, and the New Age* (Mill Spring, N.C.: Wild Flower, 1997), makes similar charges concerning mainstream science's rejection of UFOs.

32. Adrian Desmond, *The Politics of Evolution: Morphology, Medicine, and Reform in Radical London* (Chicago: University of Chicago Press, 1989).

33. Peter J. Bowler, *The Non-Darwinian Revolution: Reinterpreting a Historical Myth* (Baltimore: Johns Hopkins University Press, 1988), chap. 2.

34. See *International Journal of Modern Physics B* 12, no. 29–31 (1999). Though subject to revision, experiments *do* probe reality; see Noretta Koertge, ed., *A House Built on Sand: Exposing Postmodernist Myths About Science* (New York: Oxford University Press, 1998), pt. 3.

35. Paul Thagard, *Conceptual Revolutions* (Princeton: Princeton University Press, 1992). Kitcher, *The Advancement of Science*, also emphasizes how science is continuous *as a practice* through revolutions.

36. This method is not always a prescribed practice; for example, C. S. Peirce's view that a critical community might guarantee progress. Harris, *Against Relativism*, chap. 7.

37. Paul Thagard, *Computational Philosophy of Science* (Cambridge: MIT Press, 1988), chap. 7.

38. Kai Nielsen, *God, Scepticism and Modernity* (Ottawa: University of Ottawa Press, 1989), chap. 1. Note that we need only deny *substantive* foundations. Some sort of logic, for example, seems necessary to talk about exactly how beliefs support or undermine one another, but this says nothing about the content of our web of belief.

39. For variations on this objection to coherentism and some replies, see Keith Lehrer, *Theory of Knowledge* (Boulder, Colo.: Westview, 1990), chaps. 5–7.

40. Roger G. Newton, *The Truth of Science: Physical Theories and Reality* (Cambridge: Harvard University Press, 1997); N. David Mermin, "What's Wrong with This Sustaining Myth?" *Physics Today* 49, no. 3 (1996): 11.

41. C. G. Prado, *The Limits of Pragmatism* (Atlantic Highlands, N.J.: Humanities Press International, 1987). We achieve, in effect, correspondence through coherence.

42. I do not worry about evil demons who make no difference in the world; see Elliott Sober, "Contrastive Empiricism," reprinted in *From a Biological Point of View: Essays in Evolutionary Philosophy* (Cambridge: Cambridge University Press, 1994).

43. David Hume, *An Abstract of a Treatise on Human Nature*, in *An Enquiry Concerning Human Understanding*, ed. Antony Flew (La Salle, Ill.: Open Court, 1988), p. 35. Other philosophers with an empiricist bent have also been tempted by inductive skepticism; e.g., Karl Popper, in *The Philosophy of Karl Popper*, ed. Paul Arthur Schilpp (La Salle, Ill.: Open Court, 1974), pp. 1015–19.

44. Arguments that science needs a transcendent justification are common; e.g., Roger Trigg, *Rationality and Religion* (Oxford: Blackwell, 1998); Mariano Artigas, *The Mind of the Universe: Understanding Science and Religion* (Philadelphia: Templeton Foundation Press, 2000). For a physicalist alternative, see Nicholas Maxwell, *The Comprehensibility of the Universe: A New Conception of Science* (Oxford: Clarendon, 1998).

45. Lehrer, *Theory of Knowledge*, pp. 121–24.

46. There are many questions about classification I ignore here. See Eli Hirsch, *Dividing Reality* (New York: Oxford University Press, 1993); Jerrold L. Aronson, Rom Harré, and Eileen Cornell Way, *Realism Rescued: How Scientific Progress Is Possible* (Chicago: Open Court, 1995).

47. Thagard, *Computational Philosophy of Science*, chap. 4; David B. Leake, "Abduction, Experience, and Goals: A Model of Everyday Abductive Explanation," *Journal of Experimental and Theoretical Artificial Intelligence* 7 (1995): 407.

48. Colin Howson and Peter Urbach, *Scientific Reasoning: The Bayesian Approach* (La Salle, Ill.: Open Court, 1989), chap. 3; R. T. Cox, *The Algebra of Probable Inference*; Paul Snow, "An Intuitive Motivation of Bayesian Belief Models," *Computational Intelligence* 11, no. 3 (1995): 449. See Deborah G. Mayo, *Error and the Growth of Experimental Knowledge* (Chicago: University of Chicago Press, 1996), for a critique and alternative.

49. This is a variant of the celebrated "grue paradox"; Nelson Goodman, *Fact, Fiction, and Forecast*, 4th ed. (Cambridge: Harvard University Press, 1983), chap. 3.

50. I.e., for all hypotheses h concerning the color of norks on the other side only, $P(h|e) = P(h)$. David Stove, *The Rationality of Induction* (Oxford: Clarendon, 1986), p. 40, takes the inductive skeptic to claim $P(h|e) = P(h)$ for *all* e and h. This should be qualified to exclude cases where $e \vdash \neg h$.

51. "Subjective Bayesianism" keeps priors arbitrary because posterior probabilities become near-identical in the long run (Howson and Urbach, *Scientific Reasoning*, pp. 235–36). However, the "long run" can be arbitrarily long. Take C = "creation," E = "evolution," and let $P(C) = 1- \varepsilon$, $P(E) = \varepsilon \ll 1$. Say $P(e|C) = \delta$ and $P(e|E) = 1$, where $\varepsilon \ll \delta \ll 1$. Then $P(C|e) \approx 1 - \varepsilon/\delta$ and $P(E|e) \approx \varepsilon/\delta \ll 1$. So a perverse enough creationist could legitimately say creation is most probable after any amount of evidence. See Henry E. Kyburg Jr., *Epistemology and Inference* (Minneapolis: University of Minnesota Press, 1983).

52. E. T. Jaynes, *E. T. Jaynes: Papers on Probability, Statistics and Statistical Physics*, ed. R. D. Rosenkrantz (Dordrecht: D. Reidel, 1983); James O. Berger, *Statistical Decision Theory: Foundations, Concepts, and Methods* (New York: Springer-Verlag, 1980); Henry E. Kyburg Jr., *Science & Reason* (New York: Oxford University Press, 1990), pp. 120–21.

53. Stove, *The Rationality of Induction*, chap. 6.

54. Bayesian inference suffers from false precision in probabilities, but there are more modest approaches which might do better. Peter Walley, *Statistical Reasoning with Imprecise Probabilities* (London: Chapman and Hall, 1991); Mark Kaplan, *Decision Theory as Philosophy* (New York: Cambridge University Press, 1996).

55. Alan Sokal and Jean Bricmont, *Fashionable Nonsense: Postmodern Intellectuals' Abuse of Science* (New York: Picador, 1998); M. J. Devaney, *'Since at Least Plato . . .' and Other Postmodernist Myths* (New York: St. Martin's Press, 1997); Paul R. Gross and Norman Levitt, *Higher Superstition: The Academic Left and Its Quarrels with Science* (Baltimore: Johns Hopkins University Press, 1994).

56. Prado, *The Limits of Pragmatism*, p. 145.

57. Many defenders of religion would consider this "scientism"; e.g., Huston Smith, *Why Religion Matters: The Fate of the Human Spirit in an Age of Disbelief* (San Francisco: Harper, 2001); Ian G. Barbour, *When Science Meets Religion* (San Francisco: Harper, 2000). They define science too narrowly.

58. Ernest Gellner, *Postmodernism, Reason and Religion* (London and New York: Routledge, 1992), p. 75.

59. Patrick Grim, "Scientific and Other Values," in *Philosophy of Science and the Occult*, 2d ed., ed. Patrick Grim (Albany: State University of New York Press, 1990).

60. Sidney Hook, "The New Failure of Nerve," *Partisan Review* 9, no. 1 (1943): 2.

Chapter Nine

The Knowledge of Good and Evil

The belief in a supernatural source of evil is unnecessary; men alone are quite capable of every wickedness.
—Joseph Conrad, *Under Western Eyes* (1911)

Moral debates today can seem especially frustrating—little more than a clash of incompatible opinions. One reason is that it is becoming increasingly clear there is no divine command or constant of human nature which is the key to our moral puzzles. Without such a key, we seem left with too many irreconcilable convictions and too much confusion. Even when we agree on good and evil, there is no transcendent moral reality to underwrite these agreements. Unfortunately, this also means that a sober naturalist cannot offer a brilliant moral vision, only an effort to muddle through.

MORAL CERTAINTY

Debating whether God exists is a fascinating exercise. But it also has a rather academic flavor. Science be damned—at the end of the day, some sort of God seems *morally* necessary. God is too deeply embedded in our sense of good and evil, of guilt and hope.

With traditional religions, skepticism is not merely an intellectual error; it threatens the social order. We grow up knowing that our world depends on a moral ideal, even though our earthly societies can realize it only imperfectly. Our way of life, from abstract notions of justice to deep-seated emotions about proper behavior, revolves around the God who keeps this moral reality in place. Priests help us approach the ideal; prophets castigate us when our rituals become empty gestures. Doubting God removes us from the moral life of our community; we become infidels, not to be trusted.

Of course, communities based on old-time religion are notoriously stifling. But they do provide stability. Religion, as anthropologist Clifford Geertz observes, "supports proper conduct by picturing a world in which such conduct is only common sense."[1] Cheating a neighbor may seem to be smart if there is no penalty, but in a moral world it is ultimately foolish. In the Abrahamic traditions, this means evildoers always get punished by God. The promise of eternal life not only soothes our fear of death, but assures us justice will prevail.[2]

Today it is no longer as easy to believe in a cosmic moral order. One reason, of course, is that our sciences keep revealing an impersonal, accidental world. But the greater threat to our moral certainty is modern social life. Historically, societies required a common religion; dissent led to social unrest and moral chaos. With the Protestant Reformation and English Civil War, something strange happened. The usual methods of repression and massacre were not able to establish one party over another, and this stalemate expanded the social space for groups which cared more about worldly concerns like making money. A secular morality spread—not because belief in God disappeared, but because religious plurality became a fact people had to live with. They came to depend on everyday dealings with others who did not share the same theological views.[3] Modern life emerged; we ended up in open societies, with fluid social roles and loyalties. Today, we can adopt any of a thousand faiths, but we can no longer count on a shared religious morality.

For many of us, however, modern society means a moral scene dominated by conflicting claims and endless ambiguities, with no final authority to resolve our anxieties. This can be liberating, as when women start to question whether barefoot-and-pregnant really is their divinely ordained place. But even then, many wonder if modern life liberates us from serving a greater good, leaving everyone free to assert their individual interests. Our morality no longer refers to anything beyond social conventions and the need to negotiate conflicts. There is a certain shallowness to this life, leading to a moral nihilism where nothing *really* matters. So perhaps the first job for religion is to restore our moral certainty. Alleviating moral anxiety is, after all, more important than satisfying idle curiosity about how the world works. Especially when, lost in a fluid social world, many of us feel a strong need for absolute standards.

So religious thinkers most keenly feel the need to defeat modern skepticism when moral issues are at stake. When infidels deny anything beyond the natural world, they also reject any fixed, objective foundation for moral judgments. As a result, anything goes. Ravi Zacharias explains:

> Antitheism provides every reason to be immoral and is *bereft of any objective point of reference* with which to condemn any choice. Any antitheist who lives a moral life merely lives *better* than his or her philosophy warrants. All denunciation implies a moral doctrine of some kind, and the antitheist is forever engaged in undermining his own mines.[4]

This is not just a rerun of runaway doubt in the moral sphere. Zacharias asks, "If life is pointless, why should ethics serve any purpose except my own?"—without transcendent principles, what can guide an infidel but self-interest? There is nothing to keep her from harming others, if she sees an advantage in so behaving. Religious people, in contrast, acknowledge a higher authority: "If, on the other hand, I am fashioned by God for *His* purpose, then I need to know Him and know that purpose for which I have been made, for out of that purpose is born my sense of right and wrong."[5]

Zacharias voices conservative anxieties. But more sophisticated thinkers also associate unbelief with nihilism.[6] And religious faith can undergird resistance to power as well as an authoritarian morality. God's love helps religious liberals stand up against the almighty marketplace, defending human dignity in a world where everything seems to have a price tag. They can step outside of the amoral world of conflicting interests and proclaim what *ought to be*. Such a transcendent focus seems necessary to truly care about anything, even if only about science. Otherwise, our values are nothing but arbitrary preferences. It appears naturalism is self-defeating: it leaves us with no moral reason to live as naturalists. In fact, naturalism seems to have no explanation for moral values at all. We *know* harming people is different from damaging mere objects, and yet we see nothing in the natural world but material things with no intrinsic value.

Religion, in contrast, promises to make sense of our moral convictions. God is the source of all value; creation and revelation are infused with moral guidance. And even if much about our Gods is intellectually suspect, it seems religion works in practice. By the way it pictures the world, and by the way it organizes us around a common purpose, religion is a very effective way of giving us a moral identity. If we want to live in a community with shared values rather than in a collection of competing individuals, we might need some sort of benign religion. Unbelief would leave our society morally paralyzed.

To many secularists, such claims are preposterous. After all, someone saying we should bring back God to stop our moral decay is very likely demanding a return to old-time religion—not some sweetness-and-light liberal mush. Though old-fashioned monotheism gives us secure identities in stable communities, it does this through closed societies excluding those who do not conform. Our priests, rabbis, and mullahs insist on running our lives; they especially like to inspect between our legs. When institutions like the Catholic church still obsess about matters like the evil of masturbation,[7] it is hard to see godly morality as social salvation. Religious modernists, of course, sidestep the authoritarian side of their tradition. But their morality differs very little from secular people. When a church declares that women may now be ordained, due to "the Holy Spirit's gift of ministry so evidently bestowed on women and men,"[8] it looks very much like religion is playing catch-up to the modern world. "God" becomes a word used to sanctify attitudes which have already arisen in secular society.

All this is correct. Even so, moral arguments for religion raise some real questions. First of all, we have to see whether morality makes sense without God. And then, even if we can explain good and evil within nature, we have to ask if we can live with the secular moral options we find. Enlightenment Rationalists were once confident that by vanquishing superstition, we would remove the greatest barrier to shaping a better life for all of us. Nowadays, much that is important has become secularized, but it also seems that modern life falls short of our moral aspirations. God or no God, we have to wonder where we go from here.

HIGH WEIRDNESS BY THEOLOGY

Though religions claim good and evil as their special domain, there is something peculiar about moral arguments for God. After all, historically the biggest nuisance for theology has been the need to reconcile God and evil. Long before modern science, philosophers were spending sleepless nights trying to see how a perfect God could create a world overabundant in its cruelty. As Mark Twain put it, "With fine sarcasm, we ennoble God with the title of Father—yet we know quite well that we should hang His style of father wherever we might catch him."[9] Our Creator is supposed to be morally perfect and all-powerful, and *this* world was the best it could manage?

Evil is a famously powerful reason to disbelieve. Most everyone can understand the pain of a parent who loses her child to disease, and hear God's maddening silence in response. We need not master arcane technical skills to see how theologians' excuses for the silence are absurd in the face of suffering. And philosophers, of course, can spend careers refining these intuitions into a formidable argument against God. God and evil do not mix; one of them has to go—and evil is too obviously here.

Of course, somewhere in the far reaches of divine unknowability there may yet be a greater good served by evil. The trouble is, it is difficult to even hint at this larger purpose. Many theologians admit this; some even use it to attack their more rationalist colleagues. For example, a classic dilemma in Muslim theology concerns three brothers. One led a life of good works and went to heaven on his death, another was bad and was sent to Hell. The third brother died before the age of consent and was allowed into heaven. Knowing he would have grown up to be a bad man, God had mercy, taking him at an early age and sparing him from Hell. But if this is just, why did God not arrange for the evil brother to die young as well, knowing he would become a bad adult? Such dilemmas, Muslim conservatives argued, showed we could not constrain God by our merely human concepts of justice. God is inscrutable, and totally sovereign. Moral action and salvation are beyond human initiative.[10] Many Christians show a similarly perverse brilliance. Everyone is worthy of Hell, but an arbitrary elect will be allowed into heaven. Augustine, Catholic saint and Protestant idol, almost sounds like an atheist while arguing that there is excess evil in the world, but his examples of suffering infants are intended to show we are all depraved and deserving of punishment from the very beginning.[11]

Such high weirdness might motivate an abject surrender to faith, but it is quite crazy otherwise. Why do evils occur in the first place? We still expect that with a benevolent God in charge, ours should be the best of all possible worlds. If this is like asking for the largest number, or if only God can be so perfect, creation should at least be very good—certainly not the vale of tears too many of us live in. The easy solution is to declare this *is* a very good world; our suffering, for example, might be ennobling, even necessary to forge high-quality souls. But then, religious morality gets infected with a bizarre conservatism. If someone cures a disease, she may well do evil by reducing the greater good such ills produce. Using anesthesia in operations becomes rebellion against God. Our world is as good as it gets, so we should keep it as it is.

So it might work better to claim that ours is a very good *history*. Our moral striving to improve our world is itself a greater good. We make progress in achieving God's purpose. From our original fall from grace to the final defeat of evil, history is a cosmic drama: a tale enriched by evil and culminated by redemption. Everything is better for going through this process, in which evil has an integral role.[12]

This gets us somewhere, but not yet far enough. For we still wonder if all the gratuitous suffering in history is excused by the greater good. Something like the Jewish Holocaust, for example, not only casts doubt on our progress, but makes the cost too high. A God still in control of events cannot escape responsibility. So the next step is to limit God by yet another greater good. Whatever moral drama we achieve will be all the more valuable if we do it through free will, genuinely risking failure. As Hans Jonas speculates on God after Auschwitz,

> in order that the world might be, and be for itself, God renounced his being, divesting himself of his deity—to receive it back from the odyssey of time weighted with the chance harvest of unforseeable temporal experience: transfigured or possibly even disfigured by it.[13]

This is not the classical, all-powerful God; it is a God in the process of becoming rather than an eternally perfect reality. A God, in fact, we might forgive. Still, such a God can perhaps be blamed for abdicating responsibility. Worse, while these metaphysical pyrotechnics seem to exonerate God, they also obscure how God should be worthy of worship. What is the point, if the result of excusing evil is that we are left to our own devices?

Maybe yet one more greater good, bringing in a touch of mystical love, will do the trick.[14] God limits itself, knowing full well evil will result. But this is because God seeks *unconditional* love, untainted by any calculation of merit. Loving someone when they are obviously worthy is all too easy. Only when there is real evil in creation can we freely love God, not compelled by reason any more than by fear. A mystic does not judge God; he only gives and accepts love.

This works. High weirdness though they may be, there are ways to fiddle with God and evil until they can coexist. The catch is that we achieve this harmony by progressively hiding God from the world, until the deity is lost in a metaphysical never-never land.[15] As usual, we start with a relatively concrete idea of God, which fails to explain the world we see. Then we rework God so it is compatible with everything and explains nothing—like composing a plausible story about why the tooth fairy is very shy, so she takes extra care never to show herself. As a result, Gods like that of Jonas are pointless, since the same excuses work just as well when we replace God with an all-powerful demon. After all, our metaphysical freedom may be necessary for the greater evil: only in a world where we are free, and where there is considerable good, can pure unconditional malice manifest itself. In the end, our metaphysical scenarios are useless. We need a way to explain good and evil as we find them. None of the excuses theologians have fashioned over the centuries help us with that task, and *this* is the enduring difficulty evil poses for God.

If theologians must explain rather than excuse evil, then religion needs a supernatural theory of good and evil instead of defensive gamesmanship. This should not be too difficult; after all, conventional wisdom says only a religious perspective can grasp both the darkest depths of evil and the heights of self-sacrificing good. Scientific naturalism is fine for a handle on everyday self-interested behavior, but it is too superficial to take us further. If this is true, there is hope for God.

We can start with the commonsense belief in objective moral truths.

We recognize morality to be binding, going beyond any individual preference. Slavery is wrong—period. It is just as wrong in a society based on slavery, even when the slaves are resigned to their fate and few can imagine a world without slavery. Moral reasoning is not about negotiating a way to live together with our various preferences; it is about discovering objective rights and wrongs. The moral truths we discover, moreover, are not like ordinary facts. They have *authority* over us. This suggests that morality has its source in a transcendent authority. Moral imperatives are divine commands. We get to know good and evil, develop a moral identity, and sustain it by proper practice, all as part of our relationship with God. We are created weak, so we err. But even when we go wrong, we sense that we transgress an objective moral law.

This is a fairly typical theist view, so skeptics have a standard response.[16] We ask if something is good solely because God so decrees. If so, morality is disturbingly arbitrary. If God's whim was to permit slavery, as various holy books seem to say, would it not still be wrong? Of course, a *loving* God would presumably command differently. But then we begin to introduce criteria for goodness other than God's will.[17] So perhaps God commands something because it is good. But in that case, goodness is independent of God, and we can hope to reach it without God.

The problem, however, is not as insurmountable at it seems. God's commands do not have to define good and evil to be vital to morality.[18] Even if we have an independent concept of good, God could still be the source and end of all that is good. After all, our idea of power does not come from theology, yet we can imagine an all-powerful God who is the source of all power in creation. Though there must be more to morality than divine commands, they may yet be an important part of the story. We can build on different intuitions about God's relation to morality, explaining our moral life by a series of weaker links that lead to God.

To begin with, we can tie moral commands to creation. God creates such properties as the redness of blood. Good and evil may also be objective properties attached to certain objects and events. Values, of course, are peculiar. They differ from natural properties like color, shape, and so forth; physics cannot begin to account for them. They bind us—they are intrinsically prescriptive. They hint at what should be, and thus are not entirely part of this world. So they may have a supernatural source.[19] Only moral agents created in God's image perceive them. We do not, of course, always agree about what is good. But we make mistakes about natural properties as well, and even if many of us became colorblind, blood would stay red.

Of course, this would mean good and evil remain somewhat arbitrary, since God could have created slavery as good, just as God could have made blood green. But in that case, our moral intuitions would also change; many more of us would see slavery as an obvious good.[20] More importantly, it makes no sense to be a hyperrationalist, demanding sufficient moral reason for everything. At some point, theists have to say God willed it so and that is all. As long as the overall theory works, there is nothing wrong with this.

If morals are rooted in God's will, and God is not capricious, morality should proceed from a divine purpose for the universe and for our lives. There is a natural moral law which reflects this purpose. Our knowledge of the law is dimmed by our insensitivity to revelation and corrupted by our sin, yet it is not completely lost. Most cultures, especially those illumined by the great world religions, affirm that lying or stealing is wrong, that a woman's natural role is fulfilled as a wife and mother, and that homosexuality is an abomination. God's will, manifested in nature's boundaries, invests creation with moral truth.[21]

Nonnatural properties, divine purpose, and moral law are still rather ethereal concepts. To flesh them out, we have to connect them to our interests and emotions as they develop within a religious way of life. We are all motivated to seek certain things, but our appetites are not etched in stone. We can reflect on our interests and adjust them to increase our overall satisfaction. Going further, we also attend to *potential* interests, as when we observe the pleasures of a sophisticated appreciation of good food and company, and decide to develop our own tastes accordingly. The religious life is no different; it involves a certain development of interests and promises fulfillment. A Christian might say that as his personality is transformed by the Holy Spirit and the example of Jesus, he approaches a disinterested love of others, even a boundless love surpassing all finite satisfactions.[22] A Muslim might say she finds true peace by submitting to the One God.

The various strands of religious morality work best together. At one end, religion ties morality to our deep-felt convictions and our desire for happiness; at the other, to deep principles which underlie creation itself. It unifies our self-interest and our concern for others. Best of all, religion ties morality to a working social structure and provides an environment where believers strengthen their moral convictions and reproduce them in others. As we develop our moral life, we realize there is a Perfect Good which is the source and object of morality. Just as we trust our senses, we can trust in our moral intuitions, and we hope to refine and strengthen this trust through a process of criticism. Once we enter the circle of correct moral reasoning, we grow more and more proficient in approaching God, the source and end of all morality.

God, then, can help us construct a solid picture of morality. Philosophical difficulties with evil or with divine commands are real enough, but they are not decisive. Religious morality has more than one leg to stand on. And so Enlightenment Rationalists cannot take good and evil as something obvious, and then attack religion in a way suggesting that it fails transcendent moral norms. The problem is to explain what morality *is*, including its slippery relationship with reason, its anchor in passionate feelings, and its echoes of transcendent obligations. Religions portray moral worlds which make sense of many of our intuitions, and are tested in practice. A better alternative cannot just rely on conceptual criticism; it must provide a better explanation of our moral life.

THE MORALITY OF A SOCIAL ANIMAL

Philosophers talk of virtue, utility, duty, and more, proposing one or the other as the core of morality. Such analysis, however, can obscure as much as it reveals. Our modern moral language, for example, is full of inalienable rights. But we cannot then conclude that morality is all about negotiating rights, still less that it is about discovering rights which are Platonically real. This would elevate a folk theory of morality to a conceptual necessity. It would also fail to capture the complexity of our moral behavior. Though we have many elegant moral philosophies to choose from, to cope with everyday life, we rely on social conventions and on a motley collection of principles which little resembles a tidy philosophical system.

So let us start less ambitiously. Within science, for example, rules like "use clean test tubes" are easy to understand. If we want reliable results, we should be careful about our methods. In basketball, players ought not to put the ball in their own basket—that is, if they want to win; otherwise, why play? Our social enterprises generate norms of conduct, since not every behavior advances an enterprise's purpose. And as we take on various social roles, we adopt their purposes and join debates about how best to achieve them. We ask whether "if you want *X*, do *Y*" is good advice. If so, and if *X* is our goal, we will be motivated to do *Y*; it is not some bloodless principle we can shrug off. We also expect others to act according to their role. We trust a doctor, for example, to practice a professional ethic according to which care for the patient comes first.

Norms make no sense apart from the purposes they serve. However, we cannot just point to our various social roles and leave it there. Our purposes are not immutable or immune to criticism. In a military context, a soldier ought to obey orders. But a soldier also has other social roles, and there are nonmilitary purposes in life. We regularly have to weigh different roles against one another, and even ask whether some purposes are worthy at all. On top of that, we often find ourselves outside of well-established social enterprises, struggling to fashion new norms. Morality seems to reach beyond all ordinary roles, becoming crucial exactly when a soldier wonders if an order to burn a village is morally acceptable. Religious morality claims just such a transcendent reach. Our lives have a divine purpose, however nebulous; the fullest development of our interests lies in seeking God. We are creatures of God—all other roles are subordinate.

It is much harder to find a role above all roles if we take a secular perspective. We develop our interests, certainly, and learn to care about others. We can try to harmonize our various purposes. But without the kind of transcendent moral reality religions imagine, we need not converge on anything like "the good of humanity" or any such overarching purpose. Still, we can negotiate our conflicts in the light of common interests; if all goes well, we may invent a morality which helps us in our overall enterprise of living together. As a bonus, we can also explain our moral life as we find it, instead of trafficking in mysteries.

To understand morality, we must start with the interests which underlie our purposes. And to see where interests come from, we can look to biology. Some things are good for self-replication and some are not, so organisms have evolved to seek or avoid them. Interests came into the world as *reproductive* interests.

At first, this seems an unpromising beginning. After all, morality is supposed to raise us above our self-interest and animal appetites. Evolution sets us in a world of reproductive selfishness, competition for scarce resources, and a relentless tide of suffering. As biologist Richard Dawkins describes,

> The total amount of suffering per year in the natural world is beyond all decent contemplation. During the minute that it takes me to compose this sentence, thousands of animals are being eaten alive, many others are running for their lives, whimpering with fear, others are slowly being devoured from within by rasping parasites, thousands of all kinds are dying of starvation, thirst, and disease. It must be so. If there ever is a time of plenty, this very fact will automatically lead to an increase in the population until the natural state of starvation and misery is restored.[23]

With evolution, selfishness and morally pointless suffering are natural; they are no longer mysteries which religions can only hint at darkly. But now, the brighter side of nature is puzzling. Dogs do not eat other dogs; at least, not very often. Though the lion never quite lies down with the lamb, animals do help one another.

Cooperation, however, is hardly a mystery when it is good strategy.[24] Just simple reciprocity goes a long way. A social animal will groom her neighbor and get her own parasites removed in return. Plus, reproductive interest is not focused on direct benefits to the organism, but on propagating genes. Even an early death can serve an organism's genetic interests, if the genes that shorten its lifespan also increase its reproductive capability earlier in life. So reproductive interest goes beyond narrow selfishness; the most obvious example is parental care. We should find nurturing, even self-sacrificing behavior benefiting genetic relatives. Indeed, we do. Female lions, for example, take care of each others' cubs. If lions were highly preyed upon, this might have made sense as insurance, but they are not. Lions in a pride are, however, closely related; on average, the females share more than a quarter of their genes. By caring for cubs besides their own, lions look after their own reproductive interests.[25]

This gene-centered approach, coupled with game theory, gives biologists a sophisticated, mathematically rigorous way to explain behavior. Naturally we wonder how it applies to us; from the crude start of 1970s sociobiology to today's evolutionary psychology, there has been much effort to pin human nature down in our biology. However, a focus on genes tends to play down their complex interdependence with their environment, and it discounts culture. Though very illuminating in areas such as basic sexual behavior, gene-centered approximations become increasingly strained with the more complex variations of human culture.[26] We are not ants. But neither do our genetic interests disappear in a cloud of culture. Biologist Richard D. Alexander argues that our moral systems, in all their variety, serve basic reproductive interests. Our nonselfish acts do not drop out of the sky—they are continuous with reciprocal and kin-favoring behavior in other animals. However, our intelligence and social plasticity lets us explore a much wider variety of ways to further our interests. In particular, we form elaborate systems of *indirect reciprocity*. We keep track of how people help or harm others, and decide whether they are reliable reciprocators. We try to figure out other people's character and intent, while presenting ourselves as persons of good moral character—as dependable reciprocators. In all this, our purposes center on social and biological reproduction, including looking out for kith and kin. Public morality helps us negotiate our conflicts, often by concealing our interests.[27]

This is a realistic description of human nature. We still, however, need more to give culture its due, recognizing that it is not just a veneer over our genetic interests. After all, genes are not the only things we pass on to the next generation. We communicate a lot of complex information, not only to our offspring, but to many other people. Beliefs, ideas, and so forth—"memes," in Richard Dawkins's Darwinian metaphor[28]—are quite different than genes, but we do, in a sense, reproduce them. Our relevant interests, then, are not confined to propagating genes, but also beliefs and social enterprises with which we identify—even grandiose moral systems. We sacrifice to defend our faiths as well as our children.

This complicates matters. Still, though we have only begun to connect culture and biology, an interesting picture is emerging. Through our evolutionary heritage and our social enterprises, we have interests in life. Morality itself is an enterprise to resolve conflict, one of our many self-reinforcing enterprises which instill purposes and emotions in their participants. Our purposes are not arbitrary, since not every pattern of interests can find a stable niche in our social ecology. For example, we often find that genetic and cultural reproduction work best in concert. Our successful religions build upon sexual and parental emotions, and in turn sanctify a social environment which encourages reproduction. On the other hand, our interests are flexible, changing in history; they cannot be simply read off any theory of human nature.

This is not how we usually think of morality. In fact, there seems to be no true morality at all in this turbulent sea of replicating interests. But this is not surprising. There is no "true warmth" when we explain temperature in terms of molecular motion, no "true intelligence" among neurons. Concepts like a moral law or weird prescriptive properties which attach to our choices come from our folk theories of morality; they may well disappear from a better theory. But our purposes and our passionate commitments will remain, and we will understand them better. Evolution takes us a good part of the way. We need to do more to explain our moral perceptions, and for that, we must look at the biology of the brain.

Though evolution is rarely perfect, it can cobble together some amazing contraptions—in our case, a brain which can handle an immensely complex and ambiguous social environment. We have to act on our purposes in the company of others like us, finding occasions to cooperate without being exploited. We have to keep track of other people—their characters, purposes, and social connections. And in a fast-moving world, we need a quick and effective way to choose a good course of action. Ponderously sifting through a list of moral principles will not do—any set of rules will be too simple and sharp-edged for the complications of real life. So we have to *perceive* the good or evil in a situation, without waiting to deduce its place in our purposes. And of course, we must respond: we need to feel right and wrong in our guts.

The mass of neural networks in our skulls does just this. We are very good at pattern recognition. When we see a face we recognize it quickly, even effortlessly. We do this under varying lighting conditions, when we see only part of the face, even when we look at a stylized cartoon. Furthermore, recognition does not just evoke a name, but also emotions appropriate to the face and the circumstances. A neural network can represent a vast array of subtle possibilities, judge how closely they match "prototypes," and quickly activate appropriate response patterns.

Similar processes of pattern recognition and generalization underlie our moral perceptions as well.[29]

Neural networks are not programmed by feeding them a list of instructions; we mostly train them through examples. Our moral training also depends more on examples than on explicit rules. We learn that we need to trust people, and that we should inspire others to trust us.[30] We also see that some people might exploit us, and so reserve our fullest trust to a circle of relatives and tried-and-true friends. We learn such things by interacting with people; if Sunday school lectures work at all, it is less because of their "thou shalt nots" than because of the moral examples in their stories.

As we mature, we acquire moral knowledge, embodied in prototypes structuring the way we perceive social circumstances. This is real knowledge about the social world. There are objectively good ways to pursue our interests in this world—interests which are themselves shaped by society—and bad ways. Furthermore, this is the kind of knowledge which engages us. Mere high-sounding principles or well-meaning emotions we might shrug off, but for most of us, those stable patterns of purposes in a viable society are central to our identity. Moral knowledge helps us achieve those purposes. As Paul Churchland explains,

> Attempting to portray either accepted rules or canonical desires as the basis of moral character [invites] the skeptic's hostile question: "Why should I follow those rules?" in the first case, and "What if I don't have those desires?" in the second. If, however, we reconceive strong moral character as the possession of a broad family of perceptual, cognitive, and behavioral skills in the social domain, then the skeptic's question must become, "Why should I acquire those skills?" To which the honest answer is "Because they are easily the most important skills you will ever learn."[31]

Morality, then, is a practical skill, depending on useful knowledge. In that case, we may wonder how it compares to natural science. Churchland argues that they are very similar. Morality, like science, is a collective enterprise of critically revising knowledge. Both draw upon our ability to imagine different possibilities and test them. Science, it appears, helps us make our way in the natural world, while moral expertise lets us navigate the social world.

Churchland is right; we can find ways in which moral reasoning is continuous with science. However, we should not lose sight of some important differences. Science is a way of learning, not a device for coping; it is remarkable exactly in that it produces knowledge independent of our social contexts. Moral skills are much more tightly connected to particular purposes and social settings. Furthermore, morality does not just map our social world; it is also a way to contemplate changing it. In this sense, morality is closer to engineering than to science. So we should not see good and evil as objective in the way an analogy with science might suggest. If morality is about navigating our social world, it is much more solid than a whim—not all patterns of perception and purposes can successfully reproduce themselves. But also, there is no single correct pattern, and so different people may legitimately have differing moral perceptions of the same situation.

This lack of full-blown objectivity is bound to cause some discomfort. We too often demand something be either *subjective* or *objective*, and especially when anything significant is at stake, we panic if it looks tainted by emotion or subjectivity. But take, for example, the color red. That blood is red seems about as objective a fact as there can be. Different observers consistently agree on it, and we can see that exceptions like colorblind people fail to perceive something real. However, redness is not a physical property of any object, though we call a certain range of wavelengths of light "red." Red happens when that light interacts with receptors in an eye hooked up to a normally functioning brain. If our brains were significantly different in the way they implement color vision, we would not agree so easily that blood is red. In our species of primate, color perception is pretty basic, handled the same way in different individuals. We have evolved to lock onto an objective feature of the world, but even so, color does not exist apart from brains with a particular evolutionary history.[32]

Now consider a quality like attractiveness. We often say a person *is* attractive, expecting others to share this judgment. Indeed, our evolutionary history locks us on to the physical world in this case as well. We tend to admire symmetric features and other indications of health and reproductive fitness. But now, variation becomes more important. To begin with, attractiveness depends on the sex of the observer. Social status also affects attractiveness, and we have to learn to recognize status in a particular culture; not only genes, but also the cultures we propagate come into play. So even more than with color, it does not make sense to call attractiveness objective or subjective. It is more complicated.

Good and evil are *a lot* more complicated. We perceive an objective social world, and good ways to navigate it according to our interests. In pursuing our interests together, we cannot avoid coming upon ideas of cooperation and fairness. Moral perceptions lock on to real features of our social world. They are not, however, fully determined. Much of our moral enterprise is about variation, about open possibilities, and about efforts to collectively decide how to live together, conflicting interests and all. To insist good and evil are out there—the same for everyone, to be recognized if only a God opened our eyes—is, in our world, a mistake. It can only make us misunderstand our lives as social animals.

TALKING MORALS

We find the roots of morality in our biology. Still, there must be more to the story. We negotiate conflicts and find ways to cooperate by talking to one another. In the process, we add layers of rules and even philosophical systems on top of our perceptions of good and evil. This is no window dressing. Though we can be stubborn in our gut-level convictions, every now and then our perceptions change through moral debate. So even if morality is not rules and systems all the way down, we need to better understand what is going on when we talk morals.

A good way to flesh out this aspect of morality is to answer some criticisms from a more traditional perspective. So let us start with a conceptual objection. Biology and cognitive science have great promise for describing what people do. But they will never, a critic might say, tell us what people *ought to do*. We cannot derive prescriptions from human nature without smuggling in some moral premises. Take a Muslim example. In the Quran, 2 Al-Baqarah 228 grants women lesser rights, asserting that men are superior. Theologian Süleyman Ateş justifies this, saying:

> It is true that as a whole, the male sex has been created superior to the female. Even the sperm which carries the male sign is different from the female. The male-bearing sperm is more active, . . . the female less. The egg stays stationary, the sperm seeks her out, and endures a long and dangerous struggle in the process. Generally in nature, all male animals are more complete, more superior compared to their females. . . . Man, being more enduring at work, and superior in prudence and willpower, has been given the duty of protecting woman.[33]

Such Aristotelian biology is, of course, ridiculous for anyone paying attention to modern science. But say it was correct, and that men and women had natural roles in a created order like Ateş describes. A critic can still ask whether we *ought to* conform to these roles. In answering yes, Ateş makes the moral assumption that the created order is good. This is unwarranted conservatism. Someone with Gnostic inclinations, for example, could easily have different moral intuitions. She might agree that men are active, women passive, and so on, but she might also declare that nature is the work of an incompetent demiurge. She would then seek to free female spirits from their social bondage. It seems that even if Ateş had his facts straight, he could not derive moral prescriptions from them.

If we cannot connect fact and value, a correct biology will also fail to capture the essence of morality. Indeed, the history of "evolutionary ethics" is a long story of making pronouncements on human nature, but never quite being able to say why we should approve of this nature.[34]

Unwarranted conservatism has been especially common, from Social Darwinists who held working-class suffering necessary to improve the race to those sociobiologists who claimed that female subordination was in our genes and we should live with it. Even a careful theory like Churchland's falls under this shadow. After all, if good moral character consists of the skills to successfully operate in society, the conventional morality of any culture is the safest way to go.[35] It certainly is what most of us learn. Is then a feminist who rebels against a traditional Muslim society deficient in her moral perception?

These are serious problems. The way to overcome them is to realize that the chasm between fact and value is not a conceptual necessity. It is instead a claim about the nature of morality—and a false one at that. Imagine, once again, an Aristotelian world. For Ateş, value is not something external to the world but part of God's creation. The universe could have embodied a moral order, manifested in the natural roles we gravitated toward. Our moral intuitions would have told us such roles were right; even women would find themselves happiest as doormats. Of course, even in such a world a Gnostic or a philosopher could legitimately ask if this order was really good. But then all the Muslim would have to do would be to say yes, according to the best explanation of our moral lives, it *is* good. The fact that we can always question this conclusion does not disprove an Aristotelian view any more than the fact that we can always ask "but is it really true?" invalidates scientific theories. Ateş's mistake is that he has his biology wrong—not only the biology of the sexes, but the biology underlying morality.

What happens, then, when we get our biology right? Evolution is a story of historical accidents; it does not settle on Platonic essences or eternally optimal ways of life. Populations vary. Even within tight-knit cultures there will be shifting, unstable alliances of interests. So variety and disagreement are in the nature of social thinking. People will have different moral perceptions, sometimes because they lack skills or because they misperceive the social world, but also because they correctly see ways to follow different purposes. There is more. Our brains do not just passively represent the social world and act on deep-seated interests. Moral thinking draws on our ability to envision different actions and their consequences—it is an *imaginative* activity.[36] Acquiring skills to navigate the social world is only half the story. The same biology also underlies our ability to imagine even radically different ways to conduct our collective affairs.

Possibilities for change, then, are a natural part of our moral lives, and in open societies, so are raging public debates about the direction we should take. A radical vision like feminism starts with people who are not satisfied and who imagine a world more to their liking. They use existing ideals like fairness, and try to convince others that it is in their broader interest to live differently. For many, seeking equality for women becomes a social enterprise which is itself a source of interests. Feminism then has a chance to become a successfully self-reproducing way of life, establishing itself in a subculture and perhaps influencing a wider swath of society.

Of course, there are real-world constraints on our imagination and on what we can achieve. In an agrarian, peasant society dominated by a male warrior caste, feminism could never become more than a cluster of female resentments. But societies change. And for a technological animal, even straightforwardly biological limits are not as permanent as they seem. Women's reproductive biology has often been an obstacle to public participation, but in a world of birth control pills and breast pumps, the limits have shifted. As we learn more about biotechnology, we will have the option of redesigning human nature more directly. To control our population, we might overhaul our reproductive systems to make pregnancy a matter of explicit choice. If male violence gets to be too risky, we might engineer our genes to tone it down. Choosing any such path will change our societies, bring forth new interests, and prompt us to recreate good and evil. We will play God. These are science-fiction themes,[37] but then, fiction has always been a good way to exercise our moral imagination.

A critic might say this still misses the point. Maybe naturalizing morality does not mean unwarranted conservatism. But having more options only sharpens the problem of what we ought to do. In a society where both feminism and sex-role conservatism are live options, not to mention all the variations and shades of grey, what is the right choice? We still have lots of facts and no values.

This is true. But it also is not a problem, as there is no such thing as stand-alone value—only things which people value.[38] There could have been some value, some "oughtness" residing in a Platonic realm, ready for Reason or Revelation to apprehend. Alternatively, ours could have been an Aristotelian world where we had intrinsic roles, framing the morality of any situation. But our best explanation for how people value things sets us in a world with no fundamental moral order. Good and evil become complicated. We do, of course, confront choices. But we cannot choose by stripping ourselves of all purposes and looking at life like an external judge. We join the debate over our options, put our purposes up for negotiation, and make a choice when we have to. If we are open-minded enough, we revisit the debate from time to time. And that is all. Morality happens only inside our social enterprises; it does not reflect an eternal ideal.

Such an account is bound to raise fears of moral laxity, even of anything-goes relativism. But this exaggerates the problem. Good and evil may be more complicated than the fully objective realities our religions often demand, but they hardly collapse into arbitrariness, even less indifference. There is, however, a genuine practical concern behind such fears. We can muster up plenty of passionate commitment from within the circle of our interests. But there are many patterns of interests in our societies and hence many occasions for conflict. We cannot just leave it at that; holy wars become too much of a burden after a while. So we also have an interest in finding a way to regulate our conflicts, to set some standards which stand above the fray.

In regulating such conflicts, rules and systems come into their own. Practical reasoning is thin, bloodless, far from the rich network of interests which undergird our gut-level morals. But in large social groups, or when people with significantly different moral perceptions negotiate, we have to rely on explicit rules. Now our philosophers can run free; debating, reasoning, constructing moral systems. Some will even argue we can derive morals through reason alone. And though not so powerful as that,[39] practical reason can in fact do a lot for us. It will affirm that we need a society-wide way to resolve conflicts. Though any one neighbor may gain from plundering another's house, everyone will be more secure if they take an interest in cooperation and mutual trust. A rational social order will arrange things to minimize the opportunity for plunder, and institute sanctions against cheaters. And of course, a rational order will not systematically undermine itself, so it will ensure that new generations are trained to follow the rules.[40] This way, rules become signposts for instilling new moral perceptions. These perceptions will deepen, and people will be able to handle the grey areas and complications which strained the original rules. Neither moral perception nor moral talk are enough on their own; to solve our problems together, we need to deliberate and revise our perceptions.[41]

Our actual moral enterprises are less tidy than ideals of practical reason suggest. Into the mix go moral traditions from many communities and subcultures, along with rival proposals for regulating society. In a complex, open society we will have an unholy mess of traditions claiming our attention. Since the business of life is not to create elegant moral systems but to muddle through, we will end up using a patchwork of moral languages.[42] Untidy though it is, we have a moral enterprise. Within it, we can engage new ideas, starting from our old convictions and revising them in debate. We can go back and forth between reasons, rules, and perceptions, keeping everything open to criticism, hoping to agree on a direction. For those of us committed to this overall moral enterprise—and many of us *will* be—the norms we devise will have force, since morality itself will be one of our purposes.

This gives us most of what we want from morality. We can do without fundamental moral principles to endlessly quarrel over; we start with our purposes and perceptions, and critically revise them. We can hope to reach decisions which rational people will agree fit their interests, including their interests in maintaining a healthy moral enterprise. We cannot step out of all social roles; "participant in moral debate" is itself a role we invent. Morality is, nonetheless, a way to move beyond our more narrowly self-interested purposes, and even, if all goes well, to begin to care about others.

This is a much more solid picture than fears of relativism suggest. So once again, the moral enterprise invites comparison to science. Like science, our moral talk is also enabled by our biology, letting us start from our perceptions and folk theories, critically revising them to the point where we can even contemplate changing our nature. But again, we should remember that good and evil are more complicated than the facts which science converges upon. Our moral debates often foster the illusion that when we reach consensus, we apprehend an objective moral reality. It would be nice, for those of us committed to talking morals, if there was a moral truth we could find—if only we were rational enough, if only we knew all the facts. So far, it appears we can lock onto the mass of the earth or the wavelengths of light, but there is nothing in the world which corresponds to good and evil. We do not have the kind of solid reality checks which would prevent unresolvable moral disagreements.[43] And there will be rational people with little interest in morality, content to let others do the work.

Our moral agreements are of our own devising, though they are no less agreements for that. A God who was the source of morality probably would have arranged matters differently. But it is hardly news at this point that the world has no moral purpose.

BELIEVING THE ABSURD

Morality is important to *us*—to how we want to live and how we hope others will treat us. Abrahamic religion raises the stakes, as God, the personal reality at the bottom of everything, has an intensely moral relationship with its creation. Our modern knowledge, however, removes good and evil from the center. Nothing is transcendent anymore.

It would now seem the argument is over—we should pack our books and get on with our lives in a godless world. We might still wonder whether, without God, we must become hopeless nihilists, or if we can sustain a public morality knowing there is nothing transcendent backing it up. After all, plenty of religious thinkers argue that life is absurd without a God, asking us to believe for the sake of therapy if nothing else.[44] *If* our knowledge were negligible, and the debate over God produced nothing but the usual philosophical stalemate, we could very well believe even without knowledge. After all, religion has the advantage here. With God, harmony reigns and truth is the best therapy. But there is no stalemate. We might still be influenced by morose meditations on the absurdity of a world with no ultimate values; but to dispel the mood, sitting down with some chocolate chip cookies and a good science-fiction adventure works well enough.

There is, however, also a real issue here. Skepticism, no less than faith, is not an ethereally intellectual position. Serious piety comes from a life lived as if we have nowhere to hide from the awesome reality of God. Serious disbelief—not just indifference—also emerges from a way of life, even if only that of an intellectual subculture. And we argue about God not just out of curiosity but because different ways of life are live options for us. We decide, we defend, we try to persuade others. Arguing about God, in other words, helps advance our interests as embodied in our social enterprises. This is not mere apologetics; in an open-ended debate, beliefs are at risk and interests can shift. Nevertheless, our broader debate over God does not just concern correct knowledge, since knowledge is but one of our interests in life. In fact, for most of us outside of an intellectual community, knowledge is valuable only to the extent it helps us accomplish other purposes. So now we confront a perverse consequence of naturalizing morality. With religion, everything really important rides on getting it right with God—it is our duty to find truth and believe it. Without any transcendent moral reality, there is no such imperative. If believing there is a God is helpful, then, pragmatically speaking, it appears rational to believe even though it is false.

This grates against our intuitions. In a pragmatist mood, we think all knowledge is practical know-how and suspect that at some level truth has to be more useful than falsehood. But something is useful only in the context of a stable pattern of interests. And though our moral enterprise can help us negotiate between different interests, it is not powerful enough to demand that we seek truth above all. This now offends our commitment to intellectual integrity; it would seem "beneficial reasons" justify belief only as a tiebreaker or in extreme circumstances. Michael Martin, for example, grudgingly concedes that "an eighty-nine-year-old Black Muslim and former Catholic" on her deathbed, fearing she will never see her dead husband again unless she returns to the fold, should send for a priest—provided there is no good reason that one faith is closer to truth than the other, that her return would set no precedent, and few would ever know of her decision.[45] But though this regard for truth makes sense for scientists or philosophers, in more ordinary social circumstances, the benefits of a belief carry a lot more weight. In Ernest Gellner's words,

> 'Ideas', 'solutions' to 'problems', adaptive devices of all kinds, are normally complex, many-stranded and multi-purpose things. Their survival depends on their satisfaction of a wide variety of human and social requirements, amongst which theoretical truth is not very prominent. This is of course the central fallacy of pragmatism: the failure to see that truth cannot be assimilated to usefulness (or some circumlocution thereof), because things are useful for a wide variety of reasons other than their 'truth', and may indeed be useful because they are *not* true.[46]

We can state the same observation with a biological metaphor. Our faiths and our varieties of skepticism compete for cultural reproduction. Some false beliefs are a handicap—a sect which believes celibacy is God's law for all will not last long; a UFO cult whose members commit suicide believing they will translate to a waiting spaceship will not become the next world religion. True beliefs, however, may also suffer a reproductive disadvantage. Skeptics cannot save souls; so unless the local religion is allied too closely with the local tyrants, infidels are liable to lack in missionary zeal. Serious skeptics cannot offer the moral clarity of an Abrahamic faith. In a world with no white-hot contrast of cosmic holiness and human sin, and quite likely no single form of rational moral life, our communities still need a common purpose. A sanctified falsehood might provide this purpose, but a runaway iconoclasm will not.

So religion *might* enjoy a genuine moral advantage—whether this is actually the case for people in our circumstances is a different matter. When we pose this more concrete question, the first thing we notice is that for those of us in the Western world, old-time religion is in trouble. Visiting a Muslim country, we can still be awed by the social power of faith, but back home, God has vacated the public square. The post-Christian West is not a land of rationalists; we have a thousand faiths, we very often hold personalized, diffuse supernatural beliefs. But religion has become irrelevant to much of our public life, and vast numbers of people are simply indifferent toward God.[47]

Why, then, have our religions faded? First of all, our lives are fragmented. Instead of having well-defined places in an overall social order, we go through life occupying a bundle of roles. We might be a plumber by occupation, but we are also a parent, the president of the local bird-watching society, a union activist, a softball coach. Our varied enterprises develop internal norms which do not depend on a unified cosmic purpose in life. With such social differentiation, tight-knit communities become difficult to sustain. Religion retreats to a private sphere, serving more as individual therapy than as a legitimator of earthly order.

We are not all happy with this fragmentation; many of us would like to integrate our lives, to be a Christian plumber instead of a Christian and a plumber. However, social differentiation turns out to be an unexpectedly brilliant solution to communal conflict. Modern society arose by accident, not by philosophical design. To political thinkers before the European Enlightenment, the obvious alternative to an ideologically united polity was chaos. But then, we found out that instead of igniting a war of all against all, "differentiation often enables us to shunt off explosive conflicts into other spheres of society where they can be handled more easily."[48] A plumber could fix a heretic's pipes without violating community boundaries, and still rage against heresy in an appropriate setting. In our secularized world, we expect a Jewish doctor to treat a Christian patient the same as a Jew, a Protestant customer not to care if a shopkeeper is Catholic. It was not always like this.[49]

This is a cold peace, not the brotherhood of man in the Kingdom of God. Yet it unleashes a rational science and an autonomous economic life, both which turn out to be extremely powerful. This combination further fuels secularization, despite the social handicap of intellectual naturalism. As Ernest Gellner explains,

> This new learning respects neither the culture, nor the morality, of either the society in which it was born, or of those in which it makes itself at home by diffusion. It is, most emphatically, "beyond culture and morality." Alas, often it is not only *beyond*, but *against*. One of the bitterest and most deeply felt, and alas justified, complaints against science is, precisely, that it disrupts morality. It does not, as the previous, technologically impotent (or very nearly so) learning had done, serve to underwrite social and political arrangements, and to make men feel more or less at home in the world and at ease with it, or indeed in awe of it, whilst significantly failing to help control it. Past belief systems were technically spurious and morally consoling. Science is the opposite. Science markedly fails to perform such social services, and the attempts to enlist it to and oblige it to perform them have failed abysmally (the year 1989 witnessed the final and dramatic collapse of the most elaborate and ambitious of such attempts). Its failure to legitimate social arrangements, and to make men feel at home in the world, is the commonest charge levelled at science. The charge is entirely valid.[50]

Science undermines the myths which impose moral order on an accidental world, but it can compensate because it is inextricably linked with technology. Our modern, religiously disunited culture can enjoy better weaponry, better medicine, and improved material comforts. This is a formidable pragmatic advantage in cultural reproduction. As many a Muslim knows, a soulless secular culture with superior weapons can colonize even the most pious. Worse, the wavering faithful may see that the infidels are better off materially, and that even some of their more disreputable desires are satisfied without causing social ruin. In a backhanded way, the religious critics are right: scientific materialism has breathing space only because it supports materialism in the crass sense.

Material progress is nothing to sneeze at. Modern technological society can protect us against many of life's sufferings, it makes unprecedented individual freedom possible, and it dampens the murderous authoritarian enthusiasm monotheistic faiths occasionally ignite. There is a price: modern life thins out our social networks, making it difficult for many of us to have satisfying human relationships. And this is no small loss. So faith and skepticism end up in a social stalemate. Our public morality concerning religion is, naturally, a compromise. We do not criticize each other's beliefs unless they interfere with our lives. And religions become therapies, Band-Aids for the social isolation of modern life.[51] To committed believers and unbelievers alike, this compromise may seem shallow, even a refusal to confront reality unevasively. So it is. It also is a solution which embodies much social experience, and a good deal of moral wisdom. Modern life has serious problems, enough to keep social thinkers employed indefinitely. But if we are going to negotiate a better compromise, we would best start from what already more-or-less works.

In that case, what are our options? We could, of course, embrace religion again. Islam is reasserting itself against alien Enlightenment influences with some success. But Muslims will always struggle with technological inferiority. They often hope to import technical skills while keeping out degenerate secular culture. Unfortunately, sex, drugs, and rock music are side issues. The deeper problem for theocracy is that a secular, fragmented society is the better environment for producing sustained technological progress. Even if citizens of an Islamic republic should feel at peace with the universe and with their fellow believers, they cannot remain isolated from modern cultures which enjoy more material power.[52]

We might still want to inject more religion into our lives, while stopping short of theocracy. It might shore up our embattled communities. Plus, modern life is very often mind-numbingly superficial, and religion promises to restore a depth which comes from a focus on enduring values. Not a few secular conservatives argue that Westerners need the moral self-discipline which a healthy public religion sustains.[53] All this may be true; at least, it is something we can argue about. We might do better with a tame religion which gives us a hint of common purpose, restrains our antisocial interests, and does not overly obstruct the development of serious knowledge. But to stop here would only mean we lack moral imagination. After all, if religion has good side effects, there may be other ways to secure these benefits.

To cure some of our social ills, the left wing of the Enlightenment proposes restoring some solidarity to our economic life. Of course, we can no longer accept an irresponsibly utopian socialism which imposes a single vision on society. Nor can we do without commercial markets. The autonomous enterprises which make up civil society depend on markets for independence from the power of the state or that of a dominant faith. Eliminating them risks too much.[54] However, it is also clear that markets can get out of hand. Enterprises like medicine or science depend on their internal ethics, valuing patient care or an open exchange of information. The market subverts these values, putting doctors under the thumb of insurance executives, turning scientists into corporate researchers who have to conceal profitable discoveries. As Christopher Lasch says,

> the market . . . puts an almost irresistable pressure on every activity to justify itself in the only terms it recognizes: to become a business proposition, to pay its own way, to show black ink on the bottom line. It turns news into entertainment, scholarship into professional careerism, social work into the scientific management of poverty. Inexorably it remodels every institution in its own image.[55]

As the internal ethics of our enterprises weaken, so does our ability to conduct our collective affairs with the help of an overall moral enterprise. We do not have a common purpose dictated by God or by Dialectical Materialism. But we *do* have an opportunity to maintain a moral enterprise which allows us to forge some common goods without regimenting society. If we leave everything to markets, we can no longer do this. Our relationships take on an economic character, and we are left with an impoverished moral language centering on rights and contracts. And of course, markets notoriously create great inequalities of power. So what remains of public moral debate—politics—serves the interests of the commercially powerful. If once we were peasants ruled by priests, we have now become consumers ruled by investors.[56] We do not know exactly how to change this and construct a more democratic social world. But this, along with more immediate problems like preventing environmental degradation and nuclear war, is the greatest challenge to our moral imagination.

None of our options lead to heaven on earth. Even if, perchance, we find a way to realize the woefully half-baked vision of democratic socialism, it is hard to say religion will wither away. In an accidental world, we will always have needs which cannot be satisfied either by technology or by social experiments. Yet we will hope, and where there is hope there will be faith. And even if we cure many of our social ills, *knowing* that this is an accidental world cannot but take the edge off the moral clarity of our religious traditions, a clarity which helps many of us muddle through life. In that case, it appears we will never decisively conclude the social contest between faith and skepticism, but just move on to another pragmatic compromise.

BEYOND PRAGMATISM

This is a dismal conclusion. Moral arguments are often the most engaging part of the debate over God; we discuss weighty matters like the problem of evil, bring in exciting science on the evolution of morality, and tie it all to concrete concerns about our life together. To end with a social stalemate is almost an insult to the seriousness of these matters.

Religious people certainly cannot be satisfied with pragmatic reasons to believe, even if we found a compelling set of such reasons. If it turned out that religion makes us healthier, happier,[57] and morally more trustworthy, these are still *secular* reasons to be religious. To believe because a church lets us express our most sublime feelings, or to accept Islam because it can see us through a crisis in life, only cheapens religion. We try to use God for human ends, instead of surrendering to an awesome reality even when this goes against our worldly interests. Faith may be good therapy, but when religious apologetics start to emphasize how faith is good for us, this is a sign that religion is in decline.[58]

A dismal pragmatism will not impress the committed unbeliever either. Without God, we are on our own, and the humanist freethinking tradition takes this as an opportunity to reason our way to a better world. Many would agree with G. A. Wells when he says, "But surely it is better to find the truth and proclaim it. In the long run, it may be impossible to hide it; and instead of shutting our eyes, we more wisely use our ingenuity to exploit it, for every real remedy for our troubles can come only from knowledge and not from ignorance."[59] If anything is obvious, this is. Speaking of useful falsehoods may seem like a sober pragmatism, but it is really a capitulation to postmodern despair.

There is some truth here. However, it is also true that humanists have been naive in claiming we know of a completely godless way to live satisfying lives together. For some subcultures this may be true; certainly, many of us today live faithless and fancy free. But we need more. We have to face the fact that magical thinking is not a tragic streak of irrationality in our species, but a natural outcome of the way we perceive the world—a way which has served us fairly well. It is scientific thinking, not religion, which is profoundly unnatural for us; no matter how science progresses, most of us will be most comfortable explaining the world through the acts of personal agents.[60] And outside the realm of some exotic questions far from the concerns of everyday life, this way of thinking works fine. For most people, learning to go without a God is a costly undertaking for no clear benefit.

Even as a matter of personal therapy, we need to give religion its due. Unbelievers often, with varying degrees of psychological insight, claim religion exists because it consoles us when we encounter death and other ravages of a contingent existence.[61] Consolation, however, is no bad thing; and at some point, heroic talk about having the courage to face reality without comforting illusions becomes just posturing.

To get beyond both a dismal pragmatism and a romanticism of Reason, we again have to put our moral imagination to work. For better or worse, we cannot fully predict our social future, only make it. This occasions a sober, conservative prudence. But it is also a spur to radical thinking: today's pragmatic compromise does not bind us forever. A few centuries ago feminism, for example, would not have appeared on our short list of pragmatic moral options. Given our history, it was—and often still is—a radical, even bizarre moral vision. It is also up to us, if we so desire, to explore new ways to harmonize cognitive reason with our moral life. The words of G. A. Wells do not express some eternal metaphysical truth, but it just might be in the realm of possibility to make them a social reality.

How to start? First of all, humanists need to look beyond trying to convince a religious public that unbelievers are decent people. It is not enough to argue that we can make sense of good and evil without a God, and that interests rooted in the material world are sufficient to keep us from becoming moral monsters. If we reject myths of absolute morality, it is not only because they are false, but also because they illegitimately restrict our moral imagination. Next, and more difficult, humanists need to address the isolated individualism of modern life, even though religious dissent has found social space in just such an environment. We have to take the ideal of open, inclusive moral communities more seriously. As in science, we need to steer between letting individual opinion rule and deferring to authority. The *marketplace* of ideas is a singularly inappropriate metaphor; we want to develop standards of merit to weigh ideas with, not approach them like a consumer choosing wallpaper. Humanists should also recover some of the radicalism of the freethought tradition; secularists today are all too often complacent about the status quo.[62] There is no point complaining about a dismal compromise when humanists not only have nothing better to offer, but are not even thinking about fashioning something new.

Maybe some form of humanism can free scientific naturalism from its intellectual ghetto. Maybe not. At least, we might have a more interesting moral debate between believers and skeptics. Meanwhile, those of us interested in knowledge can still pursue it, and appreciate the stars even with no God to set them in the heavens. If thereby we can also make ourselves useful, so much the better.

NOTES

1. Clifford Geertz, "Ethos, World View, and the Analysis of Sacred Symbols," reprinted in Clifford Geertz, *The Interpretation of Cultures: Selected Essays* (New York: Basic Books, 1973), p. 129.

2. Judaism might seem an exception, but see Neil Gillman, *The Death of Death: Resurrection and Immortality in Jewish Thought* (Woodstock, N.Y.: Jewish Lights Publishing, 1997).

3. Jeffrey Stout, *Ethics After Babel: The Languages of Morals and Their Discontents* (Boston: Beacon, 1988), pp. 80–81.

4. Ravi Zacharias, *Can Man Live Without God?* (Dallas, Tex.: Word, 1994), p. 32.

5. Ibid, p. 40. For a more sophisticated defense of such metaphysical intuitions, see Franklin I. Gamwell, *The Divine Good: Modern Moral Theory and the Necessity of God* (Dallas, Tex.: Southern Methodist University Press, 1996).

6. E.g., Hans Küng, *Does God Exist? An Answer for Today*, trans. Edward Quinn (Garden City, N.Y.: Doubleday, 1980); Jeffrey Burton Russell, *Mephistopheles: The Devil in the Modern World* (Ithaca, N.Y.: Cornell University Press, 1986).

7. *Catechism of the Catholic Church* (English translation, Boston: Libreria Editrice Vaticana, 1994), ¶2352; Uta Ranke-Heinemann, *Eunuchs for the Kingdom of Heaven: Women, Sexuality, and the Catholic Church*, trans. Peter Heinegg (New York: Doubleday, 1990).

8. William A. Norgren and William G. Rusch, eds., *Implications of the Gospel: Lutheran-Episcopal Dialogue, Series III* (Minneapolis: Augsburg Fortress, 1988), p. 50.

9. Mark Twain, 1906, in Howard G. Baetzhold and Joseph B. McCullough, eds., *The Bible According to Mark Twain: Writings on Heaven, Eden, and the Flood* (Athens: University of Georgia Press, 1995), p. 326.

10. Majid Fakhry, *A History of Islamic Philosophy*, 2d ed. (New York: Columbia University Press, 1983), pp. 204–205.

11. Elaine Pagels, *Adam, Eve, and the Serpent* (New York: Random House, 1988), p. 135.

12. Jeffrey Burton Russell, *Satan: The Early Christian Tradition* (Ithaca, N.Y.: Cornell University Press, 1981), pp. 194–95. The dramatic necessity of evil is even clearer in an artificial mythology like that of J. R. R. Tolkien. His tale depends on evil with cosmic significance; appreciating his cosmology probably does more than any philosophical argument to justify evil. See *The History of Middle Earth* series, particularly vol. X, *Morgoth's Ring: The Later Silmarillion* (Boston: Houghton Mifflin, 1993), pt. 5.

13. Hans Jonas, *Mortality and Morality: A Search for the Good after Auschwitz*, ed. Laurence Vogel (Evanston, Ill.: Northwestern University Press, 1996), p. 134. Also see Hans Schwarz, *Evil: A Historical and Theological Perspective*, trans. Mark W. Worthing (Minneapolis: Fortress, 1995), pp. 181–84.

14. Michael Stoeber, *Evil and the Mystics' God: Toward a Mystical Theodicy* (London: Macmillan, 1992). Alternatively, a specific religious goal,

such as redeeming individual lives, might be the overarching good, though I find this much less promising. Marilyn McCord Adams, *Horrendous Evils and the Goodness of God* (Ithaca, N.Y.: Cornell University Press, 1999).

15. Philosophical atheists can still make a good case based on evil; e.g., A. M. Weisberger, *Suffering Belief: Evil and the Anglo-American Defense of Theism* (New York: Peter Lang, 1999); Michael Martin, *Atheism: A Philosophical Justification* (Philadelphia: Temple University Press, 1990), chaps. 14–18. The result is to drive theists to make increasingly strange excuses.

16. Kai Nielsen, *Ethics Without God* (Amherst, N.Y.: Prometheus Books, 1973), chap. 1; Michael Martin, *The Case Against Christianity* (Philadelphia: Temple University Press, 1991), app. 1.

17. Richard Garner, *Beyond Morality* (Philadelphia: Temple University Press, 1994), pp. 215–16.

18. Stout, *Ethics After Babel*, chap. 5; Michael Peterson et al., *Reason and Religious Belief: An Introduction to the Philosophy of Religion* (New York: Oxford University Press, 1991), pp. 239–41.

19. J. L. Mackie, *The Miracle of Theism: Arguments For and Against the Existence of God* (Oxford: Oxford University Press, 1982), pp. 114–18.

20. Phillip L. Quinn, *Divine Commands and Moral Requirements* (Oxford: Clarendon, 1978), pp. 58–60.

21. *Catechism of the Catholic Church*, ¶1960; J. Budziszewski, *The Revenge of Conscience: Politics and the Fall of Man* (Dallas, Tex.: Spence, 1999). For a critique, see Kai Nielsen, *God and the Grounding of Morality* (Ottawa: University of Ottawa Press, 1991), chap. 3.

22. John Gallagher, *The Basis for Christian Ethics* (New York: Paulist Press, 1985).

23. Richard Dawkins, *River Out of Eden: A Darwinian View of Life* (New York: Basic Books, 1995), pp. 131–32.

24. Robert Axelrod, *The Evolution of Cooperation* (New York: Basic Books, 1984); Patrick Grim et al., *The Philosophical Computer: Exploratory Essays in Philosophical Computer Modeling* (Cambridge: MIT Press, 1998), chap. 4.

25. Henry Plotkin, *Evolution in Mind: An Introduction to Evolutionary Psychology* (Cambridge: Harvard University Press, 1998), chap. 3.

26. Philip Kitcher, *Vaulting Ambition: Sociobiology and the Quest for Human Nature* (Cambridge: MIT Press, 1985); John Cartwright, *Evolution and Human Behavior* (Cambridge: MIT Press, 2000).

27. Richard D. Alexander, *The Biology of Moral Systems* (New York: Aldine de Gruyter, 1987). See also James Rachels, *Created from Animals: The Moral Implications of Darwinism* (Oxford: Oxford University Press, 1991), chap. 4.

28. Richard Dawkins, *The Selfish Gene*, 2d ed. (Oxford: Oxford University Press, 1989), chaps. 10–13; Susan Blackmore, *The Meme Machine* (Oxford: Oxford University Press, 1999).

29. Larry May, Marilyn Friedman, and Andy Clark, eds., *Mind and Morals: Essays on Cognitive Science and Ethics* (Cambridge: MIT Press, 1996).

30. Annette C. Baier, *Moral Prejudices: Essays on Ethics* (Cambridge: Harvard University Press, 1994), chaps. 6–9.

31. Paul M. Churchland, "The Neural Representation of the Social World,"

in May, Friedman, and Clark, *Mind and Morals*, p. 107. Moral relativists and absolutists share a common mistake about the nature of morality; that it consists of bedrock rules. John W. Cook, *Morality and Cultural Differences* (New York: Oxford University Press, 1999).

32. Daniel C. Dennett, *Consciousness Explained* (Boston: Little, Brown, 1991), pp. 375–83.

33. Süleyman Ateş, *Gerçek Din Bu*, vol. 1 (İstanbul: Yeni Ufuklar Neşriyat, 1991), p. 37. My translation.

34. Paul Lawrence Farber, *The Temptations of Evolutionary Ethics* (Berkeley: University of California Press, 1994); Alexander Rosenberg, *Darwinism in Philosophy, Social Science, and Policy* (New York: Cambridge, 2000), chap. 6.

35. James P. Sterba, in May, Friedman, and Clark, *Mind and Morals*, pp. 252–53.

36. Mark Johnson, *Moral Imagination: Implications of Cognitive Science for Ethics* (Chicago: University of Chicago Press, 1993).

37. To explore the ambiguities of, for example, dampening our aggression, see Stanislaw Lem, *Return from the Stars* (San Diego: Harvest/HBJ, 1980).

38. J. L. Mackie, *Ethics: Inventing Right and Wrong* (London: Penguin, 1990); Garner, *Beyond Morality*.

39. Edward Regis Jr., ed., *Gewirth's Ethical Rationalism: Critical Essays with a Reply by Alan Gewirth* (Chicago: University of Chicago Press, 1984).

40. Kurt Baier, *The Rational and the Moral Order: The Social Roots of Reason and Morality* (La Salle, Ill.: Open Court, 1995). Though Baier concedes that "complete success" is impossible for the moral enterprise (pp. 250–62), and that different moral orders can be rationally sound (p. 274), his argument that rationality just about requires an egalitarian social order is rather strained—see Kai Nielsen, *Why Be Moral?* (Amherst, N.Y.: Prometheus Books, 1989), pp. 295–98, for a discussion of a "classist amoralist" who is very inegalitarian but also seems quite rational.

41. Andy Clark, "Connectionism, Moral Cognition, and Collaborative Problem Solving," in May, Friedman, and Clark, *Mind and Morals*.

42. Stout, *Ethics After Babel*, pp. 218–19; Michael Luntley, *The Meaning of Socialism* (La Salle, Ill.: Open Court, 1989).

43. Gilbert Harman, *The Nature of Morality* (New York: Oxford University Press, 1977), chap. 1; Nielsen, *Why Be Moral?* chaps. 1, 2.

44. E.g., John Carmody and Denise Lardner Carmody, *Interpreting the Religious Experience: A Worldview* (Englewood Cliffs, N.J.: Prentice-Hall, 1987), which, inspired by Eric Voegelin, tries to read history as a prescription for mental health. Also see Shabbir Akhtar, *A Faith for All Seasons: Islam and the Challenge of the Modern World* (Chicago: Ivan R. Dee, 1990). He insists that an authentic ideology must overcome the crises of life, and asks, "Isn't every authentic philosophy a search for a vision that transcends the indifferent ravages of time and contingency?" (p. 122). Maybe so, but then an authentic philosophy must embrace falsehood.

45. Martin, *The Case Against Christianity*, p. 21; Martin, *Atheism*, chap. 9. For a contrary view, see Taner Edis, "The Rationality of an Illusion," *Humanist* 60, no. 4 (2000): 28.

46. Ernest Gellner, *Relativism and the Social Sciences* (New York: Cambridge University Press, 1985), p. 53.

47. Steve Bruce, ed., *Religion and Modernization: Sociologists and Historians Debate the Secularization Thesis* (Oxford: Clarendon, 1992); Steve Bruce, *Religion in the Modern World: From Cathedrals to Cults* (Oxford: Oxford University Press, 1996). In contrast, Andrew M. Greeley, *Religion as Poetry* (New Brunswick: Transaction, 1995); Rodney Stark and Roger Finke, *Acts of Faith: Explaining the Human Side of Religion* (Berkeley: University of California Press, 2000), cite poll data indicating high levels of supernatural belief as evidence against secularization; however, continuing supernaturalism can accompany a sharp decline in the *public* role of religion.

48. Don Herzog, *Happy Slaves: A Critique of Consent Theory* (Chicago: University of Chicago Press, 1989), p. 173.

49. *All* orthodox, premodern Abrahamic religions divide the people of God from the infidels and promote separation. This is no less true for Judaism, though its history of being persecuted obscures this; Israel Shahak, *Jewish History, Jewish Religion: The Weight of Three Thousand Years* (Chicago: Pluto, 1997).

50. Ernest Gellner, *Postmodernism, Reason and Religion* (London and New York: Routledge, 1992), pp. 59–60.

51. This is particularly clear in an open, diverse religious marketplace, where many take a consumer-like, pragmatic approach to faith. Richard Cimino and Don Lattin, *Shopping for Faith: American Religion in the New Millennium* (San Francisco: Jossey-Bass, 1998), argue that American religion is such an environment, where experience—particularly psychologically therapeutic experience—is decisive for spiritual consumers.

52. Don Peretz, in Don Peretz, Richard U. Moench, and Safia K. Mohsen, *Islam: Legacy of the Past, Challenge of the Future* (New York: New Horizon/North River, 1984); Fatima Mernissi, *Islam and Democracy: Fear of the Modern World*, trans. Mary Jo Lakeland (Reading, Mass.: Addison-Wesley, 1992), chap. 6.

53. E.g., Guenter Lewy, *Why America Needs Religion: Secular Morality and Its Discontents* (Grand Rapids, Mich.: Wm. B. Eerdmans, 1996). Capitalist secular conservatives in particular have reason to support religion; in the United States, popular religion often imposes an authoritarian discipline on a personal sphere while leaving the rest of life to the market. Linda Kintz, *Between Jesus and the Market: The Emotions That Matter in Right-Wing America* (Durham, N.C.: Duke University Press, 1997); Michael Lienesch, *Redeeming America: Piety and Politics in the New Christian Right* (Chapel Hill, N.C.: University of North Carolina Press, 1993).

54. Ernest Gellner, *Conditions of Liberty: Civil Society and Its Rivals* (London: Penguin, 1994), chap. 22.

55. Christopher Lasch, *The Revolt of the Elites, and the Betrayal of Democracy* (New York: W. W. Norton, 1995), p. 98; Stout, *Ethics After Babel*, pp. 283–84. Historically, Western capitalism plays down social responsibility and results in moral language receding from everyday use. See Alan MacFarlane, "The Root of All Evil," in *The Anthropology of Evil*, ed. David Parkin (New York: Basil Blackwell, 1985).

56. Luntley, *The Meaning of Socialism*; Michael Harrington, *The Politics at*

God's Funeral: The Spiritual Crisis of Western Civilization (New York: Penguin, 1983).

57. The evidence for religion being associated with well-being is ambiguous. Theodore J. Chamberlain and Christopher A. Hall, *Realized Religion: Research on the Relationship Between Religion and Health* (Philadelphia: Templeton Foundation Press, 2000) defends the link; Richard P. Sloan, E. Bagiella, and T. Powell, "Religion, Spirituality, and Medicine," *Lancet* 353, no. 9153 (1999): 664 disputes it.

58. Bruce, *Religion in the Modern World*, chap. 6.

59. G. A. Wells, *Religious Postures: Essays on Modern Christian Apologists and Religious Problems* (La Salle, Ill.: Open Court, 1988), p. 187.

60. Robert N. McCauley, "The Naturalness of Religion and the Unnaturalness of Science," in *Explanation and Cognition*, ed. Frank C. Keil and Robert A. Wilson (Cambridge: MIT Press, 2000); Pascal Boyer, *Religion Explained: The Evolutionary Origins of Religious Thought* (New York: Basic, 2001)

61. John F. Schumaker, *Wings of Illusion: The Origin, Nature, and Future of Paranormal Belief* (Cambridge: Polity, 1990), chap. 5; Shelly E. Taylor, *Positive Illusions: Creative Self-Deception and the Healthy Mind* (New York: Basic, 1989). For an exploration of the philosophical issues, see Mike W. Martin, *Self-Deception and Morality* (Lawrence: University Press of Kansas, 1986). Against the pragmatic "vital lie" tradition which affirms some self-deception, he endorses a muddled "truth-centered" pragmatism similar to the will-to-believe argument of William James.

62. Peter Marin, *Freedom and Its Discontents: Reflections on Four Decades of American Moral Experience* (South Royalton, Vt.: Steerforth, 1995), pp. 195–265.

Conclusion

The God of Song and Story

Oh, threats of Hell and Hopes of Paradise!
One thing at least is certain—This *life flies;*
One thing is certain and the rest is Lies;
The Flower that once has blown for ever dies.

—Edward FitzGerald,
The Rubáiyát of Omar Khayyám (1889)

I once spent two weeks in a remote Turkish village as a translator for a group of young Europeans. We parked our sleeping bags in the local schoolhouse, helped dig ditches and did some construction work, and got a taste of rural life in a different culture. We loved the food the villagers gave us, but even more, we came to appreciate the times we gathered at one of the coffeehouses, our conversations full of curiosity about each other. An Englishman was very impressed with the villagers' sense of community, the way they all helped one another. Modern life was encroaching, though. The coffeehouses had television. On a break from ditch digging, I helped the Englishman ask a villager what he thought of change, and whether he worried if their community values would last. He wasn't sure, he replied, but he certainly would like to own a tractor.

Another difference which fascinated us was the way in which the sexes were segregated. No Turkish women came into the coffeehouses; they invited the European women over for a separate gathering one night. And, of course, we talked about religion. Once, an Irish woman expressed interest in Islam, and our hostess brought her Quran out to show us. It was nothing special—printed on cheap paper, with bilious green arabesques on the page margins. But she kept it in a place of honor and tried to memorize some scripture from time to time. She knew no Arabic, so she had little idea what her memorized passages meant. She was sure this was a good deed noted in Heaven.

To most of the villagers, our differences did not matter. We guests were not Muslims, and the European women mixed with men in a way which would cause scandal in the village, if not bloody murder. But this was fine; we just had different ways of life, and in a few weeks, none could influence the other too greatly. The village schoolteacher was less tolerant. He had been educated in a city, and as many people from conservative rural backgrounds do, he had responded to his new environment by adopting a more austere, scripturally purist Islam. He had found, I imagine, that he had to defend his beliefs. They were no longer woven into everyone's way of life, part of the way things are always done. The one night he joined our coffeehouse conversation, he argued that the sexes *should* be segregated. Islam and nature required this—"fire and gunpowder could not be kept together." The European way, in other words, was wrong. Some villagers agreed, but most did not care.

The teacher, because of his very fundamentalism, was probably closer to myself in his attitude toward religion than most of the villagers. I am, after all, a stereotypically intellectual skeptic. I cannot just lose myself in a ritual; I certainly can't imagine memorizing holy phrases I do not comprehend. Fundamentalists like the teacher know what their prayers mean. In their strident embrace of holy writ, in their need to defend the faith against an apostasizing world, they show they are touched by modern doubt. Muslim apologists like to say that everyone, down deep, knows there is a God; atheists fool themselves even about their own beliefs. I think otherwise. If my arguments add up to anything, there is, in all likelihood, no ghost in the universe. There is no divine reality which theologians are imperfectly trying to tell us about. We have many questions, some which we may never answer, but no reason to invent a God. And even fundamentalists are influenced by skepticism, not so much because of intellectual arguments, but because doubt runs deep in modern culture.

Our more liberal religions with their sophisticated Gods are more obviously wracked by doubt. They more honestly recognize modern skepticism, even as they protest against it. Recently, during a Reform Jewish service, I started reading the rather curiously philosophical introduction to the prayer book. It was interesting that the editors of a prayer book felt a need to address unbelief, and also interesting that they did this in the form of a wistful complaint:

> If God is not, then the existence of all that is beautiful and . . . good, is but the accidental . . . by-product of blindly swirling atoms, of the equally unpurposeful . . . mechanisms of present-day physics. A man may believe that this dreadful thing is true. But only the fool will say in his heart that he is glad that it is true. For to wish there should be no God is to wish that the things which we love and strive to realize and

> make permanent should be only temporary and doomed to frustration and destruction. . . . Atheism leads not to badness but only to an incurable sadness and loneliness.[1]

I don't cite this as yet another argument. The arguments are over. But after all is said, there is still room for a more personal sort of reflection. Not all atheists wander through life drowning in pathos. True, many who were once religious do feel a sense of loss—often they miss an assurance of belonging in the world, though they are no longer able to believe. Nevertheless, they do not have the joy sucked out of their life; and there is something to be said for being free of threats of Hell and hopes of Paradise. Many do quite well with impermanent satisfactions, and say, like Janet Brazill writing in the membership newsletter of the American Humanist Association:

> rejection of religion led me to atheism. But that alone was not completely satisfying to me, and it wasn't until I discovered Humanism, with its positive approach, that I felt I had found something worthwhile. Humankind needs these principles of compassion and respect for others (even if we can't always respect their actions). By devotion to reason, science, and naturalism, and by holding strong ethical standards, we can find meaning in our lives free from religious superstition.[2]

Both the loss and the freedom which accompanies falling out of faith are foreign to me. I have never known faith from the inside—like an exotic country it fascinates me, and I feel compelled to visit it repeatedly, yet I keep coming home. However much I work up sympathy for hopes of permanence, I can't help asking whether the promises religions make are real. It appears not, and I confess I am not especially devastated by this. Still, I don't want to conclude by saying religion is false and leaving it there. Skeptic though I am, I do not live by reality alone. Our Gods do not belong in our explanations, perhaps not even in our hopes, but they should be at home, I think, in our stories and songs.

DIVINE FALSEHOODS

Take the book of Revelation, that dream of dragons and woes. It is a demented rant. More, it is a hate-filled demented rant. Many conservative Christians passionately believe it gives a detailed history of the End Times to come, which may be just about to begin. And this, more than anything else in popular Christianity, scares me. When apocalyptic theists begin naming the Antichrist, skeptics are bound to be a target.[3] Yet strangely enough, I also find Revelation captivating. As a prediction it is crazy; for anyone with humane feelings, it is often disgusting. But there is something compelling about its psychedelic imagery, its cosmic drama, even the sheer incoherent lunacy of its vision. So maybe, I am tempted to think, even those of us who are incurably skeptical can reclaim religion in a fashion. We can take religion as an art, and not just as a myth from a bygone age, but as something to weave into the stories we tell today.

Speaking of compelling stories, of course, immediately attracts liberal apologetics. Maybe myth is but a deeper truth. Andrew Greeley, for example, says "the story *is* the truth," claiming that "many of science's most fruitful terms are metaphors that tell stories: black holes, great attractors, big bangs, double helix, survival of the fittest, $E=MC^2$."[4] Nonsense. Stories are most apparent in the persistent *mistakes* about science, like the belief in the purposeful nature of evolution. Natural science conspicuously lacks stories infused with personality and purpose; theories of, say, superconductivity are not stories which enrapture any beyond a few physicists.

Reading myths as moral tales makes a bit more sense. But there, too, apologetics distorts the stories. A God-man killed by humans is a wonderfully tragic drama, but it is also an excuse to persecute Jews. In Turkish schools, we learned to celebrate the conquering march of Muslim armies—a beautiful tale of faith and triumph. It took me long to realize I was being taught to feel pride in pillage and enslavement. If I still am touched by religious myths, it is not because I share their moral ideals.

It is hard to appreciate a story when surrounded by people insisting it is the gospel truth. So let me use an explicitly fictional example: J. R. R. Tolkien's artificial mythology.

Tolkien wrote of quests set in a world of medieval technology, magic, near-human creatures such as elves and dwarves, and of course orcs and demons. Like many a swords-and-sorcery fantasy universe, this is a place of stark contrasts between good and evil, beauty and corruption. Tolkien, however, did much more than stage a conflict where the good guys could righteously slaughter deserving villains. A good story has depth—hints of what is left unsaid, tragic failures, ambiguities in even the clearest contrast. "Middle-Earth" has an intricate geography and a many-layered history, where the stones and trees conceal untold legends. Tolkien took a special delight in inventing languages, tracing their evolution through the expanses of time which he imagined. So the quests take place against a rich historical background, and the reader catches glimpses of ages past when magic was stronger and the gods walked the hills of a young world.

In fact, the bulk of Tolkien's creative work, never completed and unpublished during his lifetime, concerns these background legends. He started out intending to compose a specifically English mythology, and the tales took on a life of their own, occupying him through his life.[5] His conception, especially at the beginning, was much like Northern European myths with some Christian elements. The one God, Ilúvatar the all-father, creates a host of angel-like beings, and through their song, brings the physical world into being. After creation, however, Ilúvatar disappears into the narrative background. Great powers among the angelic beings enter the world, shape it, and preside over it as a pantheon. The mightiest of their number, Melkor, rebels and falls into evil. But even Melkor is destined to serve the greater good, since the conflict he causes brings forth drama and tragic beauty.

The world is at first flat, and the gods abide in an earthly paradise west of the ocean. There, they create the Two Trees to give light. The Elves, the firstborn Children of Ilúvatar, awake under the stars, and the gods make war on Melkor and capture him. Most Elves then make the great journey to the West, the land of light. Their bliss in the blessed land ends, however, when Melkor regains his freedom and destroys the Trees. He also steals the Silmarils, jewels in which the Elves have caught the light of the Trees. Many Elves then depart for the lands abandoned by the gods, defying the angelic powers, to make war on Melkor. The gods, meanwhile, fashion the Sun and Moon from the dying Trees, and they light the lands of Melkor just as the Elves appear to challenge the Dark Lord.

This is the stage for the earliest legends—tales of desperate heroism and fleeting victories, magical quests to regain the Silmarils, and betrayals of a noble cause. Melkor proves too strong for the Elves and their few allies among Men, overwhelming and defeating them. Against hope, finally, the forces of the gods emerge out of the West, and cast Melkor into the Void. The Elves leave the lands where they fought for an island close to the western paradise, and the dominion of Men begins.

Tolkien would always keep working on these early layers of his mythology. He also expanded it to tell of the Men who were faithful against Melkor, who were rewarded with an island home and lifespans far exceeding ordinary mortals. Yet there is a Shadow upon Men from their earliest days. Even the best who were rewarded eventually try to claim immortality by conquering the western paradise. The gods call on Ilúvatar, who makes the world round and removes the realm of the gods from its confines. But still, all this was only background in the few books Tolkien completed and published. The completed stories take place thousands of years later, referring to the deeds of the older Ages only in fragments and lore.

After Tolkien's death, his son began editing the large amount of unfinished material and publishing it, giving readers like me a vast new territory to explore. The main reason the mythology remained incomplete was Tolkien's constant changes, so the legends kept mutating into new forms. There were aesthetic reasons for these changes, but it also turned out to be impossible to find a satisfactory overall framework for everything, and efforts to achieve this coherent vision sent Tolkien in many different directions. As the legends developed, the poetic pagan-style myths began to look unsatisfactory. Toward the end of his life, Tolkien was trying to Christianize his creation, even to find what was in effect a rational theology which would make the whole mythology hang together smoothly. Tales like the making of the Sun and Moon and the rounding of the world made more sense as the poetic fancies of Men who did not know the truth as the Elves did. Tolkien even toyed with the idea of developing a Christian-style legend of a Fall, and having Ilúvatar become a Man in order to redeem the morally ever-ambiguous race of Men.

So finally, what Tolkien fanatics like myself have is not a cosmic drama with a coherent logic to it, but an ever-changing myth which stubbornly resists converging to a final form. I peruse partial narratives in many different stages of composition, with a fascination akin to that of scholars uncovering layers of scripture. Before the series was complete, I would anxiously await publication of the next volume of incomplete Tolkien texts, not to finally give me the "true" story, but to see what possibilities he hinted at. In fact, I am glad Tolkien never came up with a fully coherent story, some kind of rational theology for his universe. Then it all would "make sense" internally, but I fear the tales would also become desiccated in the process. Consistency is all very well for science, but it is not important for myth.

Perhaps we can think of religions the same way. At their best, they are stories we can appreciate regardless of whether they are remotely true, morally uplifting, or practically significant.[6] After all, human hopes and desires are an incoherent mess, so to consistently speak to us, a myth must be able to generate many different, contradicting levels of meaning. So even the strange, disreputable corners of religion—Gnostic visions and mystic cosmologies, demented apocalyptic fantasies, legends of magic and mystery set in distant times—are wonderful stories. For the same reasons, I find myself drawn to the incubating grounds of new religions, where believers discuss UFO sightings and government conspiracies, creating a new folk "science" outside respectable institutions.[7] Our myths are false, and they are sometimes dangerously paranoid. Yet I respond to these stories not despite, but partly *because* of their falsehood.

NEW STORIES

This is, of course, at best a backhanded compliment to religion. Fascination with God as a figure of imaginative literature is a poor echo of allegiance to a God who is central to all existence. And the skepticism which makes us see God as fiction often diminishes the beauty of religion. In a secular world, Sufi ceremonies become museum pieces, cathedrals tourist attractions. Liberal religions like Reform Judaism become social clubs to mark the passages of life, and places where people gather to appreciate Jewish culture and sacred literature.

So be it. To create new stories, we need freedom to tamper with the old. Liberal theologians are right; we have to bring our traditions up to date, to make them respond to modern needs. Perhaps the best way to do this is not to seek some eternal Truth beneath the magical beliefs and the faded rituals, but to treat our religions as fictions, and keep rewriting the story as we always do with our myths. Perhaps God should become like Zeus or Isis—no longer something we believe to be real, but nevertheless part of our history, still influencing the stories we tell. We can, or so I hope, inject poetry into our lives without necessarily believing that poetry tells us how deep down things really are.

I don't know if this is any help to someone who feels an "incurable sadness" without God. For those of us who do not care for existential despair, however, taking religion as a fiction which is not *merely* false may be a good idea. After all, it is not so much an indifferent universe which drives us up the wall as it is other people. Skepticism is difficult, and after a long day of fruitless argument, philosopher David Stove's words express the frustration of many an infidel:

> From an Enlightenment or Positivist point of view, . . . there is simply no avoiding the conclusion that the human race is mad. There are scarcely any human beings who do not have some lunatic beliefs or other to which they attach great importance. People are mostly sane enough, of course, in the affairs of common life: the getting of food, shelter, and so on. But the moment they attempt any depth or generality of thought, they go mad almost infallibly. The vast majority, of course, adopt the local religious madness, as naturally as they adopt the local dress. But the more powerful minds will, equally infallibly, fall into the worship of some intelligent and dangerous lunatic, such as Plato, or Augustine, or Comte, or Hegel, or Marx.[8]

Yes, we *are* mad; and yes, philosophy is often the disease rather than the cure. So we might as well enjoy our madness. By all means, let us argue about our beliefs. When we conclude they are not true, or when we think they are dangerous, let us say so without evasion or apology. But let us also play with our myths. Unlike fundamentalists, we don't have to canonize our stories. We can rewrite them, blaspheme them, turn them upside down. Our new stories will, in the long run, be as ephemeral as a night out with good food and good friends. They will be worth telling just the same.

NOTES

1. Robert I. Kahn et al., eds., *Gates of Prayer: The New Union Prayerbook* (New York: Central Conference of American Rabbis, 1975), p. 9. Ellipses in original.

2. From a letter to the editor by Janet Brazill, *Free Mind* 39, no. 3 (1996): 9. See also Judith Hayes, *In God We Trust: But Which One?* (Madison, Wis.: Freedom From Religion Foundation, 1996), chap. 10.

3. Robert C. Fuller, *Naming the Antichrist: The History of an American Obsession* (New York: Oxford University Press, 1995).

4. Andrew M. Greeley, *Religion as Poetry* (New Brunswick, N.J.: Transaction, 1995), p. 38.

5. See *The History of Middle-Earth* series, ed. Christopher Tolkien. The earliest material is in J. R. R. Tolkien, *The Book of Lost Tales, Part One* (Boston: Houghton Mifflin, 1984).

6. Tolkien's mythology also stands against the Enlightenment ideals I share; Patrick Curry, *Defending Middle-Earth: Tolkien, Myth, and Modernity* (New York: St. Martin's Press, 1997). I like it just the same.

7. See Douglas Curran, *In Advance of the Landing: Folk Concepts of Outer Space*, 2d ed. (New York: Abbeville, 2000). Sometimes a false cosmic story reveals an uncomplicated sense of wonder, uncorrupted by grant proposals and patent applications.

8. David Stove, *The Plato Cult, and Other Philosophical Follies* (Cambridge: Basil Blackwell, 1991), p. 184.

Index

accidents, 27, 61, 69, 75, 77, 85, 103–107, 109, 137–39, 190, 217, 220, 235, 238, 249. *See also* uncaused events
  frozen, 75, 106
  historical, 55, 137, 154, 220, 236, 290, 296
Alexander, Richard D., 285
Anderson, Philip W., 196
angels, 38, 39, 125, 137, 175, 180, 215, 217, 219–21, 225, 231
anomalies, 186, 187, 196, 203
anthropic principle. *See* cosmology, anthropic
anthropomorphism, 34, 58, 74, 76, 157, 237
antirationalism
  mystical, 172, 212–13, 215, 226, 230
  postmodern, 246, 248
apocalypticism, 116, 121–23, 140, 148, 151–54, 156, 157, 160, 309
Apollonius, 161, 163, 167
Aristotelian biology, 289–91
Arkoun, Mohammed, 140, 249
Armstrong, Karen, 127
artificial intelligence, 70, 71, 96, 98, 184, 203, 230–35
artificial life, 42, 50
astrology, 77, 84, 97, 195, 203, 218
Ateş, Süleyman, 289, 290
atheism, 24, 31, 46, 108, 185, 279, 302, 308, 309
Atwater, P. M. H., 199
Augustine, 50, 116–18, 279
Ayer, A. J., 34

background theory, 36, 48, 237, 256, 258–61, 264–67
Barth, Karl, 249
basic belief. *See* presuppositions
Becker, C. B., 199
Behe, Michael J., 59
Beloff, John, 185
Berger, Peter L., 39
Bernhardt, Stephen, 222
Bible, 13, 14, 54, 116–22, 125, 131, 133, 136, 139, 180, 218–20, 249, 250
  New Testament, 19, 149, 151–54, 156
  Old Testament, 19, 116, 155, 158, 161, 162, 174
big bang, 68, 85, 87, 88, 91–96, 112
Blackmore, Susan, 200, 201

Bohm, David, 81, 101, 113
boundary conditions, 66, 69, 73, 80, 87, 89, 91
Bradlaugh, Charles, 31
Bray, Gerald, 22, 24
Brazill, Janet, 309
Buckley, Michael J., 108
Buddhism, 55, 140, 181, 185, 220

Calvin, William H., 71
Camping, Harold, 117, 118
capitalism, 253, 255, 277, 297–98, 304
Carmody, D. L. and J. T., 213
Catholicism, 13, 23, 38, 77, 135, 148, 181–83, 185, 186, 189, 217, 246, 278
causality, 96, 97, 107, 138
  nonphysical, 64, 185
chance, 55, 61, 70, 72, 73, 81, 101. *See also* randomness
channeling, 129, 164, 201
chaotic dynamics, 64, 66, 73, 80, 137, 145
Christian Science, 32, 129
Christianity, 22, 53, 116, 117, 121, 125, 129, 131, 135, 137, 147–49, 152, 153, 166, 167, 172, 216, 231, 236
  Catholic. *See* Catholicism
  gnostic. *See* gnosticism
  origins of, 151–65, 167, 175, 177
  Pentecostal, 164, 181
  Protestant. *See* Protestantism
Churchland, Patricia S., 204, 232
Churchland, Paul M., 204, 228, 237, 287, 290
Clark, Austen, 228
Cohen, Edmund D., 103
common sense, 12, 36, 37, 54, 69, 75, 76, 85, 95, 96, 138, 181, 184, 185, 206, 236, 248, 250, 251, 259, 276, 280
complex order, 36, 37, 52, 55, 57, 58, 62–65, 67–70, 72, 76, 90
complex systems, 63, 66, 69, 90, 99, 137, 138
computation, 40, 105, 230, 231
consciousness, 54, 63, 71, 184, 187, 202, 203, 229, 231–32
  altered state of, 164
  cosmic. *See* quantum mysticism
  measurement and, 99, 100
  unity of, 98, 222
conspiracy theory, 43, 264
Constantine, 131, 137, 139
Cook, Michael, 126
cosmology, 37, 80, 85, 87, 92–96
  ancient, 84
  anthropic, 87–90, 92
  big bang. *See* big bang
  Darwinian, 90
  inflationary, 89–91, 95, 110
  quantum, 89, 90, 95, 96
Cox, Harvey, 181
Craig, William Lane, 93, 95
creation, 15, 39, 41, 77, 94, 95, 217
  from nothing, 42, 85, 92, 111
creationism, 14, 37, 51–54, 57–59, 63, 74–78, 85, 244, 245, 247, 253, 254, 256, 257, 259, 260, 273
  intelligent design, 58, 59, 61–64, 70, 71, 76, 79, 92. *See also* design argument
  Islamic, 53–54
creativity, 36, 55, 57, 58, 61, 62, 64, 70–72, 74, 76, 184, 205, 226, 234

Crone, Patricia, 126, 130
Crossan, John Dominic, 149, 159, 162

Darwin, Charles, 24, 51–53, 65, 81, 85, 95, 234, 235, 252, 253, 255–57
Davies, Paul, 90, 91
Davies, Stevan L., 164, 165
Dawkins, Richard, 36, 76, 284, 285
decoherence, 100, 112, 231
deism, 62–64, 77, 84, 85
Dembski, William A., 61–64, 70, 71
Dennett, Daniel C., 57, 71, 225, 234
Descartes, René, 84, 214
design argument, 23, 24, 35–41, 49, 50, 52, 53, 57, 58, 61, 64, 70–72, 74, 75, 77, 84, 85, 87, 88, 90–92, 95, 108, 115, 205, 250, 252
Desmond, Adrian, 255
determinism, 73, 101, 113, 232
Dharmasiri, Gunapala, 185, 220
Dyson, Freeman, 88

Eddy, Mary Baker, 32, 33, 129
Edelman, Gerald L., 71
Edwards, Denis, 72, 74
Einstein, Albert, 84, 87, 101–104
empiricism, 214, 224, 260, 266, 272
energy, 33, 85, 87, 104
  conservation of, 14, 33, 37, 89, 96, 104
  gravitational, 89
Enlightenment, 24, 25, 32, 117, 134, 136, 181, 214, 243, 247, 248, 251–53, 267, 269, 296, 297, 313, 314
Enlightenment rationalism, 244–47, 251, 267–69, 278, 283
entropy
  information, 48, 73
  physical, 37, 63, 68, 73, 80
epilepsy, 128, 224
EPR paradox, 102–103
esoteric experience, 161–63, 165
Eucharist, 157, 163
Eusebius, 129
evil. *See* problem of evil
evolution, 15, 36, 46, 51–53, 55, 57, 58, 60–63, 65–68, 72, 75–78, 92, 95, 106, 137, 138, 234, 237, 248, 249, 253–57, 259, 260, 284, 286, 288, 290, 310
  chemical, 75
  Darwinian, 52, 54, 55, 57, 58, 63, 72, 76, 90, 105, 234, 253, 255
  guided, 37, 58, 63–65, 69, 72, 79, 81
  symbiotic, 60
evolutionary progress, 65–69
evolutionary psychology, 285
explanation
  bottom-up, 36, 37, 57, 64, 65, 187
  personal, 34, 37, 39, 58, 59, 111, 134, 299
  supernatural, 43, 164, 188, 190, 201, 204, 252
  top-down, 36–39, 49, 54, 55, 57, 64, 183

fact and value, 289–91
faith, 46, 135, 139, 182, 243, 245, 248, 249, 279, 295, 297–99, 309
  act of, 34, 178, 244, 249, 251

faith healing, 158, 161, 162, 164, 176, 181
Fatima miracle, 189
feminism, 290, 291, 300
Findlay, J. N., 31
fine-tuning, 85, 87–90, 92
first cause. *See* proofs of God, cosmological
Flanagan, Owen, 228
Flew, Antony, 27, 186
folk theory, 164, 202–204, 218, 237, 258, 283, 286, 293
Forman, Robert K. C., 222
foundational principles. *See* presuppositions
foundationalism, 250
fraud, 122, 129, 188, 189, 192–95, 207
Fredriksen, Paula, 152, 166
free will, 40, 72, 78, 84, 232, 279, 280
fundamental commitment. *See* presuppositions
fundamentalism, 15, 39, 54, 58, 64, 142, 180, 270, 308

Gale, Richard M., 26
game theory, 234, 285
Gardner, Martin, 191
Gauquelin, Michel, 195
Geertz, Clifford, 276
Gellner, Ernest, 268, 294, 296
Gilkey, Langdon, 134
Gimello, Robert, 230
gnosticism, 154, 163, 170, 176, 239
God
  Abrahamic, 92, 118, 247
  attributes of, 21–23, 26, 31, 40–42, 45, 50, 58
  changeability of, 22, 32
  intelligibility of, 25, 31, 34, 40, 43, 217
  mystery of, 13, 22, 24, 29, 39, 40, 44, 49, 171, 212, 217, 263
  necessity of, 23–27, 30–32, 35, 44, 49, 244
  omnipotence of, 22, 40, 42, 57, 278, 280, 281
  omniscience of, 42, 57
  perfection of, 22, 23, 25, 26, 28–32, 183, 216, 217, 263, 278–80
  proofs of. *See* proofs of God
  simplicity of, 22, 29–31, 36, 216, 217, 228
God of the gaps, 54, 55, 61, 108, 167, 233
God of the philosophers, 16, 21, 24, 29–32, 34, 35, 40, 44, 53, 116, 179, 183, 216
Gödel's theorem, 70, 232, 235
Goodman's paradox, 273
gospel, 148, 149
  John, 148, 149, 165, 171, 173, 177
  Luke, 148, 150, 155, 158, 161, 171
  Mark, 148–50, 152, 155, 158, 160, 162–64, 170, 171
  Matthew, 148, 150, 155, 158, 161, 171
  Peter, 171, 176
  Q, 148, 158, 159, 167, 176
  Thomas, 158, 167
Gould, Stephen Jay, 67
Greeley, Andrew M., 310
Grenz, Stanley J., 247, 248
Grin, David Ray, 185
Grof, Stanislav, 224
Grush, Rick, 231
Guthrie, Stewart Elliott, 82

Habermas, Gary R., 168
hallucinations, 200, 221, 225
Hartle, James B., 89, 95
Haught, John F., 63, 64
Hawking, Stephen W., 89, 91, 95
heaven, 43, 120, 121, 148, 169, 199, 279, 309
  ascent to, 148, 155, 161–63, 168, 171, 215, 219, 225
Hell, 43, 133, 139, 199, 279, 309
Helms, Randel, 155
Hempel, Carl, 136
Henderson, Charles P., Jr., 46
hierarchical reality, 54, 55, 63, 64, 97, 183, 199, 217, 218, 230, 268
Hinduism, 54, 140, 181, 185, 219–21
historical contingency, 137
history
  critical, 117, 122, 134, 136, 138, 139, 141, 149, 171
  mythic. *See* myth, historical
  narrative, 117, 134–36, 138
  salvation, 39, 55, 116, 124–26, 138, 249
Holocaust, 136, 279
Home, D. D., 188
homeopathy, 195, 203
Hook, Sidney, 269
humanism, 133, 136, 299, 300, 309
Hume, David, 52, 53, 77, 186, 263–65
Humphrey, Nicholas, 190
Hyman, Ray, 194
hypnosis, 48, 169, 176, 198, 201

Ibn Sina, 231
identity, personal, 244, 246, 269, 277, 287
induction. *See* problem of induction
infinite numbers, 27, 93
inflation. *See* cosmology, inflationary
information
  loss of, 62, 66, 96, 100
  origin of, 61–64, 68
information theory, 61
intelligent design. *See* creationism, intelligent design
intermediate forms, 59–61
interpretation
  historical, 134, 136, 138
  quantum, 81, 113
  scriptural, 53, 54, 122, 123, 125, 126, 135, 140
irreducible complexity, 59
irreversibility, 66–68, 96, 99, 105, 111, 113, 114
Islam, 11, 14, 38, 53, 54, 110, 116, 124–33, 135, 137, 139, 142–44, 150, 156, 161, 215, 236, 250, 279, 289, 290, 297, 307, 308
  heterodox, 132, 133
  mystical. *See* Sufism
  origins of, 125–31
  Shii, 132
  Sunni, 132
Islamic empire, 126, 129, 131, 132, 156
Islamic philosophy, 15, 216, 231, 236, 263, 279

Jackson, Frank, 227
James, William, 224
Jastrow, Robert, 92, 93
Jennings, Theodore W., 226
Jesus, 117, 125, 141, 147–73, 175, 176, 217, 282

birth of. *See* Virgin Birth
divinity of, 147–49, 166, 171
existence of, 156–57
historical, 149–66, 172, 246
miracles of, 148, 152, 155–57, 161–63, 166, 167, 185
resurrection of. *See* resurrection, Jesus'
second coming of, 153, 154, 170
John of Damascus, 22
John the Baptizer, 150, 152, 164
Jonas, Hans, 279, 280
Jones, Richard H., 214, 227
Josephus, 156, 175
Judaism, 15, 19, 116, 125, 129, 131, 137, 152, 175, 219, 301, 304, 313
early, 119–21, 157
rabbinic, 121, 135, 136

Kant, Immanuel, 25
Kingdom of God, 148, 151–54, 158–62, 164
Kitcher, Philip, 255
Kuhn, Thomas S., 203
Kyburg, Henry E., Jr., 45

Lane Fox, Robin, 122
Laplace, 107
Lasch, Christopher, 298
law
Islamic, 132, 215
Jewish, 118, 153, 163, 168, 219
moral, 281, 282, 286
natural, 69, 84, 85, 90, 103, 134, 136, 137, 181, 182, 186, 231, 241
physical, 29, 33, 37, 41, 61, 66, 69, 73, 77, 87, 91, 94, 104–107, 113, 197, 232
Lawrence, Bruce B., 247, 248, 253
Lee, T. D., 106
Linde, Andrei, 89, 95
literalism, 51, 53, 136
Lüdemann, Gerd, 153

Maccoby, Hyam, 153
Mack, Burton L., 122, 158–60, 168
Mack, John E., 215
magic
conjuring, 188, 191
ritual, 121, 127, 161–64, 219
marginal effects, 194–97
Mars Effect, 195
Martin, Michael, 294
materialism, 64, 98, 182, 184, 185, 187, 203, 242, 297
McClenon, James, 255
mediums, 188, 191, 192
Meister Eckhart, 222
Mellon, John C., 149
memes, 71, 285
Merkley, Paul, 119
messianic beliefs, 126, 129, 132, 135, 151, 153, 161
meta-analysis, 193, 208
metaphorical speech in religion, 13, 15, 40–44, 49, 53, 123, 138, 140, 213, 219, 221
metaphysics, 15, 22–25, 27, 29, 31– 36, 39, 42–45, 47, 52, 53, 58, 63–65, 72, 74, 81, 85, 87, 91, 93, 95–97, 107, 111, 164, 173, 216, 218, 222, 226, 231, 235–36, 238, 242–44, 246, 280, 301
mind-body dualism, 36, 38, 39, 49, 98, 184, 185, 187, 203, 206
Miosi, Frank T., 32

miracles, 11, 38, 42, 84, 116, 127, 148, 151–54, 158, 161, 167, 170, 173, 179–95, 197, 198, 205, 217, 218, 235, 249
modernity, 49, 136, 139, 140, 269, 276, 278, 296, 297, 300, 307, 308
Monod, Jacques, 140, 141
monotheism, 22, 55, 84, 119, 128, 129, 131–33, 277, 297
moral imagination, 290, 291, 297, 298, 300
moral perception, 286–88, 290, 292
moral principles, 286, 289, 292, 293, 303
morality
  conflict resolution and, 276, 284–86, 288, 289, 292
  divine command, 281, 282
  evolution and, 65, 284–86, 289
  objectivity of, 276, 280–82, 288, 292, 293
  religious, 54, 128, 129, 250, 269, 275–79, 282–84, 295, 297, 299
  secular, 133, 246, 276, 278, 284, 297, 298
  self-interest and, 276, 277, 280, 282, 284, 285, 293
Mormon origins, 129, 167
Morris, Henry M., 14, 37, 59
Muhammad, 13, 116, 118, 124–28, 130, 131, 137, 143, 150, 161, 162, 174, 215
multiple universes, 89–90, 95
mystical experience, 15, 38, 127, 162, 165, 201, 205, 212–15, 218–25, 227, 238, 240
mysticism, 97, 140, 199, 205, 212–16, 218–21, 226, 230, 233, 236, 238, 248, 249, 280
  Christian, 219, 236
  introspective, 221–22, 225
  Islamic. *See* Sufism
  Jewish, 215, 219
  nontheistic, 219–21
  psychedelic, 223–24, 240
  quantum. *See* quantum mysticism
  visionary. *See* visions
myth, 118, 121–24, 126, 127, 132, 133, 135, 136, 138, 140, 141, 147–49, 153–59, 161, 164, 165, 169–73, 218, 296, 310–13
  historical, 119, 121, 123, 126, 155

Nasr, Seyyed Hossein, 54, 64
natural law. *See* law, natural
natural selection, 57, 58, 60, 62, 65, 66, 69, 71, 72, 75, 78, 105, 234, 235, 253, 259
naturalism, 15, 27, 30, 36, 38, 55, 57, 58, 86, 92, 117, 123, 134, 177, 188, 204, 212, 224, 237, 244, 250, 277, 280, 296, 300, 309
near-death experience, 198–202, 204, 209, 210, 223, 224
Neoplatonism, 15, 216–18, 222, 231. *See also* Platonism
neural networks, 202, 204, 286, 287
neuroscience, 98, 184, 203, 204, 224, 228, 231, 232, 234, 237, 240
New Age, 38, 86, 97, 109, 112, 129, 164, 172, 180, 181, 186, 188

Newton, Isaac, 85, 86, 107
Newtonian physics, 62, 64, 84–86, 88, 95, 97–99, 101, 103, 107, 108, 252, 257
Nickell, Joe, 191
Nielsen, Kai, 27, 34
nihilism, 276, 277, 293
nonalgorithmicity, 71, 232–35

objective knowledge, 244, 246–48, 252–54, 258, 268, 271
occult, 84–86, 97, 98, 101, 182, 195, 199, 204, 218, 223
Olthius, James H., 246
Otto, Rudolf, 212
out-of-body experience, 198–200, 202

Pagels, Elaine, 170
Pagels, Heinz R., 88
Pahnke, Walter N., 223
Pannenberg, Wolfhart, 141
paradigm, 203, 252, 254–57
paradigm shift, 187, 203, 252, 254, 256, 272
paradoxical speech, 40, 165, 215, 216, 226
paranormal experience, 169, 180, 181, 188, 190, 198–202, 215, 218, 223, 225, 238
parapsychology, 182, 183, 185–88, 191–97, 199, 200, 202, 203, 237, 240, 255
  ganzfeld, 193–94
  REG, 193–95, 208
  remote viewing, 193
Parrinder, Geoffrey, 232
particle creation, 33, 96, 111
past-life memories, 198–202, 209, 218
Paul, 13, 149, 153, 154, 156–58, 167–70, 172, 175, 177, 215
Payne, Steven, 220
Peacocke, Arthur R., 64
Penrose, Roger, 231–32, 235, 242
Peters, F. E., 126, 130
physical law. *See* law, physical
physical theory, 33, 34, 73, 87, 88, 91, 105
Plantinga, Alvin, 250
Platonism, 15, 28, 52, 103, 120, 214, 217, 218, 230–33, 235, 241, 258, 260, 266–68, 291. *See also* Neoplatonism
Playfair, John, 61, 62, 72
Polkinghorne, John, 64, 74, 108
positivism, 32–34, 48, 244
postmodernism, 242, 246–54, 256, 267–69, 271, 299
Prado, C. G., 268
pragmatism, 248, 268, 269, 294, 295, 299, 300, 304, 305
presuppositions, 13, 45, 107, 134, 138, 214, 226, 237, 238, 244–46, 248–52, 257–59, 261–63, 266–68, 272
Price, G. R., 192
probabilistic inference, 28–30, 48, 91, 96, 194, 195, 264–66
  Bayesian, 28, 47, 48, 264–67, 273
problem of evil, 16, 26, 27, 40, 43, 78, 92, 183, 236, 278–80, 301, 302
problem of induction, 249, 262–67
Proclus, 216, 217
proofs of God, 12, 16, 23–24, 31, 39, 46, 85, 250
  cosmological, 11, 23, 24, 27–30, 48, 74, 85, 93, 95, 97, 111, 115, 216, 263

kalam, 93–97
ontological, 23, 24, 26, 27, 31
teleological. *See* design argument
prophecy, 86, 117, 118, 121–23, 140–42, 151–55, 165
prophets, 15, 38, 109, 116, 118, 119, 124, 126–29, 137, 141, 143, 148, 150, 155, 157, 158, 162–67, 169, 170, 185, 212, 219, 221, 231, 275
Protestantism, 54, 115, 123, 134–36, 139, 179–81, 183, 186, 215, 246, 249, 251, 276
Provine, William B., 76
Pseudo-Dionysius, 216, 217, 222
psychic phenomena, 182, 185, 190, 196, 218, 255
psychic power, 97, 110, 173, 182, 185–88, 190, 192, 194–97, 200, 203, 204
psychical research. *See* parapsychology
psychology
evolutionary. *See* evolutionary psychology
folk, 98, 138, 184, 202, 204, 206, 218, 237
Jungian, 98, 209

qualia, 227–30, 241
quantum mechanics, 24, 73, 81, 87, 93, 95, 96, 98–103, 105, 111–13, 231, 237, 252, 256, 257
quantum mysticism, 86, 97–98, 100–103, 112, 184, 196, 209
Quran, 11, 13–14, 18, 21, 53, 54, 116, 124–29, 131, 132, 137, 142, 162, 219, 257, 262, 289, 307

Randi, James, 191
randomness, 61, 62, 69–76, 81, 104, 110, 233–35
quantum, 64, 73, 81, 96, 99, 101, 105, 111, 235
randomness in physics, 24, 66, 68, 73, 85, 89, 94, 96, 107, 137
rationalism. *See* Enlightenment rationalism
Rawlings, Maurice S., 199
Ray, John, 52, 77
reality. *See* hierarchical reality; supernatural reality
reality checks, 34, 45, 76, 135, 260, 261, 293
reason and faith, 13–14, 243–51
relativism
cognitive, 139, 247–49, 251, 270
moral, 292, 293, 302
relativity
general, 73, 87–89, 93–95
special, 96, 101, 104, 252, 257
religion
Abrahamic, 18, 28, 38, 54, 92, 112, 116, 117, 125, 129, 135, 139, 199, 215, 245, 276, 293, 295, 304
benefits of, 294–95, 297–99
conservative, 14–15, 49, 51, 54, 58, 59, 64, 79, 84, 85, 116–18, 123, 136, 138, 139, 148, 149, 161, 166, 199, 220, 246, 248, 251, 253, 276, 277, 279, 295, 309
individualist, 84, 86, 139, 143, 157, 162, 165, 170, 181, 199, 295

liberal, 14–16, 46, 49, 50, 52, 53, 58, 63, 64, 70, 76, 112, 117, 122, 123, 127, 129, 136, 140, 141, 143, 147, 149, 160, 166, 171, 179, 269, 277, 278, 308, 310, 313
popular, 11, 23, 38, 39, 54, 76, 179, 181, 189, 199, 205, 216, 221, 236, 304
violence and, 128–33, 276, 296, 297
women and, 129, 276, 278, 282, 289, 290
religious diversity, 46, 50, 135, 140, 238, 247, 276, 295, 304
religious experience, 15, 24, 34, 108, 123, 139, 140, 180, 181, 212–14, 221, 250, 304
religious theory, 34, 35, 37–40, 44, 59, 74, 87, 92, 94, 107, 118, 205, 220
repeatability, 186, 187, 194
replicators, 55, 56, 68, 71, 75
representative sampling, 29, 48, 267
resurrection, 116, 151, 169
Jesus', 134, 141, 148, 149, 153, 158–60, 162, 163, 165–71, 177
revelation, 24, 38, 39, 42, 49, 123, 127, 131, 133–35, 138–41, 148, 152, 165, 170, 172, 173, 180, 213, 215, 218, 231, 236, 277, 282
general, 115, 141
special, 109, 115–19, 124–29, 143
revolution
conceptual. *See* paradigm shift
scientific, 84, 136, 182, 236
Rhine, J. B., 183, 192
Ruse, Michael, 76
Russell, Jeffrey Burton, 134
saints, 132, 133, 170, 180, 181, 183, 185, 188, 189, 217, 218, 221, 223
salvation, 23, 116–18, 139, 157, 173, 180, 199, 215, 251, 279. *See also* history, salvation
Sanders, E. P, 151–53
Satan, 128, 141, 148, 199, 231
Satanic verses, 128
Scharfstein, Ben-Ami, 236
Schmeidler, Gertrude, 196
science
Enlightenment, 149, 248–50, 254
ideology and, 246–48, 253, 255–57
progress of, 251–52, 256–59
social mechanisms of, 253–56, 271
theistic, 250
science and religion
compatibility of, 53, 63, 76
conflict of, 53, 244, 245
independence of, 59, 76
scientific method, 245, 252, 257–60, 264, 268, 269, 272
scriptural inerrancy, 14, 15, 118, 125, 142, 180, 245
secondary causes, 72, 74, 77
secularization, 296, 304
self-replication, 55, 68, 71, 284, 286
Shibley, Mark A., 179
simplicity of hypotheses, 29, 30, 48
sin, 22, 116, 164, 168, 245, 248, 250, 251, 282, 295
original, 53

skepticism
  paranormal, 182, 186–88, 190–96, 199, 202, 203, 207
  philosophical, 246, 248, 250, 267, 272, 273
  religious, 11–13, 15, 18, 24, 31, 32, 34, 36, 39, 42, 76, 97, 140, 213, 220, 243, 275, 276, 294, 295, 297, 299, 308–10, 313
Smith, Jonathan Z., 158
Smith, Joseph, 129, 167
Smith, Morton, 162–65
Smolin, Lee, 90
Soal, Samuel G., 192, 193
soul, 15, 23, 39, 43, 77, 84, 98, 173, 182, 184, 185, 199, 218, 222, 230, 279, 295
  animal, 78, 184, 203
  rational, 84, 173, 217
Spanos, Nicholas, 201
speciation, 65–67
specified complexity, 61, 62, 69
spirit possession, 157, 158, 163–69, 171, 181, 198, 201
spiritual reality, 12, 13, 15, 16, 38, 39, 44, 182, 184, 202, 205, 212, 213, 223, 226, 230
Spiritualism, 181, 185, 188
Spong, John Shelby, 34, 171, 172
statistics. *See* probabilistic inference
Stenger, Victor J., 96
Steven T. Katz, 219, 240
Stove, David, 313
string theory, 95
Sufism, 133, 180, 215, 219
supernatural beings, 38, 49, 124, 129, 157, 158, 161, 163
supernatural claims, 34, 37–39, 44, 190
supernatural powers, 162, 179, 181
supernatural reality, 35, 38, 40, 43, 45, 49, 58, 84, 85, 117, 121, 129, 162, 165, 166, 173, 181, 183, 185, 199, 225
supernaturalism, 39, 49, 84, 181, 190, 240, 295, 304
survival of death, 182, 184, 198, 199, 206, 209, 276
Swinburne, Richard, 27, 29
symmetry, 66, 80, 83, 103–105, 113, 197
  broken, 83, 105, 106
  violations of, 197

Teilhard de Chardin, Pierre, 64, 65
testimony, 183, 186, 188, 258
  eyewitness, 148, 167, 168
  multiple, 190
theology
  Jewish restoration, 151–54
  natural, 52, 249
  negative, 215–18, 220, 226, 239, 248
  process, 32, 280
theory-experiment interaction, 106, 196, 231, 236, 256, 260, 261
thermodynamics, 66–68, 73, 100, 105, 111
  second law of, 36–37, 63, 67, 68, 75, 85
Thompson, Thomas L., 119
Tillich, Paul, 49
time-reversibility, 66, 96, 111
Tolkien, J. R. R., 301, 310–12, 314
Trinity, 22, 147, 219
Turkey, 11, 54, 132, 144, 307, 310

UFO abductions, 169, 201, 215, 218, 226
UFOs, 169–70, 172, 212, 218, 272, 295, 312
uncaused events, 24, 27, 30, 74, 93–97

value. *See* fact and value
violence. *See* religion, violence and
Virgin Birth, 141, 150, 151, 155, 166
Virgin Mary, 148, 217
  miracles of, 180, 183, 189, 190
visions, 127, 129, 143, 153, 162, 169–71, 173, 176, 180, 189, 190, 221–25, 239, 240
Voegelin, Eric, 134, 303

Wells, G. A., 156, 157, 299, 300
White, Ellen, 129
Wigner, Eugene P., 98
Wilber, Ken, 97
Wilson, E. O., 76
Wine, Sherwin, 136
Wiseman, Richard, 194
women. *See* religion, women and; feminism

Yahweh, 116, 118–22, 129, 136, 151, 157, 219

Zacharias, Ravi, 276, 277
Zurek, Wojciech H., 100